Chemical Demonstrations

Volume 1

Volume 1 Collaborators and Contributors

GLEN E. DIRREEN, PH.D.
Director of General Chemistry Laboratories, University of Wisconsin–Madison

GEORGE L. GILBERT, PH.D.
Professor of Chemistry, Denison University; Visiting Professor, University of Wisconsin–Madison, January–August, 1979

FREDERICK H. JUERGENS, M.A.T.
Lecture Demonstrator, University of Wisconsin–Madison

E. PHILIP PAGE, PH.D.
Project Associate, University of Wisconsin–Madison

RICHARD W. RAMETTE, PH.D.
Professor of Chemistry, Carleton College

RODNEY SCHREINER, PH.D.
Project Associate and Lecturer, University of Wisconsin–Madison

EARLE S. SCOTT, PH.D.
Professor of Chemistry, Ripon College; Visiting Professor, University of Wisconsin–Madison, June–December 1980, Summer, 1981, Summer 1982

MARY ELLEN TESTEN, M.S.
Project Assistant, University of Wisconsin–Madison, 1980–1982

LLOYD G. WILLIAMS, PH.D.
Assistant Professor of Chemistry, Hampshire College; Project Assistant and Lecturer, University of Wisconsin, 1973–1978

Chemical Demonstrations

A Handbook for Teachers of Chemistry

Bassam Z. Shakhashiri

VOLUME **1**

THE UNIVERSITY OF WISCONSIN PRESS

The University of Wisconsin Press
1930 Monroe Street
Madison, Wisconsin 53711

3 Henrietta Street
London WC2E 8LU, England

For permission to quote from previously copyrighted material, acknowledgment is gratefully made to the following authors and publishers:

P. W. Atkins, *Physical Chemistry*, used with permission of W. H. Freeman and Company, copyright © 1978.

I. B. Berlman, *Handbook of Fluorescence Spectra of Aromatic Molecules*, Academic Press, 1971.

F. A. Cotton and G. Wilkinson, *Advanced Inorganic Chemistry: A Comprehensive Text*, 4th ed., John Wiley & Sons, Inc., 1980.

F. S. Dainton and J. Bevington, "The Oxidation and Inflammation of Yellow Phosphorous," *Faraday Society, Transactions*, vol. 42, Royal Society of Chemistry, 1946.

J. A. Dean, *Flame Photometry*, used with permission of McGraw-Hill Book Company, copyright © 1960.

H. D. Gafney and A. W. Adamson, "Chemiluminescence: An Illuminating Experience," *Journal of Chemical Education*, vol. 52, 1975.

B. Lewis and G. Van Elbe, "Stability and Structure of Burner Flames," *Journal of Chemical Physics*, vol. 11, 1943.

F. E. Lytle and D. M. Hercules, "Chemiluminescence from the Reduction of Aromatic Amine Cations and Ruthenium (III) Chelates," *Photochemistry and Photobiology*, vol. 13, reprinted with permission of Pergammon Press, Ltd., copyright © 1971.

K. Maeda and T. Hayashi, "The Spectra of the Chemiluminescence, Fluorescence and Absorption of Lucigenin and Its Electron Spin Resonance," *Bulletin of the Chemical Society of Japan*, vol. 40, 1967.

Richard W. Ramette, "Exocharmic Reactions," *Journal of Chemical Education*, vol. 57, 1980.

B. Z. Shakhashiri and L. G. Williams, "Singlet Oxygen ($^1\Delta g$) in Aqueous Solution—A Lecture Demonstration," *Journal of Chemical Education*, vol. 53, 1976.

B. Z. Shakhashiri and G. E. Dirreen, "The Preparation of Polyurethane Foam—A Lecture Demonstration," *Journal of Chemical Education*, vol. 54, 1977.

B. Z. Shakhashiri, G. E. Dirreen, and F. J. Juergens, "Solubility and Complex Ion Equilibria of Silver(I) Species in Aqueous Solution," *Journal of Chemical Education*, vol. 57, 1980.

B. Z. Shakhashiri, G. E. Dirreen, and F. J. Juergens, "Color, Solubility and Complex Ion Equilibria of Nickel(II) Species in Aqueous Solution," *Journal of Chemical Education*, vol. 57, 1980.

B. Z. Shakhashiri, G. E. Dirreen, and L. G. Williams, "Paramagnetism and Color of Liquid Oxygen: A Lecture Demonstration," *Journal of Chemical Education*, vol. 57, 1980.

B. Z. Shakhashiri, G. E. Dirreen, and L. G. Williams, "Preparation and Properties of Polybutadiene (Jumping Rubber)," *Journal of Chemical Education*, vol. 57, 1980.

B. Z. Shakhashiri, L. G. Williams, G. E. Dirreen, and A. Francis, "Cool Light Luminescence," *Journal of Chemical Education*, vol. 58, 1981.

E. H. White and M. M. Bursey, "Chemiluminescence of Luminol and Related Hydrazides: The Light Emission Step," *Journal of the American Chemical Society*, vol. 86, reprinted with permission of American Chemical Society, copyright © 1964.

H. E. Winberg, J. R. Downing, and D. D. Coffman, "The Chemiluminescence of Tetrakis(dimethylamino)ethylene," *Journal of the American Chemical Society*, vol. 87, reprinted with permission of American Chemical Society, copyright © 1965.

Library of Congress Cataloging in Publication Data
Shakhashiri, Bassam Z.
Chemical demonstrations.
Bibliography: p.
1. Chemistry—Experiments.   I. Title.
QD43.S5   1983   540'.7'8   81-70016
ISBN 0-299-08890-1 (v. 1)

To
Gil Haight,
who taught me the art of presenting demonstrations,
and to
Odell "Tally" Taliaferro,
who expanded my knowledge of demonstrations.

B. Z. S.

Contents

2   CHEMILUMINESCENCE   **125**

Rodney Schreiner, Mary Ellen Testen,
Bassam Z. Shakhashiri, Glen E. Dirreen, and Lloyd G. Williams

3 POLYMERS   205
Glen E. Dirreen and Bassam Z. Shakhashiri

4 COLOR AND EQUILIBRIA OF METAL ION PRECIPITATES AND COMPLEXES   250
Earle S. Scott, Bassam Z. Shakhashiri,
Glen E. Dirreen, and Frederick H. Juergens

Preface

This is the first in a series of volumes aimed at providing teachers of chemistry at all educational levels with detailed instructions and background information for using chemical demonstrations in the classroom and in public lectures. Each volume will consist of a variety of demonstrations grouped topically in chapters. The demonstrations in this volume deal with thermochemistry, chemiluminescence, polymers, and metal ion precipitates and complexes. Future volumes, now in preparation, will include demonstrations on gases, acids, bases, colligative and other properties of solutions, corridor exhibits, colloids, oscillating reactions, clock reactions, cryogenics, electrochemistry, and other topics.

Late in 1974, I embarked on a project aimed at compiling lists of about 500 demonstrations suitable for use in college and high school chemistry courses. My aim was to have each demonstration include not only a set of specific directions but also a clear discussion of the phenomena displayed and the principles involved. In addition, I was concerned about safety aspects and disposal procedures. Although several sources of demonstration material were available, none, in my opinion, provided full enough explanations of the chemistry involved. Believing the project could be completed in a year or so, I enlisted the help of Glen Dirreen, the Director of General Chemistry Laboratories at the University of Wisconsin–Madison, and Lloyd Williams, who was a graduate student finishing his Ph.D. dissertation. We quickly discovered that a library search for explanations of the behavior of various chemical systems did not yield the information we wanted. As we approached colleagues at Wisconsin and elsewhere for help, I realized that major research efforts had to be made to achieve our goals. And so, beginning in mid-1975, I expanded the scope of the project by involving graduate students as project assistants. The results of their work contributed to understanding the behavior of several chemical systems which appear in this volume. I am pleased to acknowledge Bruce Siggins, Ann Francis, Mary Ellen Testen, and Mark Fink for their contributions to demonstrations in Chapter 2, and Lawrence David and David Piatt for their contributions to demonstrations in Chapter 3 and Chapter 4.

It is with much appreciation that I acknowledge the work of the University of Wisconsin–Madison lecture demonstrators, Fred Juergens and Vince Genna, who tested every demonstration procedure in this volume. I also thank Robert Olsen, an undergraduate assistant for three years in my laboratory, who checked data and references in the literature and helped improve many discussion sections in Chapter 1. I express thanks to Patti Puccio and Kay Kilcoyne, the secretaries in the General Chemistry Office, for their expertise and patience in typing several drafts of the manuscript.

The staff of the University of Wisconsin Press were helpful and considerate in all their dealings with us. In particular, I wish to express special gratitude to Elizabeth A. Steinberg, the chief editor, and to Robin Whitaker.

I am grateful for the collaboration of Glen Dirreen, Rodney Schreiner, Earle Scott, Lloyd Williams, Fred Juergens, Mary Ellen Testen, and George Gilbert, who are the coauthors of the chapters in this volume. I also thank Philip Page, who as editor contributed significantly to the clarity and coherence of the text and was of great assistance in ensuring consistency in style and format. The comments and suggestions of many colleagues from across the country were invaluable in developing the content and format of this volume.

In completing this volume I relied heavily on expert advice and assistance from Glen Dirreen, Rodney Schreiner, and Philip Page. Without their efforts this book would not have been finished.

January 1983   Bassam Z. Shakhashiri
Madison, Wisconsin   Professor of Chemistry
                                      University of Wisconsin–Madison

Exocharmic Reactions†

Richard W. Ramette

When young Humphry Davy passed an electric current through potassium hydroxide and "saw the minute globules of potassium burst through the crust of potash and take fire as they entered the atmosphere, he could not contain his joy—he actually bounded about the room in ecstatic delight; and some little time was required for him to compose himself sufficiently to continue the experiment. An entry in Davy's notebook concludes with the statement, *'Capital experiment!'* " [1]. And further, "the little metallic globules always appeared at the cathode, and these had an astonishing way of bursting into flame when thrown into water. They skimmed about excitedly with a hissing sound, and soon burned with a lovely lavender light" [2].

Clearly, these chemical changes evolved not only several forms of energy, but also a great quantity of *charm*. Derived from Middle English *charme,* charm, as the dictionaries define it, is a trait that fascinates, allures, or delights; a combination of entirely attractive and delightful traits; compelling attractiveness and appeal, dispelling any possible reserved or antagonistic feeling.

The joy of chemical experimentation has been well recognized, at least from the early days of alchemy, and our appreciation of chemical charm probably dates back to the prehistoric discovery of ways to make and control fire. Therefore, it seems useful to coin the term *exocharmic reaction* (from the Greek *exo-,* turning out) and, particularly in our role as chemistry teachers, to seek and share techniques for liberating as much charm as possible from the chemical changes our students see in the laboratory and classroom demonstrations.

THE LAWS OF CHARMODYNAMICS

Although the controlled uses of chemical charm may involve as much artistry as formal science, at least four main assumptions seem reasonable.

1. Any chemical change, particularly one that is both thermodynamically and kinetically spontaneous, is inherently exocharmic and possesses a significant amount of latent charm.

2. When the reaction occurs, the fraction of the latent charm evolved depends directly upon the technical and histrionic skills of the person performing the experiment.

3. The fraction of the evolved charm that is reabsorbed (the free charm) is directly dependent upon the intellectual and psychological preparation of the person viewing the experiment.

4. Charm is not conserved in the transfer process. Even though the amount liberated is restricted by the second law, *each* viewer absorbs according to the third law. Thus, a demonstration before a large class is particularly, well, charming.

†A revision of the article by the same title published in *J. Chem. Educ.* **1980,** *57,* 68.

SOME EVIDENCE FOR CHEMICAL CHARM

In my own case, an exocharmic reaction performed by my high school biology teacher immediately and permanently converted me from an unmotivated state to the exciting and dangerous life of a boy chemist. He was trying to teach us about sugars, and we yawned as he added glucose to some Fehlings solution. But when he heated the mixture, and the blue solution turned into a red solid, I sat up straight and recognized the moment as a turning point in my life. Within a day, I bought a Gilbert Chemistry Set and began threatening my attic and the peace of mind of my parents.

This experience has always made me especially appreciative of a reminiscence by Ira Remsen which supports the concept of exocharmic reactions:

> While reading a textbook of chemistry I came upon the statement, "nitric acid acts upon copper." I was getting tired of reading such absurd stuff and I was determined to see what this meant. Copper was more or less familiar to me, for copper cents were then in use. I had seen a bottle marked nitric acid on a table in the doctor's office where I was then "doing time." I did not know its peculiarities, but the spirit of adventure was upon me. Having nitric acid and copper, I had only to learn what the words "act upon" meant. The statement "nitric acid acts upon copper" would be something more than mere words. All was still. In the interest of knowledge I was even willing to sacrifice one of the few copper cents then in my possession. I put one of them on the table, opened the bottle marked nitric acid, poured some of the liquid on the copper and prepared to make an observation. But what was this wonderful thing which I beheld? The cent was already changed and it was no small change either. A green-blue liquid foamed and fumed over the cent and over the table. The air in the neighborhood of the performance became colored dark red. A great colored cloud arose. This was disagreeable and suffocating. How should I stop this? I tried to get rid of the objectionable mess by picking it up and throwing it out of the window. I learned another fact. Nitric acid not only acts upon copper, but it acts upon fingers. The pain led to another unpremeditated experiment. I drew my fingers across my trousers and another fact was discovered. Nitric acid acts upon trousers. Taking everything into consideration, that was the most impressive experiment and relatively probably the most costly experiment I have ever performed. . . . It was a revelation to me. It resulted in a desire on my part to learn more about that remarkable kind of action. Plainly, the only way to learn about it was to see its results, to experiment, to work in a laboratory [3].

EVOLUTION OF CHEMICAL CHARM IN THE CLASSROOM

I like to read Remsen's account to audiences while simultaneously performing the experiment. By putting the cent in a large beaker on an overhead projector, the color changes are prettily and clearly shown. The class reaction is always delightful and in complete support of the four laws stated above. I stop the reaction by adding water before the cent has fully dissolved, and then toss the thin disk back to the audience with the hope that it will later be handed around and discussed.

If you want to have a student exclaim "Oh, Wow!" through operation of the third law, try the following: first, inspect a few bits of granular aluminum using a stereo microscope at about 20×. Then, coat some granules with mercury by stirring them with a dilute solution of mercuric chloride in 0.1M HCl. Rinse the now shiny amalgamated granules and squeeze them between tissue to dry off the water. Immediately place them under the microscope and just watch the rapid and spectacular transformation to aluminum oxide "whiskers." Lots of charm there. And there are more good things to try with a microscope, especially precipitations. A fine example is to bring a small crystal of potassium dichromate just to the edge of a drop of 0.05M silver nitrate. When viewed under 20× magnification, the reaction is highly exocharmic.

Probably no one has mastered the art of charmodynamics more fully than Hubert Alyea, the world-famous wizard of molecular magic. If you have heard him sing the Princeton fight song during the progress of the "Old Nassau Reaction" [4], then you know what I mean. But if you can't or won't sing, try this variation: first, do the regular iodine clock reaction, telling it to "turn black" at the proper moment. Then call for two volunteers from the class, stating that you wish to test their inherent powers. Mix the solutions again, but this time use the "Old Nassau" mixture. Ask one student to command the change to black, and probably the command will be soft-spoken. Say, "Louder, please!", and if the timing is right the louder command will coincide with a change to *orange,* much to the amusement of the class. Then ask the second student to give a try at commanding a black color. This time, of course, it does turn black and you can observe that "some have it, and some have only part of it." Then discuss the chemistry.

If you follow these clock reactions with the oscillating clock reaction [5], which I consider the most charming demonstration ever published, you will have milked the second law to its fullest.

I once heard Bassam Shakhashiri make an eloquent plea to teachers to take more advantage of the opportunities for teaching through classroom demonstrations [6]. When he had successfully suspended some liquid oxygen between the poles of a magnet, the whole assembly perfectly displayed by an overhead projector, he said, *"That,* ladies and gentlemen, is paramagnetism! Not drawing little arrows up and down on the blackboard," and the audience burst into spontaneous applause. All four laws were working well, and one thought of the claim by Leonardo [7], "There is no higher or lower knowledge, but one only, flowing out of experimentation."

In an escape from the limitations of the classroom, I have found that an extra measure of charm can be transferred through "Backyard Experiments of the Week." Students who are interested troop out after class to an open area in back of the chemistry building. We have a very short discussion and then perform an experiment that is too risky to do inside. Examples include burning white phosphorus and watching the white cloud of the pentoxide; synthesizing zinc sulfide by igniting a stoichiometric mixture of zinc dust and powdered sulfur; comparing the reactivity of good-sized chunks of lithium, sodium, and potassium; and the photochemical chain reaction between chlorine and hydrogen using a balloon. In connection with the latter demonstration it is good to quote Gay-Lussac and Thenard (1808) as follows:

> We made new mixtures of hydrogen and chlorine gases and placed them in darkness, awaiting some moments of bright light. Two days after having made the mixtures we were able to expose them to the sun. Scarcely had they been exposed, when they suddenly enflamed with a very loud detonation, and the jars were reduced to splinters and projected to a great distance [8].

I usually let volunteers actually do the demonstration. It seems to me that these students will remember their freshman chemistry course more fondly because of these good times.

Sometimes, the quantity of evolved charm is unexpectedly greater than might be anticipated. One day I was letting hydrogen from a balloon stream out through a glass tip, the burning jet being directed upwards into a large beaker to show the condensation of the water formed by the combustion. A small leak developed by the mouth of the balloon, and suddenly I had a cubic foot of flame in front of me. The burning balloon settled onto my lecture notes and set them on fire, thus triggering sustained hilarity. Such enhancements of demonstrations can rarely be planned. As another example, I wanted to show the thermite reaction to the class, using magnesium ribbon as a starter. The pre-experiment discussion was finished, the lights were turned down, and one could sense the anticipation. The ribbon lit well, burned closer to the thermite, and then went out. What a letdown! In desperation, I told the class that I really wanted this reaction to go for them, and that I would use two strips of ribbon . . . *folded,* to make *four!* By this time, they were on the edges of their seats, and, when the mixture ignited with a shower of sparks, and they saw me leap back with fear, they cheered and applauded while the white-hot molten iron streamed out.

CONCLUSIONS

The point of all this is *not* that we should become clowns in the classroom, or that clever demonstrations should displace hard-core development of chemical principles. The point is that we teachers must realize, as Remsen did, that chemistry is relatively boring to read and work problems about, unless the student has some vivid mental images of the experimental side of the science. Good demonstrations not only spice up a class session, but they also help teach principles, and they help build up general experimental knowledge of a sort that makes chemistry seem less abstract. Teachers who feel that they have "been down that road" and are tired of demonstrations should remember that it is all new and interesting to their students. The teacher who does not take advantage of demonstrations is doing students a disservice.

Obviously, we teachers are in a highly responsible position when it comes to helping form attitudes toward science. Years later, our students will remember only small fragments from our course, but they definitely will have attitudes toward science. It may be that these attitudes will be unfavorable, for reasons that Tom Lippincott discussed in an editorial entitled "Why Students Hate Chemistry" [9]; it would be desirable, however, if our students retain a feeling that there is a certain amount of charm associated with the subject of chemistry.

REFERENCES

1. Partington, J. R. "A Short History of Chemistry," 3rd ed.; Harper Torchbooks: New York, 1960; p 184.
2. Weeks, M. E.; Leicester, H. M. "Discovery of the Elements," 7th ed.; Journal of Chemical Education: Easton, Pennsylvania, 1968.
3. Getman, F. H. "The Life of Ira Remsen"; Journal of Chemical Education: Easton, Pennsylvania, 1940; pp 9–10.
4. Alyea, H. N. *J. Chem. Educ.* **1977,** *54,* 167.
5. Briggs, T. S.; Rauscher, W. C. *J. Chem. Educ.* **1973,** *50,* 496.
6. Annual meeting of Midwest Association of Chemistry Teachers at Liberal Arts Colleges, Northfield, Minnesota, 1978.
7. Hildebrand, J. H. "Science in the Making," paperback ed.; Columbia University Press: New York, 1962; p 5.
8. Bozzelli, J. W.; Barat, R. B. *J. Chem. Educ.* **1979,** *56,* 675.
9. Lippincott, W. T. *J. Chem. Educ.* **1979,** *58,* 1.

Introduction

Bassam Z. Shakhashiri

Lecture demonstrations help to focus students' attention on chemical behavior and chemical properties, and to increase students' knowledge and awareness of chemistry. To approach them simply as a chance to show off dramatic chemical changes or to impress students with the "magic" of chemistry is to fail to appreciate the opportunity they provide to teach scientific concepts and descriptive properties of chemical systems. The lecture demonstration should be a process, not a single event.

In lecture demonstrations, the teacher's knowledge of the behavior and properties of the chemical system is the key to successful instruction, and the way in which the teacher manipulates chemical systems serves as a model not only of technique but also of attitude. The instructional purposes of the lecture dictate whether a phenomenon is demonstrated or whether a concept is developed and built by a series of experiments. Lecture experiments, which some teachers prefer to lecture demonstrations, generally involve more student participation and greater reliance on questions and suggestions, such as "What will happen if you add more of . . . ?" Even in a lecture demonstration, however, where the teacher is in full control of directing the flow of events, the teacher can ask the same sort of "what if" questions and can proceed with further manipulation of the chemical system. In principle and in practice, every lecture demonstration is a situation in which teachers can convey their attitudes about the experimental basis of chemistry, and can thus motivate their students to conduct further experimentation, and lead them to understand the interplay between theory and experiment.

Lecture demonstrations should not, of course, be considered a substitute for laboratory experiments. In the laboratory, students can work with the chemicals and equipment at their own pace and make their own discoveries. In the lecture hall, students witness chemical changes and chemical systems as manipulated by the teacher. The teacher controls the pace and explains the purposes of each step. Both kinds of instruction are integral parts of the education we offer students.

In teaching and in learning chemistry, teachers and students engage in a complex series of intellectual activities. These activities can be arranged in a hierarchy which indicates their increasing complexity [1]:

(1) observing phenomena and learning facts
(2) understanding models and theories
(3) developing reasoning skills
(4) examining chemical epistemology

This hierarchy provides a framework for the purposes of including lecture demonstrations in teaching chemistry.

At the first level, we observe chemical phenomena and learn chemical facts. For example, we can observe that, at room temperature, sodium chloride is a white crystalline solid and that it dissolves in water to form a solution with characteristic properties of its own. One such property, electrical conductivity, can be readily observed when two wire electrodes connected to a light bulb and a source of current are dipped into the solution. There are additional phenomena and facts that can be introduced: the white solid has a very high melting point; the substance is insoluble in ether; its chemical formula is NaCl; etc.

At the second level, we explain observations and facts in terms of models and theories. For example, we teach that NaCl is an ionic solid compound and that its aqueous solution contains hydrated ions: sodium cations, $Na^+(aq)$, and chloride anions, $Cl^-(aq)$. The solid, which consists of Na^+ and Cl^- particles, is said to have ionic bonds, that is, there are electrostatic forces between the oppositely charged particles. The ions are arranged throughout the solid in a regular three-dimensional array called a face-centered cube. Here, the teacher can introduce a discussion of the ionic bond model, bond energy, and bond distances. Similarly, a discussion of water as a molecular covalent substance can be presented. The ionic and covalent bonding models can be compared and used to explain the observed properties of a variety of compounds.

At the third level, we develop skills which involve both mathematical tools and logic. For example, we use equilibrium calculations in devising the steps of an inorganic qualitative analysis scheme. We combine solubility product, weak acid dissociation, and complex ion formation constants for competing equilibria which are exploited in analyzing a mixture of ions. The logical sequence of steps is based on understanding the equilibrium aspects of solubility phenomena.

At the fourth level, we are concerned with chemical epistemology. We examine the basis of our chemical knowledge by asking questions such as, "How do we know that the cation of sodium is monovalent rather than divalent?" and "How do we know that the crystal structure of sodium chloride can be determined from x-ray data?" At this level we deal with the limits and validity of our fundamental chemical knowledge.

Across all four levels, the attitudes and motivations of both teacher and student are crucial. The attitude of the teacher is central to the success of interactions with students. Our motivation to teach is reflected in what we do and, as well, in what we do not do, both in and out of the classroom. Our modes of communicating with students affect their motivation to learn. All aspects of our behavior influence students' confidence and their trust in what we say. Our own attitudes toward chemicals and toward chemistry itself are reflected in such matters as how we handle chemicals, adhere to safety regulations, approach chemical problems, and explain and illustrate chemical principles. In my opinion, the single most important purpose that lectures serve is to give teachers the opportunity to convey an attitude toward chemistry—to communicate to students an appreciation of chemistry's diversity and usefulness, its cohesiveness and value as a central science, its intellectual excitement and challenge.

PRESENTING EFFECTIVE DEMONSTRATIONS

In planning a lecture demonstration, I always begin by analyzing the reasons for presenting it. Whether a demonstration is spectacular or quite ordinary I undertake to use the chemical system to achieve specific teaching goals. I determine what I am going to say about the demonstration and at what stage I should say it. Prior to the lecture, I practice doing the demonstration. By doing the demonstration in advance, I often see aspects of the chemical change which help me formulate both statements and questions that I then use in class.

Because one of the purposes of demonstrations is to increase the students' ability to make observations, I try to avoid saying, "Now I will demonstrate the insolubility of barium sulfate by mixing equal volumes of 0.1M barium chloride and 0.1M sodium sulfate solutions." Instead, I say, "Let us mix equal volumes of 0.1M barium chloride and 0.1M sodium sulfate solutions and observe what happens." Rather than announcing what should happen, I emphasize the importance of observing all changes. Often, I ask two or three students to state their observations to the entire class before I proceed with further manipulations. In addition, I help students to sort out observations so that relevant ones can be used in formulating conclusions about the chemical system. Some valid observations may not be relevant to the main purpose of the demonstration. For example, when the above-mentioned solutions are mixed, students may observe that the volumes are additive. However, this observation is not germane to the main purpose of the demonstration, which is to show the insolubility of barium sulfate. However, this observation is relevant if the purposes include teaching about the additive properties of liquids.

Every demonstration that I present in lectures is aimed at enhancing the understanding of chemical behavior. In all cases, the chemistry speaks for itself more eloquently than anything I can describe in words, write on a chalk board, or show on a slide.

Wesley Smith of Ricks College, who was a visiting faculty member at the University of Wisconsin–Madison from 1974 to 1977, has outlined six characteristics of effective demonstrations which best promote student understanding [2]:

1. *Demonstrations must be timely and appropriate.* Demonstrations should be done to meet a specific educational objective. For best results, plan demonstrations that are immediately germane to the material in the lesson. Demonstrations for their own sake have limited effectiveness.

2. *Demonstrations must be well prepared and rehearsed.* To ensure success, you need to be thoroughly prepared. *All* necessary material and equipment should be collected well in advance so that they are ready at class time. You should rehearse the entire demonstration from start to finish. Do not just go through the motions or make a dry run. Actually mix the solutions, throw the switches, turn on the heat, and see if the demonstration really works. Only then will you know that all the equipment is present and that all the solutions have been made up correctly. Always practice your presentation.†

3. *Demonstrations must be visible and large-scale.* A demonstration can help only those students who experience it. Hence, you need to set up the effect for the whole class to see. If necessary, rig a platform above desktop level to ensure visibility.

Perhaps the most important factor to consider is the size of what you are presenting. Only in the very tiniest of classes can the students see phenomena on the milligram and milliliter scale. Many situations require the use of oversized glassware and specialized equipment. Solutions and liquids should be shown in full-liter volumes, and solids should be displayed in molar or multi-molar amounts.

Contrasting backgrounds help emphasize chemical changes. A collection of large white and black cards to place behind beakers and other equipment is a valuable addition to your demonstration equipment. These are inexpensive, easy to use, and can provide an extra bit of polish to your demonstration.

4. *Demonstrations must be simple and uncluttered.* A common source of distraction is clutter on the lecture bench. Make sure that the demonstration area is neat and free of extraneous glassware, scattered papers, and other disorder. All attention must be focused on the demonstration itself.

†As Fred Juergens likes to say, "Prior practice prevents poor presentation."

5. *Demonstrations must be direct and lively*. Action is an important part of a good demonstration. It is the very ingredient that makes demonstrations such efficient attention-grabbers. Students are eager to see something happen, but if nothing perceptible occurs within a few seconds you may lose their attention. The longer they have to wait for results, the less likely it is that the demonstration will have maximum educational value.

6. *Demonstrations must be dramatic and striking*. Usually, a demonstration can be improved by its mode of presentation. A lecture demonstration, according to Alfred T. Collette, is like a stage play. "A demonstration is 'produced' much as a play is produced. Attention must be given to many of the same factors as stage directors consider: visibility, audibility, single centers of attention, audience participation, contrasts, climaxes" [*3*]. The presentation of effective demonstrations is such an important part of good education that "no instructor is doing his best unless he can use this method of teaching to its fullest potential" [*4*].

USING THIS BOOK

The demonstrations in this volume are grouped in topical chapters dealing with thermochemistry, chemiluminescence, polymers, and metal ion precipitates and complexes. Each chapter has an introduction which covers the chemical background for the demonstrations that follow. We confine the discussion of relevant terminology and concepts to the introduction rather than repeating it in the discussion section of each demonstration. Accordingly, when teachers read the discussion section of any particular demonstration, they may find it necessary to refer to the chapter introduction for background information. For additional information teachers may wish to consult the sources listed at the end of each chapter's introduction.

Each demonstration has seven sections: a brief summary, a materials list, a step-by-step account of the procedure to be used, an explanation of the hazards involved, information on how to store or dispose of the chemicals used, a discussion of the phenomena displayed and principles illustrated by the demonstration, and a list of references. The brief summary provides a succinct description of the demonstration. The materials list for each procedure specifies the equipment and chemicals needed. Where solutions are to be used, we give directions for preparing stock amounts larger than those required for the procedure. The teacher should decide how much of each solution to prepare for practicing the demonstration and for doing the actual presentation. The availability and cost of chemicals may also affect decisions about the volumes to be prepared.

The procedure section often contains more than one method for presenting a demonstration. In all cases, the first procedure is the one the authors prefer to use. However, the alternative procedures are also effective and valid pedagogically.

The hazards and disposal sections include information compiled from sources believed to be reliable. We have enumerated many potentially adverse health effects and have called attention to the fact that many of the chemicals should be used only in well-ventilated areas. In all instances teachers should inquire about and follow local disposal practices and should act responsibly in handling potentially hazardous material. We recognize that several chemicals such as silver and mercury can be recovered and re-used and have given references to recovery and purification procedures.

The purpose of the discussion section is to provide the teacher with information for explaining each demonstration. We include discussion of chemical equations, relevant data, properties of the materials involved, as well as a theoretical framework through which the chemical processes can be understood. Again, we remind teachers that they should refer to the introduction of each chapter for background information not included in the discussion section of each demonstration. Finally, each demonstration contains a list of references used in developing procedures and providing information for the demonstration.

A WORD ABOUT SAFETY

Jearl Walker, professor of physics at Cleveland State University and editor of the Amateur Scientist section in *Scientific American,* has been quoted in newspaper stories as saying "The way to capture a student's attention is with a demonstration where there is a possibility the teacher may die." Walker is said to get the attention of his students by dipping his hand in molten lead or liquid nitrogen, or by gulping a mouthful of liquid nitrogen, or by lying between two beds of nails and having an assistant with a sledge hammer break a cinder block on top of him. Walker reportedly has been injured twice, once when he used a small brick instead of a cinder block in the bed-of-nails demonstration, and once when he walked on hot coals and was severely burned.

We disagree strongly with this kind of approach. Demonstrations that result in injury are likely to confirm beliefs that chemicals are dangerous and that their effects are bad. In fact, every chemical is potentially harmful if not handled properly. That is why every person who does lecture demonstrations should be thoroughly knowledgeable about the safe handling of all chemicals used in a demonstration and should be prepared to handle any emergency. A first-aid kit, a fire extinguisher, a safety shower, and a telephone must be accessible in the immediate vicinity of the demonstration area. Demonstrations involving volatile material, fumes, noxious gases, or smoke should be rehearsed and presented only in well-ventilated areas.

We recognize that any of the demonstrations in this book can be hazardous. Our procedures are written for experienced chemists who fully understand the properties of the chemicals and the nature of their behavior. We take no responsibility or liability for the use of any chemical or procedure specified in this book. We urge care and caution in handling chemicals and equipment.

REFERENCES

1. I have adapted many ideas from Paul Saltman's address at the Third Biennial Conference on Chemical Education which was sponsored by the American Chemical Society, Division of Chemical Education, and held at Pennsylvania State University, State College, Pennsylvania (1974); see *J. Chem. Educ.* **1975**, *52,* 25.
2. Chemical Demonstrations Proceedings, Western Illinois University and Quincy-Keokuk Section of the American Chemical Society, Macomb, Illinois (1978).
3. Collette, A. T. "Science Teaching in the Secondary School"; Allyn and Bacon: Boston, 1973.
4. Miller, R.; Culpepper, F. W., Jr. *Ind. Arts Voc. Educ.* **1971**, *60,* 24.

Sources Containing Descriptions of Lecture Demonstrations

We call attention to the following sources of information about lecture demonstrations. These lists are not intended to be comprehensive. Some of the books are out of print but may be available in libraries.

BOOKS

Alyea, H. N. "TOPS in General Chemistry," 3rd ed.; Journal of Chemical Education: Easton, Pennsylvania, 1967.

Alyea, H. N.; Dutton, F. B., Eds. "Tested Demonstrations in Chemistry," 6th ed.; Journal of Chemical Education, Easton, Pennsylvania, 1965.

Blecha, M. T., Ph.D. Dissertation, "The Development of Instructional Aids for Teaching Organic Chemistry," Kansas State University, Manhattan, Kansas, 1981.

"Chemical Demonstrations Proceedings"; Western Illinois University and Quincy-Keokuk Section of the American Chemical Society, Macomb, Illinois, May 5–6, 1978.

"Chemical Demonstrations Proceedings"; Western Illinois University and Quincy-Keokuk Section of the American Chemical Society, Macomb, Illinois, May 4–5, 1979.

"Chemical Demonstrations Proceedings"; Western Illinois University and Quincy-Keokuk Section of the American Chemical Society, Macomb, Illinois, May 1–2, 1981.

"Chemical Demonstrations Proceedings"; Fifth Annual Symposium at 16th Great Lakes Regional Meeting of the American Chemical Society, Normal, Illinois, June 8, 1982.

Chen, Philip S. "Entertaining and Educational Chemical Demonstrations"; Chemical Elements Publishing Co.: Camarillo, California, 1974.

Faraday, M. "The Chemical History of a Candle"; The Viking Press: New York, 1960.

Ford, L. A. "Chemical Magic"; T. S. Denison & Co.: Minneapolis, Minnesota, 1959.

Fowles, G. "Lecture Experiments in Chemistry," 5th ed.; Basic Books, Inc.: New York, 1959.

Frank, J. O.; assisted by Barlow, G. J. "Mystery Experiments and Problems for Science Classes and Science Clubs," 2nd ed.; J. O. Frank: Oshkosh, Wisconsin, 1936.

Freier, G. D.; Anderson, F. J. "A Demonstration Handbook for Physics," 2nd ed.; American Association of Physics Teachers: Stony Brook, New York, 1981.

Gardner, R. "Magic Through Science"; Doubleday & Co., Inc.: Garden City, New York, 1978.

Hartung, E. J. "The Screen Projection of Chemical Experiments"; Melbourne University Press: Carlton, Victoria, 1953.

Lippy, J. D., Jr.; Palder, E. L. "Modern Chemical Magic"; The Stackpole Co.: Harrisburg, Pennsylvania, 1959.

Meiners, H. F., Ed. "Physics Demonstration Experiments," Volume 1 and Volume 2; The Ronald Press Company: New York, 1970.

"My Favorite Lecture Demonstrations," A Symposium at the Science Teachers Short Course, W. Hutton, Chairman; Iowa State University, Ames, Iowa, March 6–7, 1977.

Newth, G. S. "Chemical Lecture Experiments"; Longmans, Green and Co.: New York, 1928.

Sharpe, S., Ed. "The Alchemist's Cookbook: 80 Demonstrations"; Shell Canada Centre for Science Teachers, McMaster University: Hamilton, Ontario, undated.

Siggins, B. A., M.S. Thesis, "A Survey of Lecture Demonstrations/Experiments in Organic Chemistry," University of Wisconsin–Madison, Madison, Wisconsin, 1978.

Walker, J. "The Flying Circus of Physics—With Answers"; Interscience Publishers, John Wiley and Sons: New York, 1977.

Wilson, J. W.; Wilson, J. W., Jr.; Gardner, T. F. "Chemical Magic"; J. W. Wilson: Los Alamitos, California, 1977.

ARTICLES

Bailey, P.S.; Bailey, C. A.; Anderson, J.; Koski, P. G.; Rechsteiner, C. *J. Chem. Educ.* **1975**, *52*, 524–25 (Producing a Chemistry Magic Show).

Bent, H. A.; Bent, H. E. *J. Chem. Educ.* **1980**, *57*, 609 (What Do I Remember? The Role of Lecture-Experiments in Teaching Chemistry).

Castka, J. F. *J. Chem. Educ.* **1975**, *52*, 394–95 (Demonstrations for High School Chemistry).

"Chem 13 News," **1976**, *81*. The November issue contained a collection of chemical demonstrations.

Gilbert, G. L., Ed. Regular column in *J. Chem. Educ.* since 1976 (Tested Demonstrations).

Hanson, R. H. *J. Chem. Educ.* **1976**, *53*, 577–78 (Chemistry is Fun, Not Magic).

Hughes, K. C. *Chem. in Australia* **1980**, *47*, 458–59 (Some More Intriguing Demonstrations).

McNaught, I. J.; McNaught, C. M. *School Sci. Review* **1981**, *62*, 655-66 (Stimulating Students with Colourful Chemistry).

Rada Kovitz, R. *J. Chem. Educ.* **1975**, *52*, 426 (The SSP Syndrome).

Schibeci, R. A.; Webb, J.; Farrel, F. *Chem. in Australia* **1980,** *47*, 246–47 (Some Intriguing Demonstrations).

Schwartz, A. T.; Kauffman, G. B. *J. Chem. Educ.* **1976**, *53*, 136–38 (Experiments in Alchemy, Part I: Ancient Arts).

Schwartz, A. T.; Kauffman, G. B. *J. Chem. Educ.* **1976**, *53*, 235–39 (Experiments in Alchemy, Part II: Medieval Discoveries and "Transmutations").

Shakhashiri, B. Z.; Dirreen, G. E.; Cary, W. R. in "Sourcebook for Chemistry Teachers"; Lippincott, W. T., Ed.; American Chemical Society, Division of Chemical Education: Washington, D.C., 1981; pp 3–16 (Lecture Demonstrations).

Sources of Information
on Hazards and Disposal

In preparing the Hazards and Disposal sections, we used the following references. The order of listing reflects our degree of utilization.

Muir, G. D., Ed. "Hazards in the Chemical Laboratory," 2nd ed.; The Chemical Society: London, 1977. A third edition, edited by L. Bretherick, was published in 1981.

Windholz, M., Ed. "The Merck Index," 9th ed.; Merck and Co.: Rahway, New Jersey, 1976.

"Prudent Practices for Handling Hazardous Chemicals in Laboratories." *Natl. Res. Counc.* **1981**; Committee on Hazardous Substances in the Laboratory.

"Laboratory Waste Disposal Manual"; Manufacturing Chemists Association: 1975. This book is out of print, but some of the information was available in the 1981 Reagent Catalog of MCB Manufacturing Chemists, Inc., 2909 Highland Avenue, Cincinnati, Ohio 45212.

"Registry of Toxic Effects of Chemical Substances"; Dept. of Health, Education and Welfare (NIOSH): Washington, D.C., revised annually. Available from Superintendent of Documents, U.S. Government Printing Office, Washington, D.C. 20402.

"Fire Protection Guide on Hazardous Materials," 6th ed.; National Fire Protection Association: 470 Atlantic Ave., Boston, Massachusetts 02210, 1975. New editions are published at intervals.

"Safety in Academic Chemistry Laboratories," 3rd ed.; American Chemical Society Committee on Chemical Safety: Washington, D.C., 1979. A fourth edition is scheduled for publication. The bibliography lists many journal articles and books.

"Health and Safety Guidelines for Chemistry Teachers"; American Chemical Society Dept. of Educational Activities: Washington, D.C., 1979. The bibliography lists journal articles and books.

Steere, N. V. "Handbook of Laboratory Safety," 2nd ed.; CRC Press: Cleveland, Ohio, 1971.

"Guide for Safety in the Chemical Laboratory," 2nd ed.; Van Nostrand Reinhold Co., Litton Educational Publishing, Inc.: New York, 1972.

Steere, N. V., Ed., "Safety in the Chemical Laboratory"; Journal of Chemical Education: Easton, Pennsylvania, Vol. 1, 1967; Vol. 2, 1971; Vol. 3, 1974.

Renfrew, M. M., Ed. "Safety in the Chemical Laboratory"; Journal of Chemical Education: Easton, Pennsylvania, Vol. 4, 1981.

Sax, N. I. "Dangerous Properties of Industrial Materials," 3rd ed.; Van Nostrand Reinhold Co., Litton Educational Publishing, Inc.: New York, 1968.

Chemical Demonstrations

Volume 1

1

Thermochemistry

George L. Gilbert, Lloyd G. Williams,
Bassam Z. Shakhashiri, Glen E. Dirreen,
and Frederick H. Juergens

Chemical reactions are accompanied by changes in energy. When a chemical system releases heat to the surroundings, the reaction is said to be *exothermic*. When a chemical system absorbs heat from the surroundings, the reaction is said to be *endothermic*. Exothermic reactions sometimes release light energy and sound energy and sometimes produce fire and smoke.

We assume that teachers know the principles of thermochemistry and can refresh such knowledge by referring to physical chemistry textbooks. We follow the usual thermodynamic conventions in using changes in enthalpy values, ΔH. Enthalpy is a state function and, at constant pressure, ΔH is equal to the heat flow, Q. Thus, under constant pressure conditions, the thermochemical heat quantity, Q, is equal to the change in the thermodynamic state function, ΔH. When all reactants and products are in their standard state (that is, their most stable form under pressure of 1 atmosphere and at 25°C), the standard enthalpy change, $\Delta H°$, refers to the difference between the sum of enthalpies of the products and the sum of enthalpies of the reactants. In all cases we have expressed ΔH changes in terms of joules (J) or kilojoules (kJ) by multiplying tabulated values by the conversion factor: 1 calorie = 4.184 J.

Spontaneous chemical reactions occur when the free energy change for the system, ΔG, is negative at constant temperature and pressure. Under these conditions the free energy change is

$$\Delta G = \Delta H - T\Delta S$$

where ΔH is the enthalpy change,
   T is the temperature in degrees Kelvin, and
   ΔS is the entropy change.
The relative magnitude of ΔH and ΔS determines whether or not ΔG is negative. For an exothermic reaction, ΔH is negative and ΔS may be either positive or negative. If ΔS is positive, then ΔG must be negative; if ΔS is negative *and* the term $T\Delta S$ is larger than ΔH, then ΔG must be negative. For an endothermic reaction, ΔH is positive and ΔS may be either positive or negative. If ΔS is positive and the term $T\Delta S$ is larger than ΔH, ΔG must be negative; otherwise, ΔG must be positive.

In this chapter we include chemical systems in which heat is the main form of energy flowing between a system and its surroundings. A few of the reactions are endothermic, but most are exothermic. We have included several examples of exothermic reactions that are accompanied by the release of smoke and flames and a few examples of explosive reactions. In all exothermic reactions, the main characteristic is the release of heat energy. However, the rate of each reaction, the rate of energy release, and the properties of the products determine whether or not smoke, flames, or loud sounds are emitted. If the rate of energy release is very high, an explosion usually occurs along with a flash of light and a loud noise.

The demonstrations are arranged in three main categories. Demonstrations 1.1 through 1.14 take place in solution or involve liquids, 1.15 through 1.37 are solid phase or heterogeneous mixture reactions, and 1.38 through 1.44 involve explosions. At the end of the chapter we list titles of nine demonstrations whose use we do not recommend because they are extremely dangerous.

We wish to emphasize that detailed explanations of several phenomena described in this chapter are not known. This is particularly true for the solid phase and heterogeneous mixture reactions. In practically all such reactions, the mechanism by which reactants change to products is not known, and yet each reaction is characterized by evolution of heat and other forms of energy. For many of these systems, we have conducted both literature and experimental research and have included our findings. We invite readers to share with us their knowledge and findings about these systems so that we can update the information in a future revision of this chapter.

Teachers are urged to call the attention of students to two familiar and simple exothermic reactions. The first is the striking of a match. Although their composition differs, either a safety match or a strike-anywhere (SAW) match can be used. The cause and nature of combustion can be observed and discussed. The second exothermic reaction is the burning of a candle. We refer you to informative and delightful readings about candles [1, 2].

REFERENCES

1. Faraday, M. "The Chemical History of a Candle"; The Viking Press: New York, 1960.
2. Young, J. A. "Chemistry, A Human Concern"; Macmillan Company: New York, 1978; pp 23–41.

1.1

Evaporation as an Endothermic Process

An aspirator reduces the pressure in a boiling flask containing acetone. The acetone then boils, and its temperature decreases.

MATERIALS

acetone, ca. 500 ml

thermometer, $-10°C$ to $+110°C$

2-hole rubber stopper to fit boiling flask

3-way stopcock

1-liter thick-walled round-bottomed boiling flask

boiling chips

cork ring

water aspirator, with trap and vacuum tubing

PROCEDURE

Insert the thermometer in one hole of the 2-hole stopper and insert one arm of the 3-way stopcock in the other. Fill the flask approximately one-half full of acetone, add several boiling chips, and insert the stopper into the neck of the flask. The bulb of the thermometer should be below the surface of the liquid (see figure). Place the flask on the cork ring and read the temperature of the liquid.

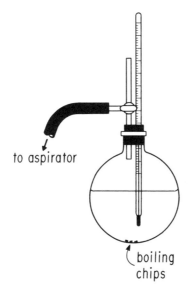

to aspirator

boiling
chips

Attach the vacuum hose from the trap to one of the open arms of the stopcock. Connect the trap to the aspirator. Turn the aspirator to maximum water flow and adjust the stopcock so that air is drawn from the flask. When the pressure inside the flask is low enough, the acetone begins to boil. As it boils, the temperature of the liquid decreases. A temperature of less than 10°C can be readily obtained. Moisture may condense on the outside of the flask.

Before disconnecting the assembly, turn the stopcock to allow air into the flask and turn off the water aspirator.

HAZARDS

Since acetone is flammable, this demonstration should be performed away from flames or other sources of heat.

The flask may implode if the pressure inside is lowered sufficiently. This risk can be minimized by using a thick-walled boiling flask rather than a standard round-bottomed flask.

DISPOSAL

The acetone should be poured into a suitable container for used solvents or flushed down the drain with water.

DISCUSSION

This demonstration illustrates that evaporation is an endothermic process. The temperature of the liquid decreases as it evaporates under reduced pressure (see table). The normal boiling point of acetone is 56.5°C.

Vapor Pressure of Acetone
at Different Temperatures [1]

Vapor pressure (mm Hg)	T (°C)
760	56.5
400	39.5
100	7.7
40	−9.4
10	−31.1
1	−59.4

With tight connections, the pressure inside the flask can be reduced to the vapor pressure of the water at the tap water temperature. For example, when the water temperature is 20°C, its vapor pressure is 17.5 mm Hg [2]. This is the lowest pressure that can be achieved by means of the aspirator at this temperature.

The heat or enthalpy of vaporization can be calculated from the table and from the relationship: $\log P = -\dfrac{\Delta H_{vap}}{2.3RT} + C$

where P is the vapor pressure in mm of Hg,

ΔH_{vap} is the heat or enthalpy of vaporization in J/mole,

R is the molar gas constant 8.31 J/K/mole, and

T is the temperature in K.

A plot of log P vs 1/T yields a straight line whose slope is equal to $-\dfrac{\Delta H}{2.3R}$.

Although other volatile liquids can be used in this demonstration, we do not recommend the use of ether [3] because of its flammability and the risk of releasing it into the room. Acetone is also flammable, but we prefer it because of its water solubility.

REFERENCES

1. Weast, R. C., Ed. "CRC Handbook of Chemistry and Physics," 54th ed.; CRC Press: Cleveland, Ohio, 1973; p D-172.
2. Weast; p D-159.
3. Alyea, H. N. *J. Chem. Educ.* **1970**, *47*, A387.

1.2

Chemical Cold Pack

First-aid cold packs demonstrate spontaneous endothermic reactions.

MATERIALS

Instant cold packs can be obtained from medical supply outlets. We use a Kwik Kold Instant Ice Pack [1] which contains ammonium nitrate and water.

PROCEDURE

The Kwik Kold Instant Ice Pack consists of an outer pouch containing solid ammonium nitrate and an inner pouch containing water with a blue dye. Firmly squeeze the inner pouch to release the water and mix it with the ammonium nitrate. Since the mixture is well sealed in the outer pouch, the pack can be passed around the class.

HAZARDS

Since the cold pack reaches temperatures significantly lower than body temperature, prolonged application to the body should be avoided.

This mixture of chemicals is not harmful. Ammonium nitrate is a very strong oxidant in dry form and can initiate combustion in flammable materials. If the outer pouch breaks, its contents should be flushed down the drain with water. Spills should be absorbed with a cloth or sponge, which should then be thoroughly rinsed with water.

DISPOSAL

The sealed pack should be discarded in a waste container. If the pack breaks, its contents should be flushed down the drain with water.

DISCUSSION

When the inner pouch containing the water is broken, the ammonium nitrate dissolves. This process absorbs heat from the materials present and causes the temperature of the system to drop. The reaction is

$$NH_4NO_3(s) \longrightarrow NH_4^+(aq) + NO_3^-(aq)$$

This reaction absorbs 26.2 kJ/mole or approximately 326 J/g of ammonium nitrate [2]. The standard heat of formation values, ΔH_f°, for $NH_4NO_3(s)$ and $NH_4NO_3(aq)$ are -365.6 kJ/mole and -339.4 kJ/mole, respectively [2]. Recorded values for the solubility of ammonium nitrate vary sharply, ranging from 118.3 g/100 ml [3, 4] to 50 g/100 ml [5] in water at 0°C. Our observations are consistent with the lower value. In the commercial cold pack system used, 214 g of NH_4NO_3 were present in the outer pouch and 220 ml of water in the inner pouch. We mixed these in a Dewar flask with a heat capacity of 42 J/deg and obtained a temperature of -7°C. Significant amounts of undissolved solid remained.

REFERENCES

1. Kwik Kold Division of Kay Laboratories, Inc., Moberly, Missouri 65270; patent numbers 2 925 719 and 3 175 558.
2. "Selected Values of Chemical Thermodynamic Properties." *Natl. Bur. Stand. (U.S.)* **1965**; Technical Note 270-1; p 67.
3. Weast, R. C., Ed. "CRC Handbook of Chemistry and Physics," 49th ed.; CRC Press: Cleveland, Ohio, 1968; pp D-38, B-176.
4. Kaufman, J. A. *J. Chem. Educ.* **1970**, *47*, 518.
5. Linke, W. F. "Solubilities of Inorganic and Metal Organic Compounds," 4th ed.; American Chemical Society: Washington, D.C., 1965; Vol. II, pp 708–9.

1.3

Endothermic Reactions of Hydrated Barium Hydroxide and Ammonium Salts

Mixing two dry solids, barium hydroxide and one of several ammonium salts, demonstrates an endothermic reaction. The odor of ammonia is emitted from the flask, a noticeable amount of liquid forms, and the flask becomes cold. Within a minute or two, the temperature of the mixture drops about 45°C below room temperature.

MATERIALS

32 g barium hydroxide octahydrate, $Ba(OH)_2 \cdot 8H_2O$

11 g ammonium chloride, NH_4Cl, or 17 g ammonium nitrate, NH_4NO_3

125-ml or 250-ml Erlenmeyer flask

thermometer, $-38°C$ to $+50°C$

small block of wood, ca. 15 cm $\times$ 15 cm $\times$ 2 cm

PROCEDURE

Place the pre-weighed amounts of solid barium hydroxide and one of the ammonium salts in the Erlenmeyer flask. Shake the flask gently to mix the reagents.

Within about 30 seconds, the odor of ammonia can be detected from the flask, and a noticeable amount of liquid forms. The flask becomes cold to the touch, and, in a humid environment, frost collects on the outside. The temperature of the mixture drops from room temperature (22°C) to about $-25°C$ or $-30°C$ within 1–2 minutes after mixing, and remains below $-20°C$ for several minutes.

The flask can be passed around the class to allow students to observe the products of the reaction and feel the cold flask. When students check for the presence of ammonia vapor, they should be warned to avoid inhaling large amounts of the pungent gas.

The temperature change can be measured directly with a thermometer. In addition, to display dramatically the decrease in temperature, wet the small wooden block with a few drops of water and set the reaction flask on it. The flask will quickly freeze to the board and both can be lifted together off the table [1].

HAZARDS

Since the reaction produces temperatures significantly lower than body temperature, the flask should be handled with care and prolonged contact with the skin should be avoided.

Soluble barium salts are poisonous if ingested. Upon contact with the skin, barium and ammonium salts may produce minor irritations or allergic reactions. If the flask is spilled or broken, its contents should be flushed down the drain with water.

Inhalation of concentrated ammonia vapor causes edema of the respiratory tract, spasm of the glottis, and asphyxia.

DISPOSAL

The contents of the flask should be flushed down the drain with water.

DISCUSSION

Ammonium thiocyanate has been used with barium hydroxide to demonstrate endothermic chemical rections [1–3]. However, the hygroscopicity of ammonium thiocyanate and the lack of thermodynamic data make this combination less preferable than the combination of barium hydroxide and either ammonium chloride or ammonium nitrate. The reaction between barium hydroxide and ammonium thiocyanate is [1–4]

$$Ba(OH)_2 \cdot 8H_2O(s) + 2\ NH_4SCN(s) \longrightarrow$$
$$Ba(SCN)_2(aq) + 2\ NH_3(aq) + 10\ H_2O(l) \qquad (1)$$

In the reactions between $Ba(OH)_2 \cdot 8H_2O$ and either ammonium nitrate or ammonium chloride, the barium salts that are formed have limited solubility [5]:

$$Ba(OH)_2 \cdot 8H_2O(s) + 2\ NH_4NO_3(s) \longrightarrow$$
$$Ba(NO_3)_2(s) + 2\ NH_3(aq) + 10\ H_2O(l) \qquad (2)$$

$$Ba(OH)_2 \cdot 8H_2O(s) + 2\ NH_4Cl(s) \longrightarrow$$
$$BaCl_2 \cdot 2H_2O(s) + 2\ NH_3(aq) + 8\ H_2O(l) \qquad (3)$$

These reactions display endothermic processes and illustrate the interaction between changes in enthalpy and entropy in spontaneous chemical reactions. For a process to take place spontaneously at constant temperature and pressure, the change in free energy (ΔG), calculated from the relation $\Delta G = \Delta H - T\Delta S$, must be negative. An endothermic reaction may thus be spontaneous at constant pressure if the positive value of ΔH (corresponding to heat absorbed) is offset by a sufficient increase in entropy (randomness). In the reactions between $Ba(OH)_2 \cdot 8H_2O$ and each of the solid ammonium salts, the large increase in entropy is related to the increase in the number of particles present and their states. Here, two solid species form a new solid as well as aqueous ammonia and water.

Thermochemical data for barium hydroxide, ammonium salts, and other relevant substances are listed in the table.

Thermochemical Data for Barium Hydroxide, Ammonium Salts, and Other Substances [6]

Substance	ΔH_f° (kJ/mole)	S_{298}°	$C_p^\circ(298)$ (J/mole·K)	ΔG_f° (kJ/mole)
$Ba(OH)_2 \cdot 8H_2O(s)$	-3342	427	—	-2793
$Ba(NO_3)_2(s)$	-992.07	214	151.4	-796.72
$BaCl_2 \cdot 2H_2O(s)$	-1460.1	203	162.0	-1296.5
$NH_4NO_3(s)$	-365.6	151.1	139	-184.0
$NH_4Cl(s)$	-314.4	94.6	84.1	-203.0
$H_2O(l)$	-285.83	69.91	75.291	-237.2
$NH_3(aq)$	-80.29	111	—	-26.6

From the data in the table, the values of the standard enthalpy, entropy, and free energy changes for reactions 2 and 3 are calculated:

	ΔH_{298}° (kJ)	ΔS_{298}° (J/K)	ΔG_{298}° (kJ)
Reaction 2	$+62.3$	406	-60.2
Reaction 3	$+63.6$	368	-47.7

For reaction 1, using ammonium thiocyanate, the corresponding calculations cannot be made because the values of ΔH_f° and S_{298}° for $Ba(SCN)_2(aq)$ and the value of S_{298}° for $NH_4SCN(s)$ are not available.

REFERENCES

1. Hambly, A. N. *J. Chem. Educ.* **1969**, *46*, A55.
2. Herstein, K. M. In "Inorganic Synthesis," Andrieth, L. F., Ed; McGraw-Hill: New York, 1950; Vol. III, p 24.
3. Burt, N. E. *J. Chem. Educ.* **1974**, *51*, A178.
4. Harris, A. D. *J. Chem. Educ.* **1979**, *56*, 477.
5. Seidell, A. "Solubilities of Inorganic and Metal Organic Compounds," 3rd ed.; Van Nostrand: New York, 1940; Vol. I, pp 154, 169.
6. "Selected Values of Chemical Thermodynamic Properties." *Natl. Bur. Stand. (U.S.)* **1965**, **1971**; Technical Note 270-1, pp 13, 62, 67, 71; Technical Note 270-6, pp 79, 82, 89.

1.4

The Nonburning Towel

After an ordinary cotton towel is immersed in a solution of alcohol and water and then held over a lighted burner, a blue flame surrounds the towel without burning it.

MATERIALS

50 ml isopropyl alcohol, $CH_3CHOHCH_3$

50 ml distilled water

1-liter beaker

cloth towel, ca. 20 cm $\times$ 40 cm

Bunsen burner

tongs

PROCEDURE

Mix 50 ml of alcohol and 50 ml of distilled water in the beaker. Immerse the dry towel in the alcohol-water mixture and thoroughly wet the towel. Squeeze the towel to remove excess liquid. Light the burner. Hold the towel at the center with crucible tongs. At arm's length, move the towel over the burner flame and allow the towel to catch on fire. Continue to hold the flaming towel with an extended arm to minimize any chance of burns.

As the flame subsides, snuff it out with a quick jerk to avoid catching the towel on fire.

HAZARDS

To avoid burns the flaming towel must be handled with care. Combustible materials should be removed from the vicinity, and a fire extinguisher should be available.

Isopropyl alcohol can damage the eyes severely.

DISPOSAL

The excess alcohol-water mixture can be saved for a future demonstration or flushed down the drain with water.

DISCUSSION

The combustion reaction of isopropyl alcohol is

$$2\ C_3H_7OH(l)\ +\ 9\ O_2(g) \longrightarrow 6\ CO_2\ (g)\ +\ 8\ H_2O(g)$$

The heat of reaction [1] for this equation is -1987 kJ/mole, which is sufficient to cause the combustion of the towel. This can be shown by using 50 ml of pure alcohol in place of the alcohol-water mixture. The function of the water is to absorb some of the energy of combustion as the water heats to its boiling point and is vaporized. The energy available from the combustion of 50 g of isopropyl alcohol (0.833 moles) is approximately 1659 kJ [2]. Heating 50 g of water from 20°C to 100°C requires 17 kJ, and the added step of vaporizing the water requires 113 kJ [3], making a total of 130 kJ absorbed in these processes.

Clearly, most of the heat produced by the combustion of the alcohol is lost to the surrounding air with modest amounts used in the vaporization of the alcohol. This vaporization is visible in the extensive flame surrounding the towel as combustion proceeds.

Attempts to reduce the ratio of water to alcohol lead to some charring of the cloth, clear evidence that the cooling function of the water is barely sufficient to cool the cloth in the ratio specified.

Other combustible materials, such as paper, can be used to demonstrate the same phenomenon [4]. Other low-boiling alcohols such as methanol and ethanol can be substituted for the isopropyl alcohol. We do not recommend using a mixture of carbon disulfide and carbon tetrachloride.

REFERENCES

1. Weast, R. C., Ed. "CRC Handbook of Chemistry and Physics," 49th ed.; CRC Press: Cleveland, Ohio, 1968; p D-184.
2. Weast; p D-186.
3. Weast; p A-10.
4. Jardin, J.; Murray, P.; Tyszka, J.; Czarnecki, J. *J. Chem. Educ.* **1978**, *55*, 655.

1.5

Heat of Neutralization

The addition of three pairs of liquids to three beakers yields varying temperatures. This variation illustrates the difference in the effects of dilution and neutralization.

MATERIALS

300 ml distilled water

200 ml 6M hydrochloric acid, HCl

200 ml 6M sodium hydroxide, NaOH

thermometer, $-10°C$ to $+110°C$

3 300-ml beakers (tall-form preferable)

3 100-ml graduated cylinders

3 stirring rods

PROCEDURE

Place the thermometer in a 300-ml beaker, add 100 ml of distilled water, and record the initial temperature. Then add 100 ml of 6M HCl, stir, and record the temperature change. In 2 minutes the temperature will typically rise about 3°C.

Remove the thermometer and rinse it with distilled water. Place the thermometer in an empty 300-ml beaker, add 100 ml of distilled water, and then add with stirring 100 ml of 6M NaOH. The temperature will increase about 1°C in 2 minutes.

Briefly rinse the thermometer again and place it in a third beaker. Simultaneously add 100 ml of 6M NaOH and 100 ml of 6M HCl. This time the temperature increase will be much greater, approximately 45°C.

HAZARDS

Either hydrochloric acid or sodium hydroxide solutions may cause severe burns. Hydrochloric acid vapors are extremely irritating to the skin, eyes, and respiratory system. Dust from solid sodium hydroxide is very caustic.

DISPOSAL

The solutions produced should be flushed down the drain with water.

DISCUSSION

The addition to water of relatively concentrated acids or bases, such as 6M HCl (reaction 1) or 6M NaOH (reaction 2), leads to the evolution of heat because of these reactions:

$$H^+(aq)_{6M} + Cl^-(aq)_{6M} \longrightarrow H^+(aq)_{3M} + Cl^-(aq)_{3M} \tag{1}$$

$$Na^+(aq)_{6M} + OH^-(aq)_{6M} \longrightarrow Na^+(aq)_{3M} + OH^-(aq)_{3M} \tag{2}$$

Standard enthalpy, ΔH_f°, values are tabulated [1] for moles of H_2O per mole of solute. For any solution, for example 6M HCl, we must first use its density [2] to calculate moles of H_2O per mole of solute in 1 liter:

mass of 1 liter of solution = $(1.10 \text{ g/ml})(1000 \text{ ml}) = 1100$ g

mass of 6 moles HCl = $(6 \text{ mole})(36.46 \text{ g/mole}) = 218.8$ g

mass of water in 1 liter of solution = $(1100 \text{ g} - 218.8 \text{ g}) = 881$ g

$$\text{moles of water in 1 liter of solution} = \frac{881 \text{ g } H_2O}{18.02 \text{ g } H_2O/\text{mole } H_2O} = 48.9 \text{ mole } H_2O$$

$$\frac{48.9 \text{ mole } H_2O}{6 \text{ mole HCl}} = \frac{8.15 \text{ mole } H_2O}{\text{mole HCl}}$$

Using the thermochemical data provided in the table, we calculate the energy changes associated with the dilutions in reactions 1 and 2 to be -2.78 kJ/mole and -0.306 kJ/mole, respectively. The observed temperature changes correspond to these values.

Enthalpies for HCl, NaOH, and NaCl [1, 2]

Substance	Density (g/ml)	$\dfrac{\text{Moles } H_2O}{\text{mole solute}}$	ΔH_f° (kJ/mole)
6M HCl	1.10	~8	-161.80
3M HCl	1.05	~17	-164.58
6M NaOH	1.22	~9	-469.227
3M NaOH	1.12	~18	-469.533
3M NaCl	1.12	~17	-409.09

The neutralization reaction between hydrogen ions (H^+) and hydroxide ions (OH^-) is significantly more exothermic than the first two reactions. For the reaction

$$H^+(aq)_{6M} + OH^-(aq)_{6M} \longrightarrow H_2O(l)$$

the energy change is -63.9 kJ/mole. The corresponding observed temperature change reflects this large energy value.

REFERENCES

1. "Selected Values of Chemical Thermodynamic Properties." *Natl. Bur. Stand. (U.S.)* **1952**; Circ. 500; pp 23, 448, 451.
2. Weast, R. C., Ed. "CRC Handbook of Chemistry and Physics," 59th ed.; CRC Press: Cleveland, Ohio, 1978–79; pp D-278, D-300, D-303.

1.6

Heat of Dilution
of Sulfuric Acid

After various quantities of concentrated sulfuric acid are added to 100-ml aliquots of water, the temperature increases are correlated with the additions of sulfuric acid.

MATERIALS

300 ml distilled water

60 ml concentrated (18M) sulfuric acid, H_2SO_4

3 9-oz (ca. 270-ml) styrofoam cups

3 400-ml beakers

3 thermometers, $-10°C$ to $+110°C$

3 50-ml graduated cylinders

100-ml graduated cylinder

3 stirring rods

rubber gloves

PROCEDURE

Place a styrofoam cup inside each beaker. (The beaker provides a stable base and a receptacle if the cup breaks.) Add 100 ml of distilled water to each cup and measure the water temperature.

Add 10 ml of 18M H_2SO_4 to the first cup, briefly stir the mixture with a stirring rod, and measure the maximum temperature attained.

Repeat the addition of acid to the second and third cups using 20 ml and 30 ml of acid, respectively. Again, record the maximum temperatures attained.

HAZARDS

Since concentrated sulfuric acid is a strong acid and a powerful dehydrating agent, it must be handled with great care. Spills should be neutralized with an appropriate agent, such as sodium bicarbonate ($NaHCO_3$), and then wiped up.

The reactions produce sufficient heat to cause burns.

DISPOSAL

The acidic solutions should be neutralized with a solution of sodium hydroxide (NaOH) and then flushed down the drain with water.

DISCUSSION

The dilution of concentrated H_2SO_4 with water releases considerable energy. The important reaction can be written

$$H_2SO_4(\text{conc.}) + n\, H_2O \longrightarrow H_2SO_4(\text{dil.})$$

where n = moles H_2O/mole H_2SO_4.

Concentrated sulfuric acid (18M) contains 95–98% H_2SO_4 and has a density of 1.84 g/ml at 25°C. Its heat capacity is 1.4 J/g·K. When 10 ml of 18M H_2SO_4 (taken to be 96% H_2SO_4) are added to 100 ml of water, the heat of dilution is -11 kJ. One ml of concentrated H_2SO_4 contains approximately 0.07 g of water and 1.77 g of H_2SO_4 which corresponds to a mole ratio, n, of 0.2. The heat of formation of H_2SO_4 in 0.2 moles of water, obtained by graphical interpolation [1], is -822.6 kJ/mole. Since 10 ml of 18M H_2SO_4 contain 0.18 moles of H_2SO_4 and since 100 ml of water correspond to 5.55 moles of H_2O, the value of n for this solution is 31 and the heat of formation of the solution is -884.00 kJ/mole [1]. The heat of dilution is thus

$$\Delta H_{rxn} = \Delta H_f^\circ\, [H_2SO_4(\text{aq}) \text{ in } 31\ H_2O] - \Delta H_f^\circ\, [H_2SO_4(\text{aq}) \text{ in } 0.2\ H_2O] =$$

$$-884.00 - (-822.6) = -61.4 \text{ kJ/mole}$$

Since 0.18 mole of H_2SO_4 is used, the enthalpy change is -11 kJ. The calculated values of the heats of dilution for the three samples are

H_2SO_4 (ml)	H_2O (ml)	n	ΔH_{rxn} (calculated) (kJ)	Typical ΔT values (°C)
10	100	31	-11	25
20	100	15.6	-26	48
30	100	10.5	-30	70

The increase in temperature is thus related to the quantity of sulfuric acid used.

REFERENCE

1. "Selected Values of Chemical Thermodynamic Properties." *Natl. Bur. Stand. (U.S.)* **1952**; Circ. 500; pp 41–42.

1.7

Reaction of Calcium Oxide and Water (Slaking of Lime)

Water added to a beaker half full of lumps of quicklime produces steam and a temperature increase.

MATERIALS

100 g calcium oxide, CaO (in lumps; must be fresh)

250-ml beaker

25 ml distilled water

thermometer, $-10°C$ to $+110°C$

rubber gloves

PROCEDURE

Wearing gloves, place the lumps of calcium oxide in the beaker. Insert the thermometer into the lumps so it touches the bottom of the beaker and note temperature. Carefully add 25 ml of distilled water to the lumps. Observe the evolution of steam and rapid temperature increase. The temperature rises to about 95°C within seconds and to almost 115°C in about 5 minutes. The volume of the solid material increases noticeably.

HAZARDS

Calcium oxide (quicklime) readily absorbs water, generating considerable heat. Since it is very caustic, it can severely irritate the skin, eyes, and respiratory system. The product of the reaction, calcium hydroxide (slaked lime), is also very caustic. A saturated solution at 25°C has a pH of 12.4 [1].

DISPOSAL

The reaction products should be flushed down the drain with water.

DISCUSSION

Slaked lime, $Ca(OH)_2$, is formed by the reaction of quicklime, CaO, and water and is accompanied by a threefold increase in volume. Because of this expansion rate, building contractors must be careful to keep their lime supplies dry.

The standard heat of formation of calcium oxide is -635.09 kJ/mole, and the standard heats of formation of $Ca(OH)_2(s)$ and $Ca(OH)_2(aq)$ are -986.09 kJ/mole and -1002.82 kJ/mole, respectively [2]. The standard heat of hydration of calcium oxide, ΔH°_{hyd}, can be calculated from these values and from the standard heat of formation of water (for water, $\Delta H^\circ_f = -285.84$ kJ/mole):

$$CaO(s) + H_2O(l) \longrightarrow Ca(OH)_2(s) \qquad \Delta H^\circ_{hyd} = -65.16 \text{ kJ/mole}$$

$$CaO(s) + H_2O(l) \longrightarrow Ca(OH)_2(aq) \qquad \Delta H^\circ_{hyd} = -81.89 \text{ kJ/mole}$$

The solubility of $Ca(OH)_2$ in water decreases as the temperature increases: 0.0218 moles/1000 g H_2O at 20°C and 0.0090 moles/1000 g H_2O at 100°C [3].

REFERENCES

1. Windholz, M., Ed. "The Merck Index," 9th ed.; Merck and Co.: Rahway, New Jersey, 1976; p 212.
2. "Selected Values of Chemical Thermodynamic Properties." *Natl. Bur. Stand. (U.S.)* **1971**; Technical Note 270-6; pp 30–31.
3. Washburn, E. W., Ed. "International Critical Tables of Numerical Data, Physics, Chemistry and Technology"; McGraw-Hill: New York, 1929; Vol. IV, p 229.

1.8

Heat of Solution
of Lithium Chloride

The dissolution of lithium chloride in water is accompanied by the release of heat.

MATERIALS

42 g lithium chloride, LiCl

50 ml distilled water

2 9-oz (ca. 270-ml) styrofoam cups

thermometer, $-10°C$ to $+110°C$

PROCEDURE

Place the two cups inside each other. Place 42 g of lithium chloride in the inner cup and add 50 ml of distilled water. Stir and measure the temperature. The temperature will increase about 65° above room temperature.

HAZARDS

The reaction produces sufficient heat to cause burns.

DISPOSAL

The solution should be flushed down the drain with water.

DISCUSSION

The standard heat of formation of crystalline lithium chloride is -408.8 kJ/mole [1]:

$$Li(s) + ½ Cl_2(g) \longrightarrow LiCl(s) \qquad \Delta H_f^° = -408.8 \text{ kJ/mole}$$

The standard heat of formation of aqueous lithium chloride is -445.93 kJ/mole [1]:

$$Li(s) + ½ Cl_2(g) + H_2O(l) \longrightarrow Li^+(aq) + Cl^-(aq)$$
$$\Delta H_f^°(aq) = -445.93 \text{ kJ/mole}$$

The standard heat of solution of lithium chloride can then be calculated:

$$LiCl(s) + H_2O(l) \longrightarrow Li^+(aq) + Cl^-(aq)$$

$$\Delta H^\circ_{soln} = -37.1 \text{ kJ/mole or } -875 \text{ J/g of LiCl}$$

The solubility of lithium chloride ranges from 63.7 g/100 ml at 0°C to 130 g/100 ml at 95°C [2]. Higher temperatures are thus clearly possible.

Table 1. Solubility and Thermochemical Data for Various Lithium Salts

Compound	Solubility[a] at °C (g/100 ml H_2O)	Calculated heat of solution (kJ/g)	(kJ/100 ml of solution)[b]
LiF	0.27[18]	+0.444	+0.12
LiCl	63.7[0], 130[95]	−0.875	−113.7
LiBr	145[4], 254[90]	−0.565	−143.5
LiI	165[20], 433[80]	−0.473	−204.7
LiNO₃	89.8[27], 234[100]	−0.039	−9.20
Li₂CO₃	1.54[0], 0.72[100]	−0.24	−0.17
LiOH	12.8[20], 17.5[100]	−0.880	−15.4

[a] The solubility data are from reference 2.

[b] Heat of solution data are calculated for highest solubility from data in reference 1.

Various lithium salts can be used instead of lithium chloride. Most evolve less heat than lithium chloride because of both lower enthalpy changes and lower solubility (Table 1). Lithium bromide and lithium iodide yield higher enthalpy changes but are much more expensive than lithium chloride.

The most important factor in the high energy of solution for lithium chloride is the very large value for the hydration energy of lithium ions. This large value is evident when the heat of solution of lithium chloride is compared with the heats of solution of other chlorides (Table 2).

Table 2. Thermochemical Data for Selected Chlorides

Compound	ΔH°_f (kJ/mole)	ΔH°_f(aq) (kJ/mole)	ΔH° (solution)[a] (kJ/mole)
NH₄Cl	−315.4	−300.2	+15.2
LiCl	−408.8	−445.93	−37.1
NaCl	−411.0	−407.1	+3.9
KCl	−435.89	−418.65	+17.24
RbCl	−430.5	−413.8	+16.7

[a] Heat of solution data are calculated from standard heat of formation data in reference 1.

REFERENCES

1. "Selected Values of Chemical Thermodynamic Properties." *Natl. Bur. Stand. (U.S.)* **1952**; Circ. 500; p 433.
2. Weast, R. C., Ed. "CRC Handbook of Chemistry and Physics," 49th ed.; The CRC Press: Cleveland, Ohio, 1968; pp B-212, B-213, D-38, D-43, D-45, D-46.

1.9

Heat of Hydration of Copper(II) Sulfate

The addition of a small quantity of water to copper sulfate powder in a test tube results in the evolution of heat.

MATERIALS

25 g copper sulfate, $CuSO_4$ (anhydrous powder)

15 ml distilled water

test tube, 25 mm $\times$ 150 mm

thermometer, $-10°C$ to $+110°C$

PROCEDURE

Place the anhydrous copper sulfate in the test tube and measure the temperature of the solid. Add 15 ml of distilled water and note the increase in temperature. Part of the solid dissolves, yielding a deep blue solution. The temperature will increase from room temperature to about 80°C.

HAZARDS

The test tube becomes hot enough to cause burns.

DISPOSAL

The mixture of reactants and products is soluble in water and should be flushed down the drain with water.

DISCUSSION

The standard heat of formation of anhydrous copper(II) sulfate is -769.9 kJ/mole [1]:

$$Cu(s) + S(s) + 2\ O_2(g) \longrightarrow CuSO_4(s) \qquad \Delta H_f^\circ = -769.9\ \text{kJ/mole}$$

The standard heat of formation of copper(II) sulfate pentahydrate is -2278.0 kJ/mole [1]:

$$Cu(s) + S(s) + \frac{9}{2}O_2(g) + 5\,H_2(g) \longrightarrow CuSO_4 \cdot 5H_2O(s)$$

$$\Delta H_f^\circ = -2278.0 \text{ kJ/mole}$$

The standard heat of hydration of copper(II) sulfate, ΔH_{hyd}°, can be calculated from these values and from the standard heat of formation of water (for water, $\Delta H_f^\circ = -285.9$ kJ/mole):

$$CuSO_4(s) + 5\,H_2O(l) \longrightarrow CuSO_4 \cdot 5H_2O(s)$$

$$\Delta H_{hyd}^\circ = [-2278.0] - [-769.9 + 5(-285.9)] = -78.6 \text{ kJ/mole or } -315 \text{ J/g}$$

The solubility of anhydrous $CuSO_4$ in water ranges from 14.3 g/100 ml at 0°C to 75.4 g/100 ml at 100°C, and the solubility of $CuSO_4 \cdot 5H_2O$ is 31.6 g/100 ml at 0°C and 203.3 g/100 ml at 100°C [2]. The highly exothermic nature of the hydration of $CuSO_4$ is due to the hydration energy of Cu(II). Various metal sulfates can be used (see table).

Solubility and Heat of Solution of Metal Sulfates

Compound	Solubility at °C (g/100 ml H_2O)	Calculated heat of solution	
		(kJ/g)	(kJ/100 ml of solution)[a]
Li_2SO_4	26.1[0], 23[100]	−0.304	−6.99
$Li_2SO_4 \cdot H_2O$	34.9[23], 29.2[100]	−0.141	−4.1
Na_2SO_4	4.76[0], 42.7[100]	−0.03	−1.3
$Na_2SO_4 \cdot 10H_2O$	11[0], 92.7[30]	+0.34	+31.4
$CaSO_4$	0.209[20], 0.162[100]	−0.123	−0.008
$CaSO_4 \cdot \frac{1}{2}H_2O$	0.3[20], sl. s.[hot]	−0.120	—
$CaSO_4 \cdot 2H_2O$	0.241[cold], 0.222[100]	−0.088	+0.000054
$CuSO_4$	14.3[0], 75.4[100]	−0.448	−33.7
$CuSO_4 \cdot 5H_2O$	31.6[0], 203.3[100]	−0.031	−6.28

[a] Heat of hydration data are calculated for highest solubility from standard heat of formation data in reference 1.

REFERENCES

1. "Selected Values of Chemical Thermodynamic Properties." *Natl. Bur. Stand. (U.S.)* **1952**; Circ. 500; pp 211–12.
2. Weast, R. C., Ed. "CRC Handbook of Chemistry and Physics," 54th ed.; CRC Press: Cleveland, Ohio, 1973; p B-89.

1.10

Reactions of Metals and Hydrochloric Acid

When a ball of steel wool is moistened with hydrochloric acid, the temperature increases. Temperature increases also occur when samples of other metals are placed in different beakers and covered with hydrochloric acid.

MATERIALS

16 g coarse steel wool, #3

500 ml concentrated (12M) hydrochloric acid, HCl

4 g fine steel wool, #0

7 g mossy zinc

2.5 g magnesium turnings

4 600-ml beakers

gloves, rubber or plastic

thermometer, $-10°C$ to $+110°C$

PROCEDURE

Fill the bottom third of one beaker with coarse steel wool. Wear gloves to protect your hands from spattering acid. Add 200 ml of 12M HCl solution and place the thermometer in the beaker. Observe the color change of the solution, the release of gas, and a temperature increase (about 15°C above room temperature). Rinse the thermometer before using it again.

In another beaker place 4 g of fine steel wool and add 100 ml of 12M HCl. Observe the color change of the solution, the evolution of gas, and a temperature increase (about 7°C above room temperature).

Place 7 g of mossy zinc in the third beaker and add 100 ml of 12M HCl. Observe the clarity of the solution, the release of gas, and the temperature increase (about 35°C above room temperature).

In the fourth beaker, place 2.5 g of magnesium turnings and add 100 ml of 12M HCl. Observe the foaming and a temperature increase (about 55°C above room temperature).

HAZARDS

Steel wool can cut the skin.

Hydrochloric acid may cause severe burns, and the vapors are extremely irritating to the skin, eyes, and respiratory system. The reactions of aqueous hydrochloric acid with zinc or magnesium produce sufficient heat to cause burns.

Since the hydrogen gas formed in each case is released rapidly, the acid solution may spatter in the immediate vicinity. Care should therefore be exercised while measuring the temperature.

Avoid having flames in the vicinity because of the potentially violent combustion of H_2.

DISPOSAL

The products of each reaction should be flushed down the drain with water. If some steel wool remains, it should be rinsed with water and discarded in a waste container.

DISCUSSION

The reactions and their respective enthalpies are [1]

	ΔH°_{rxn} (kJ/mole)
$Fe(s) + 2\,HCl(aq) \longrightarrow FeCl_2(aq) + H_2(g)$	-87.7
$Zn(s) + 2\,HCl(aq) \longrightarrow ZnCl_2(aq) + H_2(g)$	-150
$Mg(s) + 2\,HCl(aq) \longrightarrow MgCl_2(aq) + H_2(g)$	-460

The increasing temperature changes in the iron-zinc-magnesium series are consistent with the heats of formation for the aqueous chlorides of these metals.

This demonstration illustrates that heat is released when a metal reacts with a strong acid. We prefer this procedure to the one involving iron and acetic acid [2]. We do not recommend the use of oxidizing acids, such as H_2SO_4 or HNO_3, because they involve the anion in the redox reaction.

In the reaction between iron and hydrochloric acid, the iron product is Fe(II) as long as there is a reducing atmosphere due to H_2. Eventually Fe(II) will be air-oxidized to Fe(III).

REFERENCES

1. "Selected Values of Chemical Thermodynamic Properties." *Natl. Bur. Stand. (U.S.)* **1952**; Circ. 500; p 2.
2. Alyea, H. N. *J. Chem. Educ.* **1970**, *47*, A-387.

1.11

Crystallization from Supersaturated Solutions of Sodium Acetate

Exothermic crystallization from a supersaturated solution is demonstrated by "seeding" supersaturated solutions of sodium acetate trihydrate with crystals of sodium acetate trihydrate. The crystallization is displayed either by pouring a solution of sodium acetate trihydrate onto crystals on a piece of hardboard (Procedure A) or by adding a crystal of sodium acetate trihydrate to a large test tube of solution (Procedure B).

MATERIALS FOR PROCEDURE A

175 g sodium acetate trihydrate, $NaC_2H_3O_2 \cdot 3H_2O$

50 ml distilled water

2-liter beaker

hot plate

500-ml Erlenmeyer flask

100-ml beaker

piece of hardboard, ca. 30 cm × 30 cm

spatula

MATERIALS FOR PROCEDURE B

250 g sodium acetate trihydrate, $NaC_2H_3O_2 \cdot 3H_2O$

100 ml distilled water

2-liter beaker

hot plate

test tube, 50 mm × 400 mm

rubber stopper to fit test tube

thermometer, −10°C to +110°C

light source (To make a suitable light source, cut a 3.5-cm hole in the bottom of a 3-lb coffee can and invert the can over a light bulb and socket, as illustrated by the cut-away view in the figure.)

ring stand

large utility clamp to hold test tube

PROCEDURE A

Provide a boiling water bath by filling the 2-liter beaker about three-fourths full of water and heating it on the hot plate. To prepare the supersaturated solution, place 175 g of sodium acetate trihydrate and 50 ml of distilled water (3.5 g $NaC_2H_3O_2 \cdot 3H_2O$/ml H_2O) in the 500-ml Erlenmeyer flask. Heat the mixture in the boiling water bath and swirl the flask occasionally until a clear, homogeneous solution is obtained. Then, invert the 100-ml beaker over the mouth of the flask and allow the solution to cool undisturbed until it reaches room temperature (1–3 hours). Alternatively, place the flask in a large beaker and cool with running water.

To demonstrate crystallization, place a few crystals of $NaC_2H_3O_2 \cdot 3H_2O$ or $NaC_2H_3O_2$ on the piece of clean hardboard and slowly pour the solution onto the crystals. Crystallization from the supersaturated solution occurs immediately, forming a mound of white solid sodium acetate. All the water is trapped within the solid, leaving no visible trace of liquid. The solid feels warm to the touch. The shape of the mound will depend on the manner in which the solution is poured; if desired, pillars of different shapes and heights can be produced. Another method is to pour the liquid slowly to form different patterns on the hardboard. If crystallization does not occur, add a few more crystals.

After the demonstration, the solid can be cut up with a spatula and returned to the Erlenmeyer flask to be used again. The supersaturated solution is restored by heating in a boiling water bath; it can be re-used repeatedly unless contaminated. After several cycles of use, *small* amounts of water may have to be added to compensate for evaporation losses.

PROCEDURE B

Provide a boiling water bath using the large beaker and hot plate. Place 250 g of sodium acetate trihydrate and 100 ml of distilled water (2.5 g $NaC_2H_3O_2 \cdot 3H_2O$/ml H_2O) in the large test tube. Heat the mixture in the boiling water bath and swirl the tube occasionally until a clear solution is obtained. Stopper the tube loosely and allow the solution to cool undisturbed until it reaches room temperature (1–3 hours). Alternatively, place the test tube in a large beaker and cool with running water.

To demonstrate crystallization, place the test tube on the light source and support it with the ring stand and the utility clamp. Remove the stopper and suspend a thermometer into the liquid (see figure). Drop a single crystal of $NaC_2H_3O_2 \cdot 3H_2O$ or $NaC_2H_3O_2$ into the tube. Long, needle-like crystals of sodium acetate trihydrate will grow slowly throughout the test tube. The illumination from below makes the phenomenon more easily visible. Note the temperature increase. If the tube is left standing on the light source for several minutes, the solid at the bottom will begin to melt. The mixture of sodium acetate and water can be re-used repeatedly.

The supersaturated solutions in both procedures of this demonstration are sometimes difficult to prepare. All glassware should be clean and scratch-free. If, after several attempts, the solution cannot be cooled to room temperature without crystallization, it should be discarded. Clean or replace the glassware and start with another sample.

Since even a small bump may cause crystallization to occur, more than one sample of supersaturated solution should be prepared.

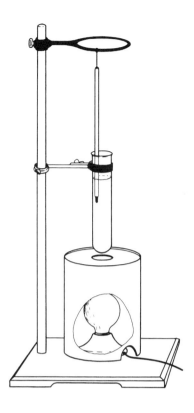

HAZARDS

The solutions are hot enough to cause burns. If the crystallized sodium acetate touches the skin, wash with water.

DISPOSAL

The materials can be recycled. When disposal is warranted, the solid should be flushed down the drain with water.

DISCUSSION

At temperatures below 58°C, solutions that are saturated with respect to anhydrous sodium acetate are supersaturated with respect to the trihydrate (see table) [1]. The compositions of the solutions prepared correspond to 88 g $NaC_2H_3O_2$/100 g H_2O (Procedure A) and 75 g $NaC_2H_3O_2$/100 g H_2O (Procedure B). The 100 g of water include both added water and water of hydration. When these solutions are cooled to 20°C, they are unsaturated with respect to $NaC_2H_3O_2$ but supersaturated with respect to $NaC_2H_3O_2 \cdot 3H_2O$. "Seeding" the solutions with $NaC_2H_3O_2 \cdot 3H_2O$ thus causes formation of crystals of hydrated sodium acetate. Since there is no residual water in Procedure A, higher hydrates than the trihydrate may be formed, but no evidence of them exists [2]. The composition of the solution in Procedure A corresponds to $NaC_2H_3O_2 \cdot 5.2H_2O$.

Composition of Saturated Sodium Acetate
Solutions[a] [1]

T (°C)	Trihydrate	Anhydrous
− 18	30.4	—
− 10	33.0	—
0	36.3	119
10	40.8	121
20	46.5	123.5
30	54.5	126
40	65.5	129.5
50	83.0	134
60	—	139.5
70	—	146
80	—	153
90	—	161
100	—	170
110	—	180
120	—	191
123[b]	—	193

[a] In each case the entry gives the composition of a saturated solution of the indicated salt expressed as grams of $NaC_2H_3O_2/100$ g H_2O.

[b] Boiling point of saturated solution.

When solid sodium acetate trihydrate is heated above 58°C, it loses its water of hydration and begins to dissolve in that water. The salt becomes completely dissolved when the mixture reaches 79°C [1], the temperature at which a saturated solution of anhydrous sodium acetate has the same composition as the trihydrate (152 g $NaC_2H_3O_2/100$ g H_2O). This solution can then be cooled to yield a solution that is supersaturated with respect to both the trihydrate and the anhydrous form. If "seeds" are excluded, the crystals deposited from these solutions are anhydrous sodium acetate [1, 2].

The heat of solution of $NaC_2H_3O_2 \cdot 3H_2O$ is 19.7 ± 0.1 kJ/mole [3]. The process is endothermic. The crystallization of $NaC_2H_3O_2 \cdot 3H_2O$ is thus exothermic, and the solid or the tube feels warm.

REFERENCES

1. Green, W. F. *J. Phys. Chem.* **1908**, *12*, 655.
2. Connor, W. W. *J. Am. Chem. Soc.* **1931**, *53*, 2806.
3. Calvet, E. *J. chim. phys.* **1933**, *30*, 140; *Chem. Abstr.* **1933**, *27*, 2868.

1.12

Crystallization of Sodium Thiosulfate

A single crystal of $Na_2S_2O_3 \cdot 5H_2O$ is added to a test tube containing a "melt" of sodium thiosulfate pentahydrate which has been cooled to room temperature. The liquid begins to crystallize immediately and the test tube becomes warm as the heat of crystallization is released.

MATERIALS

46 g sodium thiosulfate pentahydrate, $Na_2S_2O_3 \cdot 5H_2O$

test tube, 25 mm $\times$ 200 mm

Bunsen burner or 1-liter beaker and hot plate

ring stand and clamp

thermometer, $-10°C$ to $+110°C$

spatula

PROCEDURE

Fill the test tube one-half to two-thirds full of solid sodium thiosulfate pentahydrate. With a Bunsen burner or in a boiling water bath, heat the test tube until the solid "melts" and a clear, homogeneous liquid is obtained. Clamp the test tube in an upright position and allow the liquid to cool to room temperature. To save time, the solid should be melted and cooled prior to the demonstration.

To initiate crystallization, drop a single crystal of $Na_2S_2O_3 \cdot 5H_2O$ into the cooled liquid. Crystallization begins immediately, and the test tube becomes warm as the heat of crystallization is released. The exothermic nature of this process can be demonstrated in several ways: the test tube can be passed around the class to let students feel the heat produced, the change in temperature can be measured with a thermometer as the solid forms, or the test tube can be placed in a small water bath and a thermometer used to measure the temperature change in the water bath.

The sample of sodium thiosulfate in the test tube can be used for subsequent demonstrations. Samples of supersaturated sodium thiosulfate solutions stored in stoppered test tubes are stable for several days. The stability of the solution can be enhanced by filtration of the hot solution.

HAZARDS

This system produces sufficient heat to cause burns.

DISPOSAL

The potential of recycling this system many times reduces the need for disposal. When disposal is warranted, the solid should either be flushed down the drain with water or heated to the fused form and poured into running water. In the latter case, the test tube should then be rinsed.

DISCUSSION

Sodium thiosulfate exhibits several different hydrates which, in turn, take a number of different forms [1]. Under the conditions encountered in this demonstration, only the ordinary pentahydrate (called the "primary" pentahydrate [1]) is important. When $Na_2S_2O_3 \cdot 5H_2O$ is heated above 48.2°C, a liquid begins to form. This temperature, however, is not a true melting point. The pentahydrate loses water of hydration and begins to dissolve until the sample completely liquifies at about 60°C [2]. The liquid is really a solution with the same composition as the salt: 5 moles of water to 1 mole of sodium thiosulfate. The fused pentahydrate is easily cooled below 48.2°C to form a supersaturated solution of $Na_2S_2O_3 \cdot 5H_2O$. The solubility of the pentahydrate at several temperatures below the "melting" point is shown in the table.

Solubility of Sodium Thiosulfate in Water [1, 3]

T (°C)	g $Na_2S_2O_3$/100 g H_2O
0	50.15
5	54.64
10	59.66
15	64.22
20	70.07
25	75.90
30	82.45
35	91.24
40	103.37
45	123.87

The composition of the pentahydrate corresponds to 175.54 g $Na_2S_2O_3$/100 g H_2O.

Under proper conditions, fusions of $Na_2S_2O_3 \cdot 5H_2O$ can be converted to other hydrates or other forms [1, 2]. However, when the solution is seeded with the ordinary pentahydrate, crystallization occurs in this form [2]. Crystals with other structures are not effective at initiating crystallization. This ineffectiveness can be demonstrated by adding a crystal of sodium chloride to the liquid prior to initiating crystallization with a crystal of $Na_2S_2O_3 \cdot 5H_2O$.

Since formation of the liquid $Na_2S_2O_3 \cdot 5H_2O$ is endothermic, the reverse process is exothermic. The heat of crystallization of $Na_2S_2O_3 \cdot 5H_2O$ from supersaturated solutions is -128 ± 1 J/g [4], which is equivalent to -31.9 ± 0.3 kJ/mole.

One model for the behavior of supersaturated systems presumes the presence of crystalline aggregates that are not present in the fused system [2]. As the system cools, portions or fragments of these aggregates coalesce to provide sites for crystallization. The size and character of these fragments allow them to be separated by filtration, an operation that reduces the tendency towards crystallization. As these aggregates unite, they assume the pattern of the crystal form of the system used, or, if the crystal system can assume more than one shape (as with sodium thiosulfate pentahydrate), the aggregates assume one of the possible forms.

REFERENCES

1. Young, S. W.; Burke, W. E. *J. Am. Chem. Soc.* **1906**, *28*, 315.
2. Young, S. W.; Mitchell, J. P. *J. Am. Chem. Soc.* **1904**, *26*, 1389.
3. Young, S. W.; Burke, W. E. *J. Am. Chem. Soc.* **1904**, *26*, 1413.
4. Tikhomiroff, N.; Pultrini, F.; Heitz, F.; Gilbert, M. *Compt. Rend.* **1965**, *261*, 334; *Chem. Abstr.* **1965**, *56*, 15631.

1.13

Supercooling of Thymol

When a crystal of thymol is added to supercooled liquid thymol, the liquid crystallizes and the tube becomes warm.

MATERIALS

35 g thymol, $C_{10}H_{14}O$

1-liter beaker

hot plate

test tube, 25 mm $\times$ 200 mm

ring stand and clamp

spatula

thermometer, $-10°C$ to $+110°C$

PROCEDURE

Fill the 1-liter beaker approximately three-fourths full of water and heat to boiling on the hot plate. Fill the test tube one-half to two-thirds full of thymol and place it in the boiling water bath. When the thymol is completely melted, remove the tube from the water bath and clamp it in an upright position. Allow the liquid thymol to cool undisturbed to approximately room temperature. Since the liquid can require as much as an hour to cool, you may want to prepare a sample and cool it beforehand.

When the liquid is cool, add a single crystal of thymol to the test tube using the spatula. The entire sample will slowly solidify, and the test tube will become warm to the touch. The heat produced by the crystallization can be demonstrated by allowing students to feel the warm tube or by using a thermometer. However, a thermometer may aid in nucleation, making it impossible to cool the liquid thymol to room temperature. The solidified thymol can be left in the stoppered test tube and re-used.

HAZARDS

Thymol is an irritant which may cause allergic reactions in some individuals. While warming the thymol, care should be taken to avoid breathing the vapor.

DISPOSAL

Tubes containing thymol can be retained for future use. When disposal becomes necessary, thymol can be discarded as solid waste or dissolved in an organic solvent and incinerated. Thymol is only slightly soluble in water (approximately 1 g/liter) [1].

DISCUSSION

Thymol (2-hydroxy-1-isopropyl-4-methyl benzene, $C_{10}H_{14}O$) has the structure:

$$
\begin{array}{c}
CH_3 \quad CH_3 \\
\diagdown \diagup \\
CH \\
| \qquad OH \\
\bigcirc \\
| \\
CH_3
\end{array}
$$

Thymol melts at 51.5°C and boils at 233.5°C. Once melted, liquid thymol can be readily supercooled, that is, cooled below its freezing point without solidifying. One source lists the low temperature limit of thymol as 24 ± 5°C [2]; we have routinely cooled samples to 21°C without solidification.

The heat of fusion of thymol is approximately 17.3 kJ/mole [3]. When liquid thymol crystallizes, this heat is released, warming the test tube.

If crystals other than thymol, such as sodium chloride, are added to supercooled thymol, crystallization does not occur. The "seed" crystal must fit the crystal structure of thymol to provide a site for formation of the solid. This lock-and-key effect can be demonstrated by first adding a foreign crystal to the test tube of liquid thymol.

REFERENCES

1. Windholz, M., Ed. "The Merck Index," 9th ed.; Merck and Co.: Rahway, New Jersey, 1976; p 1214.
2. Akers, F.; CHEM 13 NEWS, October 1976, p 12.
3. "International Critical Tables"; McGraw-Hill: New York, 1928; Vol. V, p 134.

1.14

Chemical Hot Pack

First aid hot packs demonstrate spontaneous exothermic reactions.

MATERIALS

Instant hot packs can be obtained from medical supply outlets. We use Kwik Heat Hot Packs [1], which consist of either sodium thiosulfate as a supersaturated solution (Procedure A) or solid calcium chloride and a separate water pouch (Procedure B).

PROCEDURE A

The sodium thiosulfate system consists of an inner pouch, which contains a supersaturated aqueous sodium thiosulfate solution and some ethylene glycol, and an outer pouch containing a few crystals of a solid (analyzed as sodium thiosulfate). To release the liquid, firmly squeeze the inner pouch. The supersaturated solution crystallizes rapidly and releases heat.

PROCEDURE B

The calcium chloride pack consists of solid calcium chloride in the outer pouch and water with a red dye in the inner pouch. To release the water and mix it with the calcium chloride, firmly squeeze the inner pouch. This system generates substantially higher temperatures than the thiosulfate system and should be handled cautiously. The packet is well sealed but should be wrapped in a towel if passed around the class.

HAZARDS

Both hot packs produce sufficient heat to cause burns.
If the outer pouch breaks, its contents should be flushed down the drain with water. This should be done quickly if the material is still hot.

DISPOSAL

The sealed packs should be discarded in a waste container. If the pack breaks, its contents should be flushed down the drain.

DISCUSSION OF PROCEDURE A

When the inner pouch containing the solution of supersaturated sodium thiosulfate is broken, the crystallization reaction proceeds rapidly. The first step in this reaction is the formation of anhydrous sodium thiosulfate from the aquated ions, according to the reaction:

$$2\ Na^+(aq)\ +\ S_2O_3^{2-}(aq) \longrightarrow Na_2S_2O_3(s)$$

The heat of reaction for this step is $+8$ kJ/mole. The second step involves the hydration of the anhydrous salt, according to the reaction:

$$Na_2S_2O_3(s)\ +\ 5\ H_2O\ (l) \longrightarrow Na_2S_2O_3 \cdot 5H_2O(s)$$

The heat of reaction for this step is -56.1 kJ/mole. Thus, the overall reaction is the sum of these two steps:

$$2\ Na^+(aq)\ +\ S_2O_3^{2-}(aq)\ +\ 5\ H_2O(l) \longrightarrow Na_2S_2O_3 \cdot 5H_2O(s)$$

The enthalpy change for the overall reaction is -48.1 kJ/mole or approximately -194 J/g of product formed [2]. The standard enthalpies of formation, ΔH_f°, are -1130 kJ/mole for $Na_2S_2O_3(aq)$, -2600 kJ/mole for $Na_2S_2O_3 \cdot 5H_2O(s)$, and -280 kJ/mole for $H_2O(l)$ [2].

When we mixed the ingredients in a Dewar flask (with a heat capacity of 42 J/°C), the temperature rose from 22°C to 46°C. A sample pouch contained approximately 350 g of crystallized solid. Iodometric analysis of the solid indicated a content of 150 g $Na_2S_2O_3$. We assumed the remainder to be water and ethylene glycol. The ethylene glycol may be included to reduce the possibility of freezing in storage.

DISCUSSION OF PROCEDURE B

When the inner pouch is broken, the water released dissolves the solid calcium chloride:

$$CaCl_2(s) \longrightarrow Ca^{2+}(aq)\ +\ 2\ Cl^-(aq)$$

This reaction evolves 82.93 kJ/mole or approximately 747 J/g of dissolved calcium chloride [2]. The standard enthalpies of formation, ΔH_f°, are -795 kJ/mole for $CaCl_2(s)$ and -877.89 kJ/mole for $CaCl_2(aq)$ [2]. The solubility of calcium chloride in water ranges from 74.5 g/100 ml at 20°C to 159 g/100 ml at 100°C [3]. In the commercial hot pack we used, 218 g of $CaCl_2$ were present in the outer pouch and 170 ml of water were present in the inner pouch. When we mixed the contents of a pack in a Dewar flask with a heat capacity of 59 J/°C, we observed a temperature increase from 22°C to 72°C. In principle, temperatures above the boiling point of water can be achieved [4].

REFERENCES

1. Kwik Kold Division of Kay Laboratories, Inc., Moberly, Missouri 65270; patent numbers 2 925 719 and 3 951 127.
2. "Selected Values of Chemical Thermodynamic Properties." *Natl. Bur. Stand. (U.S.)* **1952**; Circ. 500; pp 9, 387, 459.
3. Weast, R. C., Ed. "CRC Handbook of Chemistry and Physics," 49th ed.; The CRC Press: Cleveland, Ohio, 1968; pp D-46, D-47, D-40, B-186.
4. Smith, W. L. *Chemistry* **1972**, *45* (7), 19.

1.15

Burning of Magnesium

When a flashbulb is activated, a brilliant flash of light is observed and the bulb becomes hot. When a piece of magnesium ribbon is ignited, light and heat are produced.

MATERIALS FOR PROCEDURE A

flashbulbs
clamp
ring stand
power supply (3V DC) or 2 flashlight batteries in series
wires

MATERIALS FOR PROCEDURE B

6–8 cm magnesium ribbon
Meker burner
tongs

PROCEDURE A

Attach the clamp to the ring stand and insert a bulb in the clamp (see figure). Plug in the power supply and attach the wires to the proper terminals.

Set the output of the power supply to 3V DC, hold one wire against the outer metal base of the bulb, and momentarily touch the metal tip of the base with the second wire. **Do not look directly at the bulb.** Do not touch the hot bulb for several minutes.

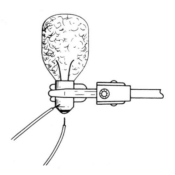

PROCEDURE B

Light the burner and adjust the flame to light blue.

Holding one end of the magnesium ribbon with the tongs, place the other end in the flame until it ignites. The burning ribbon should be held at arm's length. **Do not**

look directly at the burning ribbon. The burning magnesium yields brilliant light and intense heat.

HAZARDS

Both the flashbulb and the burning magnesium ribbon produce light of sufficient intensity to cause temporary loss of sight. Avoid looking directly at either light source.

The burning of magnesium in air produces intense heat, which can cause burns and initiate combustion in flammable materials. Since a carbon dioxide fire extinguisher will not extinguish burning magnesium, a dry-powder extinguisher must be used (see Demonstration 1.37).

DISPOSAL

Once cooled to room temperature, the used flashbulb and the magnesium oxide solid from the ribbon should be discarded in a waste container.

DISCUSSION

The oxidation or combustion of magnesium in air has long been used as a source of intense light in photography and for other photochemical reactions. The energy released in this combustion occurs as the magnesium reacts with oxygen in the air according to the equation:

$$2 \text{ Mg(s)} + \text{O}_2\text{(g)} \longrightarrow 2 \text{ MgO(s)}$$

The heat of formation for magnesium oxide is -601.83 kJ/mole [1].

Because of the high temperature achieved in this combustion, a small portion of the magnesium reacts with nitrogen in the air according to the reaction:

$$3 \text{ Mg(s)} + \text{N}_2\text{(g)} \longrightarrow \text{Mg}_3\text{N}_2\text{(s)}$$

The heat of formation of magnesium nitride is -461.24 kJ/mole [1].

The presence of magnesium nitride can be confirmed by placing a sample of the solid residue in a test tube, placing a damp piece of pink litmus across the top of the test tube, and wetting the mixture. The magnesium nitride decomposes to yield ammonia which turns the litmus blue. The reaction is

$$\text{Mg}_3\text{N}_2\text{(s)} + 6 \text{ H}_2\text{O(l)} \longrightarrow 3 \text{ Mg(OH)}_2\text{(s)} + 2 \text{ NH}_3\text{(g)}$$

When magnesium burns, approximately 10% of the energy of combustion occurs as light, a "value unapproached among known transformations of energy used in the production of light" [2].

REFERENCES

1. "Selected Values of Chemical Thermodynamic Properties." *Natl. Bur. Stand. (U.S.)* **1952**; Circ. 500; p 374.
2. Mellor, J. W. "A Comprehensive Treatise on Inorganic and Theoretical Chemistry"; Longmans, Green and Co.: London, 1928; Vol. IV, pp 259–60.

1.16

Combustion Under Water

Sparks are produced when a small amount of sulfuric acid is added to potassium chlorate and white phosphorus. These sparks are readily observed in a darkened room [1, 2].

MATERIALS

100 ml distilled water

10 g potassium chlorate, $KClO_3$

piece of white (yellow) phosphorus, ca. 3 mm in diameter

2 ml concentrated (18M) sulfuric acid, H_2SO_4

100-ml graduated cylinder

stirring rod

gloves, plastic or rubber

forceps

petri dish

knife

long-stemmed funnel or thistle tube

clamp

ring stand

PROCEDURE

Fill the graduated cylinder nearly full with distilled water. Add 10 g of potassium chlorate to the cylinder and with the stirring rod form a flat layer of crystals. Wearing gloves and using forceps, place the piece of white phosphorus under water in the petri dish. Cut the phosphorus into four or five pieces. With the forceps, place these pieces on top of the layer of potassium chlorate.

Insert the funnel or thistle tube into the graduated cylinder and clamp it so that the tip is just above the phosphorus (see figure). Add 1 ml of the concentrated sulfuric acid. Quickly darken the room. If sparks are not seen at the bottom of the cylinder within 15 seconds, add another milliliter of the concentrated acid. The solution becomes slightly yellowish green as ClO_2 forms. **Do not add more than a total of 2 ml of 18M H_2SO_4 since a violent reaction may ensue.** If no reaction occurs, follow the disposal instructions.

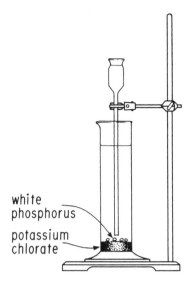

white
phosphorus

potassium
chlorate

HAZARDS

Since sulfuric acid is a strong acid and a powerful dehydrating agent, it can cause burns. Spills should be neutralized with an appropriate agent, such as $NaHCO_3$, and then rinsed clean.

White phosphorus is spontaneously flammable in air. Combustion in air produces phosphorus pentoxide. Both white phosphorus and phosphorus pentoxide are very poisonous and can cause severe burns. Chronic effects can result from continued absorption of small amounts. Exposure of phosphorus to the air should be very brief, and all cutting should be done under water.

Chlorine dioxide gas is explosive, but very dilute aqueous solutions are safe to handle.

Since potassium chlorate is a strong oxidizing agent, mixtures of $KClO_3$ with combustible materials can be flammable or explosive.

DISPOSAL

Allow the reaction to run its course, then place the graduated cylinder in a deep sink, and run water into the thistle tube. This action dilutes the reaction mixture and slowly dissolves the potassium chlorate layer. After about 5 minutes, the graduate can be removed and cleaned. Any unreacted white phosphorus should be placed on a small piece of filter paper under a hood. When dry, the white phosphorus will burn spontaneously.

DISCUSSION

The oxidizing power of potassium chlorate is significantly enhanced in the presence of sulfuric acid. This powerful oxidant mixture has long been used to demonstrate combustion of a wide variety of substances [2]. The lack of a clear-cut stoichiometry makes it difficult to specify the products in the system described, although oxidation of the phosphorus to phosphorus pentoxide seems likely.

REFERENCES

1. Alyea, H. N.; Dutton, F. B., Eds. "Tested Demonstrations in Chemistry," 6th ed.; Journal of Chemical Education: Easton, Pennsylvania, 1965; p 197.
2. Mellor, J. W. "A Comprehensive Treatise on Inorganic and Theoretical Chemistry"; Longmans, Green and Co.: London, 1930; Vol. VIII, p 289.

1.17

Combustion of Cellulose Nitrate (Guncotton)

A piece of cotton-like material is placed on the base of a ring stand and ignited with a lighted candle on a meter stick. The cotton rapidly burns leaving little residue.

MATERIALS FOR PROCEDURE A

70 ml concentrated (18M) sulfuric acid, H_2SO_4

30 ml concentrated (16M) nitric acid, HNO_3

250 ml 1M sodium bicarbonate, $NaHCO_3$

250-ml beaker

ice bath

5 g absorbent cotton

tongs

paper towels

ring stand with metal base

candle, mounted on a meter stick

ADDITIONAL MATERIALS FOR PROCEDURE B

1 g phosphorus pentoxide, P_2O_5

spatula

distilled water in wash bottle

PROCEDURE A

To prepare cellulose nitrate [1], place a 250-ml beaker in an ice bath, and add 70 ml of concentrated sulfuric acid and 30 ml of concentrated nitric acid to the beaker. Divide the cotton into pieces of approximately 0.7 g. With tongs, immerse each piece in the acid solution for 1 minute. Next, rinse each piece in three successive baths of 500 ml of water. Use fresh water for each piece. Then immerse each piece in 250 ml of 1M $NaHCO_3$. If substantial bubbling occurs, rinse the piece in water once more and

43

check for residual acid by immersing in the $NaHCO_3$ solution. Squeeze dry and spread on paper towels to dry overnight.

Place one of the pieces of the cellulose nitrate on the base of the ring stand. Light the candle and ignite the cellulose nitrate from the side to heighten the inflammation effect. The cellulose nitrate quickly burns, leaving a small amount of unburned residue.

PROCEDURE B

Prepare the cellulose nitrate as above. Place one of the pieces on the base of the ring stand. Using the spatula, place the phosphorus pentoxide in a small pile at the edge of the cellulose nitrate in intimate contact with it. From a distance of at least 1 m, squirt a stream of distilled water from the wash bottle at the phosphorus pentoxide. Do not soak the cellulose nitrate before hitting the phosphorus pentoxide. When the water reacts with the phosphorus pentoxide, sufficient heat is liberated to ignite the cellulose nitrate, which quickly burns to a fine black ash.

HAZARDS

Nitric acid and sulfuric acid are strong acids and powerful oxidizing agents, which can cause severe burns. In addition, sulfuric acid is a powerful dehydrating agent. Spills should be neutralized with an appropriate agent, such as $NaHCO_3$, and then rinsed thoroughly.

Since the mixing of concentrated sulfuric acid and concentrated nitric acid evolves considerable heat, this process must be carried out in an ice bath. This mixture is extremely hazardous.

Cellulose nitrate is extremely flammable and can explode if ignited in an enclosed space. If larger amounts are used, we recommend a longer holder for the candle.

Phosphorus pentoxide reacts violently with water, which is the basis for the method of ignition in Procedure B. Phosphorus pentoxide is very corrosive. The dust is a strong irritant and can cause severe burns to skin, eyes, and mucous membranes.

DISPOSAL

The small amount of carbonaceous residue remaining after ignition should be discarded in a waste container. We recommend burning any unused cellulose nitrate rather than storing it.

DISCUSSION

Guncotton is a cellulose trinitrate which contains 12.5–13.5% nitrogen. The equation for decomposition of cellulose trinitrate is

$$\longrightarrow \frac{9}{2}x\ CO + \frac{3}{2}x\ CO_2 + \frac{7}{2}x\ H_2O + \frac{3}{2}x\ N_2$$

The compound contains sufficient oxygen for complete conversion to gaseous products. In this demonstration, the decomposition of cellulose trinitrate is nonexplosive because of the method of ignition and the open conditions.

REFERENCE

1. Turner, E. University of Massachusetts, Amherst, Massachusetts, personal communication, 1979.

1.18

Combustion of Peroxyacetone

When a small pile of powdered peroxyacetone is ignited on a piece of paper, the powder burns instantly but the paper remains unaffected.

MATERIALS

4 ml acetone, CH_3COCH_3

4 ml 30% hydrogen peroxide, H_2O_2

4 drops concentrated (12M) hydrochloric acid, HCl

distilled water in wash bottle

gloves, plastic or rubber

150-mm test tube

9-cm filter paper

filter funnel

12-cm watch glass

ring stand with metal base

candle, mounted on a meter stick

PROCEDURE

Be extremely cautious in handling peroxyacetone. Serious injuries have been reported when this shock-sensitive and extremely flammable compound was prepared. See Hazards section.

Wear gloves while preparing the peroxyacetone. Add 4 ml of acetone and 4 ml of 30% hydrogen peroxide to a 150-mm test tube. Then add 4 drops of concentrated hydrochloric acid to the mixture. In 10–20 minutes a white solid should begin to separate from the solution. If no change is observed, warm the test tube in a water bath at 40°C to initiate the reaction. Allow the reaction to continue for 2 hours. Swirl the slurry and filter it. Rinse the solid remaining in the test tube onto the filter paper with small portions of distilled water. Open the filter paper on a watch glass and allow the peroxyacetone solid to dry for at least 2 hours.

To demonstrate the combustion reaction, place the filter paper with the peroxyacetone solid on the base of a ring stand. Light the candle on the meter stick and bring it to the solid along a horizontal path. The solid bursts into flame quickly but the filter paper does not burn.

Caution: do not ignite the peroxyacetone from a distance of less than 1 m. Do not drop or jar the solid, since it is shock sensitive. Use all the solid prepared; do **not store it.**

HAZARDS

Peroxyacetone is extremely flammable and hazardous. When the peroxyacetone is ignited, the flame expands to an area about 10 times the size of the solid. Keep the solid away from all hot surfaces or objects, since it might ignite. Do not drop or jar the solid, since it is shock sensitive. Use all the solid prepared; do not store it.

Since 30% hydrogen peroxide is a strong oxidizing agent, contact with skin and eyes must be avoided. In case of contact, immediately flush with water for at least 15 minutes; get immediate medical attention if the eyes are affected.

Avoid contact between 30% hydrogen peroxide and combustible materials. Avoid contamination from any source, since any contaminant, including dust, will cause rapid decomposition and the generation of large quantities of oxygen gas. Store 30% hydrogen peroxide in its original closed container, making sure that the container vent works properly.

Hydrochloric acid may cause severe burns. Hydrochloric acid vapors are extremely irritating to the skin, eyes, and respiratory system.

DISPOSAL

The only residue after the reaction is the filter paper, which should be discarded in a waste container. Any peroxyacetone not used in the demonstration should be burned in a safe area.

DISCUSSION

The probable structure of peroxyacetone, 2,2-dihydroperoxypropane ($C_3H_8O_4$), is

$$\begin{array}{c} OOH \\ | \\ CH_3\text{-}C\text{-}CH_3 \\ | \\ OOH \end{array}$$

It was first prepared by Milas and Golubovic [1] from a 1:1 mole ratio of acetone and hydrogen peroxide. This product undergoes spontaneous decomposition to form 2,2'-dihydroperoxydiisopropyl peroxide according to the equation:

$$2\ \begin{array}{c} OOH \\ | \\ CH_3\text{-}C\text{-}CH_3 \\ | \\ OOH \end{array} \longrightarrow 2\ \begin{array}{c} O\cdot \\ | \\ CH_3\text{-}C\text{-}CH_3 \\ | \\ OOH \end{array} + 2\ HO\cdot \longrightarrow$$

$$\begin{array}{c} CH_3 \quad CH_3 \\ | \qquad | \\ CH_3\text{-}CO\text{-}OC\text{-}CH_3 \\ | \qquad | \\ OOH \quad OOH \end{array} + H_2O + \tfrac{1}{2}\ O_2$$

In the presence of hydrogen ions and an excess of acetone, 2,2-dihydroperoxypropane converts into the cyclic trimer, 1,1,4,4,7,7-hexamethyl-1,4,7-cyclononatriperoxane, in 90% yield:

The identity of the final product of this synthesis has not been clearly established.

We have scaled down the synthesis reported by El-Awady and Prell [2] to permit preparation of sufficient material for only one or two demonstrations. These authors report that this compound is not shock sensitive, but our experience clearly demonstrates such sensitivity when a sample is struck with a hammer. For this reason, we strongly recommend against the storage of this material.

The rapid combustion and shock sensitivity of this compound [3] illustrate the general instability of organic peroxides.

REFERENCES

1. Milas, N. A.; Golubovic, A. *J. Am. Chem. Soc.* **1959**, *81*, 6461.
2. El-Awady, A.; Prell, L. J. "Chemical Demonstrations Proceedings"; Western Illinois University and Quincy-Keokuk Section of the American Chemical Society, 1979; p 47.
3. Sax, N. I. "Dangerous Properties of Industrial Materials," 3rd ed.; Reinhold Book Corp.: New York, 1968; p 1003.

1.19

Reaction of Zinc and Iodine

The addition of a few milliliters of water to a mixture of zinc and iodine results in the evolution of purple vapor.

MATERIALS

2 g 20–30 mesh zinc (Do not use finer mesh.)

7 g iodine

50 ml ethyl alcohol, C_2H_5OH

mortar and pestle, ca. 100 mm

test tube, 18 mm $\times$ 150 mm

test tube clamp

ring stand and iron ring

3–5 liter round-bottomed, Florence, or Erlenmeyer flask

10-ml graduated cylinder

PROCEDURE

To assure a finely divided and intimately mixed condition, grind the zinc and iodine together using the mortar and pestle. Attach the test tube clamp to the ring stand. Mount the test tube vertically in the test tube clamp and then pour the mixture into the test tube. Attach the iron ring to the ring stand above the test tube clamp. Invert the flask and support it in the iron ring so that the flask touches the clamp. The open end of the test tube should extend well into the flask (see figure).

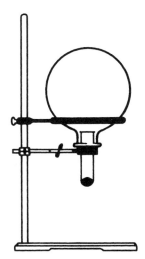

With one hand, lift and hold the flask; with the other hand, add 3 ml of water from the graduated cylinder. Immediately replace the flask. A cloud of iodine vapor will rise from the test tube into the flask, and the test tube will become too hot to touch. Adequate ventilation should be provided, since iodine vapors may escape from the flask. Upon cooling, iodine vapors will condense on the flask and test tube, and the bottom of the test tube will contain a mixture of aqueous zinc iodide (ZnI_2) and reactants.

HAZARDS

Contact with solid iodine causes burns. The vapor irritates the eyes and respiratory system. The evolution of iodine vapor in the classroom should be avoided; adequate ventilation should be provided in case such evolution occurs.

The test tube becomes hot enough to cause burns.

DISPOSAL

The products should be flushed down the drain with water. The test tube should be rinsed with ethyl alcohol to remove the excess solid iodine.

DISCUSSION

The addition of water facilitates the reaction of zinc and iodine, possibly because of its role as a reaction medium. The water also removes the soluble zinc iodide product from the zinc surface, allowing further reaction to occur. The heat of solution of zinc iodide contributes to the initiation and the overall exothermic nature of the reaction.

The energy produced increases the temperature beyond the melting point of the iodine ($113.5°C$) [1], at which point purple vapors are observed due to the sublimation of iodine.

The reactions and their ΔH values are [2]

$$Zn(s) + I_2(s) \longrightarrow ZnI_2(s) \qquad \Delta H_f° = -208 \text{ kJ/mole}$$

$$ZnI_2(s) + H_2O(l) \longrightarrow ZnI_2(aq) \qquad \Delta H_f° = -56.23 \text{ kJ/mole}$$

The high solubility of zinc iodide and its high heat of solution result in a significant contribution to the overall reaction energy [3].

REFERENCES

1. Weast, R. C., Ed. "CRC Handbook of Chemistry and Physics," 49th ed.; CRC Press: Cleveland, Ohio, 1968; p B-205.
2. "Selected Values of Chemical Thermodynamic Properties." *Natl. Bur. Stand. (U.S.)* **1968**; Technical Note 270-3; p 237.
3. Weast; p B-263.

1.20

Reaction of Zinc and a Mixture of Ammonium Nitrate and Ammonium Chloride

The addition of several drops of water to a pile of ammonium salts mixed with zinc dust yields a puff of white smoke.

MATERIALS

4 g ammonium nitrate, NH_4NO_3

0.5 g ammonium chloride, NH_4Cl

4 g zinc dust

0.5 g iodine (optional)

spatula

90-mm evaporating dish

dropper bottle of distilled water

PROCEDURE

With a spatula mix the dry ammonium salts in the evaporating dish. Add the zinc dust to the mixture and mix cautiously since the reaction may be initiated by any moisture present. If a violet smoke is desired, add 0.5 g of iodine.

To initiate the reaction add 1–2 drops of water to the pile. Keep the hands and body well away from the mixture. The reaction generates smoke, a white cloud, and a bluish flame. Afterward, a pale yellow solid will be found in the dish and in the immediate vicinity. The room should be well ventilated since the reaction disperses smoke and particulate matter.

HAZARDS

Do not use larger amounts of materials in this demonstration because of the explosive potential of ammonium nitrate. Ammonium nitrate is a very strong oxidant in dry form and can initiate combustion in flammable materials. Moreover, flammable objects should be removed from the immediate vicinity. Once the mixture is ignited, spattering occurs.

Good ventilation must be provided. The smoke should not be inhaled, since zinc oxide and iodine vapors are both irritants.

The presence of wet surfaces or high humidity may initiate this reaction inadvertently.

DISPOSAL

The solids produced in this reaction should be wiped up with a damp cloth or sponge, which should then be thoroughly rinsed with water.

DISCUSSION

The decomposition of ammonium nitrate is catalyzed by chloride ions [1]. This catalytic effect can be observed by using different sources of anhydrous chloride ions (such as LiCl, NaCl, and $SrCl_2$). We did not observe characteristic flame colors with LiCl and $SrCl_2$. The addition of a few drops of water results in the controlled (as opposed to explosive) decomposition of ammonium nitrate [2]. The decomposition reaction becomes autocatalytic as water is produced:

$$NH_4NO_3(s) \longrightarrow N_2O(g) + 2\ H_2O(g)$$

As this reaction proceeds, the temperature increase presumably causes the ammonium nitrate to melt (melting point $= 169.6°C$) [3] and enhances the dissolution and oxidation of zinc [4]. Although the nitrogen-containing products have not been clearly established, a reasonable equation for the reaction is

$$Zn(s) + NH_4NO_3(s) \longrightarrow N_2(g) + ZnO(s) + 2\ H_2O(g)$$

The energy changes for the two reactions shown above are calculated to be -36.0 kJ and -466.5 kJ, respectively [5]. The energy, observed in the form of the flame, causes the dispersal of zinc oxide as a white cloud.

In the presence of iodine, the cloud is reddish maroon because of the dispersal of sublimed iodine.

REFERENCES

1. Mellor, J. W. "A Comprehensive Treatise on Inorganic and Theoretical Chemistry"; Longmans, Green and Co.: London, 1964; Vol. VIII, Supplement I, p 544.
2. Mellor; p 543.
3. Weast, R. C., Ed. "CRC Handbook of Chemistry and Physics," 49th ed.; CRC Press: Cleveland, Ohio, 1968; p 205.
4. Mellor; p 545.
5. "Selected Values of Chemical Thermodynamic Properties." *Natl. Bur. Stand. (U.S.)* **1965**, **1968**; Technical Note 270-1, pp 13, 62, 67; Technical Note 270-3, p 233.

1.21

Reaction of Zinc and Sulfur

Insertion of a red-hot wire into a small pile of powdered zinc and sulfur results in a chemical reaction including light, a hissing sound, and a mushroom-shaped cloud.

MATERIALS

6 g zinc powder

1 g sulfur powder

combustion spoon or heavy iron wire, ca. 30 cm long

iron tripod, or ring stand and iron ring

Meker burner

spatula

ring stand with metal base

PROCEDURE

Since the reaction produces thick smoke, this demonstration should be performed in a well-ventilated room.

Before beginning the demonstration, place the combustion spoon on the tripod so that the handle is over the Meker burner. Light the burner and adjust it so that the tip of the spoon's handle is in the hottest part of the flame.

Mix the zinc and sulfur to an homogeneous color. The mixing can be done on weighing paper using a spatula. Pile the zinc-sulfur mixture in the middle of the base of the ring stand.

When the tip of the combustion spoon handle is red hot, quickly remove it from the flame and place the hot tip in the center of the pile of powder. This must be done quickly, since the tip of the wire must be as hot as possible. The reaction is quite violent, so the hands and body should be kept well away from the powder. Almost instantly, a brilliant flash of light, the emission of hot sparks, a hissing sound, and a mushroom-shaped cloud of smoke are produced. Afterward, pale yellow flakes remain on the table, and a mixture of a pale yellow solid and an orange solid remains on the base of the ring stand. The orange color fades as the mixture cools.

HAZARDS

Because of the smoke, this demonstration should be performed in a well-ventilated room. The smoke should not be inhaled. Sulfur dioxide is a toxic and irritating gas. Inhalation of zinc oxide fumes can cause chills and fever.

Since the reaction is quite violent, hands and body should be kept well away from the reaction zone.

DISPOSAL

The residues should be wiped up with a damp cloth and discarded in a waste container.

DISCUSSION

Although the molar ratio of zinc to sulfur used in this demonstration is about 3:1, all the zinc appears to be consumed. Both ZnS and ZnO are white, but ZnO becomes yellow when heated [1]. The pale yellow residue is a mixture of ZnO and ZnS. The base of the ring stand turns somewhat black at the site of the reaction and where hot sparks strike its surface. We believe that the chemical reactions are

$$Zn(s) + S(s) \longrightarrow ZnS(s)$$
$$Zn(s) + \tfrac{1}{2} O_2(g) \longrightarrow ZnO(s)$$
$$S(s) + O_2(g) \longrightarrow SO_2(g)$$

The smoke contains SO_2 as well as particulate matter consisting of ZnS and ZnO. The flash of light is predominantly yellowish orange, although a bluish edge is sometimes observed.

The three reactions are all exothermic. The heats of formation (ΔH_f°) of ZnS(s) and ZnO (sphalerite) are -206.0 kJ/mole and -348.3 kJ/mole, respectively, while the heat of formation of $SO_2(g)$ is -297.04 kJ/mole [2]. The reaction must be initiated by heat, but once the process starts, the heat evolved is sufficient to sustain the reaction.

REFERENCES

1. Cotton, F. A.; Wilkinson, G. "Advanced Inorganic Chemistry: A Comprehensive Text"; Interscience Publishers, John Wiley and Sons: New York, 1966; p 604.
2. "Selected Values of Chemical Thermodynamic Properties." *Natl. Bur. Stand. (U.S.)* **1965, 1968**; Technical Note 270-1, p 45; Technical Note 270-3, pp 177, 180.

1.22

Reaction of Iron and Sulfur

Insertion of a red-hot wire into a small pile of powdered iron and sulfur produces a glow that travels through the pile accompanied by a few sparks.

MATERIALS

10 g iron powder

5 g sulfur powder

combustion spoon or heavy iron wire, ca. 30 cm long

iron tripod, or ring stand and iron ring

Meker burner

spatula

ring stand with metal base

PROCEDURE

Before beginning the demonstration, place the combustion spoon on the tripod so that the handle is over the Meker burner. Light the burner and adjust it so that the tip of the spoon's handle is in the hottest part of the flame.

Mix the iron and sulfur to an homogeneous color. The mixing can be done on weighing paper using a spatula. Pile the iron-sulfur mixture in the middle of the base of the ring stand.

When the tip of the combustion spoon handle is red hot, quickly remove it from the flame and place it in the center of the pile of powder. A glow will move from the center of the pile toward the edges. The reaction, once initiated, continues at a moderate rate, accompanied by a red glow and occasional sparks. This effect is more visible in dim light. Afterward, the spoon will be attached to the gray solid product.

HAZARDS

The reaction produces sufficient heat to cause burns. Sulfur dioxide is a toxic and irritating gas.

DISPOSAL

The cake of solid can be removed from the handle by gentle pounding, and the solid residue should be discarded in a waste container.

DISCUSSION

The mole ratio of iron to sulfur used in this demonstration is approximately 1:1. We presume that the chemical equation for the reaction is

$$Fe(s) + S(s) \longrightarrow FeS(s)$$

The standard heat of formation, ΔH_f°, for FeS is -100 kJ/mole [1].

"Although the iron-sulfur system has probably been studied more extensively than any other sulfide system, it has not yet been completely disentangled" [2]. As this statement asserts, the identity and nature of the product(s) of this reaction are not definite. Because of the absence of clear-cut differences between the properties of the reactant mixture and the reaction product(s) [3], the demonstration should not be used to distinguish physical change (simple mixing of iron and sulfur) from chemical change [4].

REFERENCES

1. "Selected Values of Chemical Thermodynamic Properties." *Natl. Bur. Stand. (U.S.)* **1968**; Technical Note 270-4, p 80.
2. Nickless, G., Ed. "Inorganic Sulfur Chemistry"; Elsevier Publishing Co.: New York, 1968; p 715.
3. Winderlich, R. *Z. Physik. Chem. Unterricht* **1920**, *33,* 100–103.
4. Alyea, H. N.; Dutton, F. B., Eds. "Tested Demonstrations in Chemistry," 6th ed.; Journal of Chemical Education: Easton, Pennsylvania, 1965; p 5.

1.23

Reaction of Sodium Peroxide and Sulfur

Addition of a few drops of water to a pile of sodium peroxide and sulfur produces a bright flame.

MATERIALS

0.1 g sulfur powder

1 g sodium peroxide, Na_2O_2

spatula

evaporating dish, ca. 9 cm in diameter

dropper bottle of water

PROCEDURE

Mix 0.1 g of sulfur powder and 1 g of Na_2O_2 on weighing paper using a spatula to form a uniform mixture. Pile the mixture in the middle of the evaporating dish and make a slight depression in the top of the pile.

Add 1–3 drops of water to the depression and quickly step back. After a few seconds the reaction yields a bright flash and a cloud of smoke.

HAZARDS

Sodium peroxide is a strong oxidizing agent, especially when wet. Make sure that all surfaces used are dry. Avoid contact with the skin.

Sulfur dioxide is a toxic and irritating gas.

DISPOSAL

After it cools, the evaporating dish should be wetted carefully with water to decompose any excess peroxide. The surface should then be rinsed free of remaining solid.

DISCUSSION

The mole ratio of sodium peroxide to sulfur used in this demonstration is 4:1. We presume that the important reactions are

$$2 \, Na_2O_2(s) + 4 \, H_2O(l) \longrightarrow 4 \, NaOH(aq) + 2 \, H_2O_2(aq) \tag{1}$$

$$S(s) + 2 \, H_2O_2(aq) \longrightarrow 2 \, H_2O(l) + SO_2(g) \tag{2}$$

$$2 \, H_2O(l) + 2 \, Na_2O_2(s) + S(s) \longrightarrow 4 \, NaOH(aq) + SO_2(g) \tag{3}$$

For reaction 1: $\Delta H^{\circ}_{rxn} = -108$ kJ, and for reaction 2: $\Delta H^{\circ}_{rxn} = -486$ kJ. For reaction 3, which is the sum of reactions 1 and 2: $\Delta H^{\circ}_{rxn} = -594$ kJ [1]. Reaction 1 produces aqueous hydrogen peroxide [2], which in a basic solution rapidly oxidizes sulfur [3]. Reaction 2 produces water, which continues the decomposition of sodium peroxide leading to the net equation shown in reaction 3.

REFERENCES

1. "Selected Values of Chemical Thermodynamic Properties." *Natl. Bur. Stand. (U.S.)* **1952**; Circ. 500; pp 9, 10, 37, 447, 448.
2. Chernyayev, I. I.; Vol'nov, I. I.; Dobrynina, T. A., Eds. "Chemistry of Peroxide Compounds"; U.S. Dept. of Commerce, Office of Technical Services (JPRS: 27,465), 1964; p 13.
3. Cotton, F. A.; Wilkinson, G. "Advanced Inorganic Chemistry: A Comprehensive Text"; Interscience Publishers, John Wiley and Sons: New York, 1966; p 373.

1.24

Reaction of Sodium Peroxide and Aluminum

Addition of a few drops of water to a pile of sodium peroxide and powdered aluminum results in a bright flash of light.

MATERIALS

1 g sodium peroxide, Na_2O_2

0.5 g aluminum powder

glass rod

spatula

ring stand with metal base

Mohr pipette or 25-cm piece of 6-mm glass tubing drawn to a tip

50-ml beaker

PROCEDURE

Mix the sodium peroxide and the aluminum powder on weighing paper using a glass rod. Do not grind! With a spatula, pile the mixture in the middle of the base of the ring stand. The spatula and the base of the ring stand must be dry.

With the Mohr pipette draw 1–2 drops of water from the beaker and add to the top of the pile. Step back immediately. Do not look directly at the pile. In a few seconds, a brilliant flash of light occurs. The red-hot pile of products turns gray upon cooling.

HAZARDS

Sodium peroxide is a strong oxidizing agent, especially when wet. Make sure that all surfaces are dry. Avoid contact with the skin.

Since the bright light produced may be harmful, observers should be cautioned not to look directly at the pile after the reaction is initiated.

DISPOSAL

After the products cool, the base of the ring stand should be wetted carefully with water to decompose any excess peroxide. The surface should then be rinsed free of remaining solid.

DISCUSSION

The mole ratio of sodium peroxide to aluminum used in this demonstration is 1:1.5. The important reactions that occur upon the addition of water are

$$Na_2O_2(s) + 2 H_2O(l) \longrightarrow 2 NaOH(aq) + H_2O_2(aq) \tag{1}$$

$$2 Al(s) + 3 H_2O_2(aq) \longrightarrow Al_2O_3(s) + 3 H_2O(g) \tag{2}$$

$$4 Al(s) + 3 O_2(g) \longrightarrow 2 Al_2O_3(s) \tag{3}$$

For reaction 1: $\Delta H^{\circ}_{rxn} = -54.0$ kJ, for reaction 2: $\Delta H^{\circ}_{rxn} = -1820$ kJ, and for reaction 3: $\Delta H^{\circ}_{f} = -1670$ kJ/mole [1]. Reaction 1 produces aqueous hydrogen peroxide [2], which in a basic solution rapidly oxidizes aluminum [3]. The excess of aluminum is consumed in reaction 3, which is initiated by the high temperatures achieved in reaction 2.

REFERENCES

1. "Selected Values of Chemical Thermodynamic Properties." *Natl. Bur. Stand. (U.S.)* **1952**; Circ. 500; pp 9, 10, 319, 447, 448.
2. Chernyayev, I. I.; Vol'nov, I. I.; Dobrynina, T. A., Eds. "Chemistry of Peroxide Compounds"; U. S. Dept. of Commerce, Office of Technical Services (JPRS: 27,465), 1964; p 13.
3. Cotton, F. A.; Wilkinson, G. "Advanced Inorganic Chemistry: A Comprehensive Text"; Interscience Publishers, John Wiley and Sons: New York, 1966; p 373.

1.25

Reaction of Sodium and Chlorine

A drop of water is added to a piece of sodium in a flask of chlorine. The sodium ignites in the chlorine producing a bright yellow flame and a cloud of white sodium chloride.

MATERIALS

0.5 cm^3 sodium metal

chlorine cylinder, with valve and rubber tubing

ethyl alcohol, C_2H_5OH

spatula

sand, ca. 100 g

1-liter or 2-liter Florence, Erlenmeyer, or round-bottomed flask

forceps

glass stirring rod, to fit flask with 6 cm protruding

cork stopper to fit flask

ring stand with appropriate clamp to hold neck of flask

dropper bottle of water

PROCEDURE

Cut a piece of clean sodium metal with the spatula to obtain a piece approximately 0.5 cm^3. Pour dry sand into the flask to a depth of 1 cm. With forceps drop the piece of sodium on the sand and center it with the stirring rod.

In a hood, fill the flask with chlorine gas by upward displacement of air. Cork the flask. Clamp it to the ring stand about 15 cm above the bench top (see figure).

To perform the demonstration, remove the cork and add 2–3 drops of water from the dropper so that the water contacts the sodium. Do not replace the cork. When water reacts with sodium, sufficient heat is evolved to melt the sodium. The hot sodium reacts with the chlorine gas, yielding a bright yellow flame and forming white fumes of sodium chloride. When the reaction is complete, re-cork the flask to contain any unreacted chlorine.

HAZARDS

Sodium is dangerous when exposed to moisture, since it reacts to produce hydrogen with sufficient heat to cause an explosion. Fire extinguishers containing water or

61

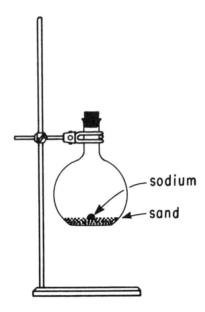

carbon dioxide should not be used on fires involving sodium metal. Sodium should be stored under nitrogen or kerosene (not chlorinated hydrocarbons), and only the amount to be used immediately should be removed from the container.

Avoid contact with sodium metal since it reacts exothermically with the moisture in body tissues to produce both thermal and chemical burns.

This demonstration should be performed in a well-ventilated room, since sodium oxide may be released when sodium is heated in air.

Chlorine gas irritates the eyes and mucous membranes and, if inhaled, can cause severe lung irritation and fatal pulmonary edema. In high concentrations, the gas irritates the skin.

Chlorine is a strong oxidizing agent, and combustible materials will burn in an atmosphere of Cl_2. Chlorine forms explosive mixtures with flammable vapors such as acetylene.

DISPOSAL

Open the flask in a hood to dissipate any excess chlorine. After the flask has cooled, it should be rinsed first with ethyl alcohol (to remove any unreacted sodium) and then with water. The wet sand should be discarded in a waste container.

DISCUSSION

The white solid that is formed is principally sodium chloride, although oxides of sodium are probably minor products. The reaction that occurs is exothermic:

$$2 \text{ Na(s)} + \text{Cl}_2(\text{g}) \longrightarrow 2 \text{ NaCl(s)}$$

The standard heat of formation, ΔH_f°, of sodium chloride is -411 kJ/mole [1].

This demonstration illustrates a typical reaction between an alkali metal and a halogen. Although the elements sodium and chlorine are extremely hazardous, the compound they form is a food seasoning.

REFERENCE

1. "Selected Values of Chemical Thermodynamic Properties." *Natl. Bur. Stand. (U.S.)* **1952**; Circ. 500; p 450.

1.26

Reaction of Antimony and Chlorine

Antimony powder shaken into a flask of chlorine gas produces sparks and white fumes.

MATERIALS

chlorine cylinder, with valve and rubber tubing

antimony powder, in salt shaker

1-liter or 2-liter Florence flask

cork stopper for flask

ring stand with appropriate clamp to hold neck of flask

PROCEDURE

Working in a hood, fill the flask with chlorine gas by upward displacement of air. Cork the flask. Fill a salt shaker with antimony powder.

To perform the demonstration, clamp the flask high on the ring stand. Remove the cork from the flask, hold the salt shaker inverted over the flask, and tap the shaker so that antimony powder falls into the flask.

HAZARDS

Chlorine gas irritates the eyes and mucous membranes and, if inhaled, can cause severe lung irritation and fatal pulmonary edema. In high concentrations, the gas irritates the skin.

Chlorine is a strong oxidizing agent, and combustible materials will burn in an atmosphere of Cl_2. Chlorine forms explosive mixtures with flammable vapors such as acetylene.

The antimony trichloride and pentachloride that are formed are poisonous and are skin irritants. Avoid breathing or contacting the product vapors or the antimony powder.

DISPOSAL

In a hood, fill the flask with water to expel the unreacted chlorine and to dissolve the antimony chlorides. The resulting solution should be flushed down the drain with water.

DISCUSSION

Antimony metal in finely divided powdered form reacts rapidly with excess chlorine to form antimony chloride. The principal product is antimony pentachloride [1, 2]. The reactions are exothermic, the trichloride evolving 382.2 kJ/mole and the pentachloride yielding 438.5 kJ/mole according to these reactions [3]:

$$Sb(s) + \frac{3}{2} Cl_2(g) \longrightarrow SbCl_3(s)$$

$$Sb(s) + \frac{5}{2} Cl_2(g) \longrightarrow SbCl_5(l)$$

REFERENCES

1. Mellor, J. W. "A Comprehensive Treatise on Inorganic and Theoretical Chemistry"; Longmans, Green and Co.: London, 1928; Vol. IX, p 486.
2. Bailar, J. C.; Emeleus, H. J.; Nyholm, R.; Trotman-Dickenson, A. F. "Comprehensive Inorganic Chemistry"; Pergamon Press: Oxford, 1973; Vol. II, p 648.
3. "Selected Values of Chemical Thermodynamic Properties." *Natl. Bur. Stand. (U.S.)* **1952**; Circ. 500; p 92.

1.27

Reaction of Iron and Chlorine

When a glowing wad of steel wool is placed in a flask containing chlorine gas, a large amount of reddish brown smoke is immediately produced, and the steel wool glows brightly.

MATERIALS

chlorine cylinder, with valve and rubber tubing

ca. 7 g steel wool

iron wire from coat hanger, ca. 40 cm long

2 corks to fit neck of 5-liter flask

5-liter round-bottomed flask

cork ring to support flask

burner

PROCEDURE

Remove the paint from the coat-hanger wire and bend one end into a small hook with a radius of about 2 cm. Pierce 1 cork and insert the wire into it so that, when the cork is in the flask, the hook will be in the center of the flask (see figure). Remove this assembly from the flask. In a hood, fill the flask with chlorine gas by upward displacement of air. Stopper the flask with the second cork.

To perform the demonstration, wrap the steel wool around the hook on the wire. Light the burner. Heat the steel wool in the flame until some of it glows brightly. Remove the cork from the flask and immediately plunge the glowing steel wool into the flask, so that the cork on the wire is seated loosely. As the hot steel wool contacts the chlorine, reddish brown smoke will form immediately and will rise from the neck

of the flask, although it will be somewhat contained by the cork. The flask will become hot. After about a minute, remove the steel wool to show that it continues to glow brightly.

HAZARDS

Chlorine gas irritates the eyes and mucous membranes and, if inhaled, can cause severe lung irritation and fatal pulmonary edema. In high concentrations, the gas irritates the skin.

Chlorine is a strong oxidizing agent, and combustible materials will burn in an atmosphere of Cl_2. Chlorine forms explosive mixtures with flammable vapors such as acetylene.

Anhydrous iron(III) chloride is violently decomposed by water with the formation of hydrogen chloride. Inhalation of fine crystals of $FeCl_3$ can irritate or burn the mucous membranes. The anhydrous salt can cause thermal and acid burns to eyes and skin by reacting with moisture in the body.

Steel wool can cut the skin.

DISPOSAL

When the flask is cool, it should be opened in a hood to draw off the acrid smoke. To decompose and dissolve the iron chloride, carefully add water to the flask and then flush down the drain with water.

When cool, the excess steel wool should be discarded in a waste container.

DISCUSSION

Iron reacts with chlorine to form $FeCl_3$:

$$Fe(s) + \frac{3}{2} Cl_2(g) \longrightarrow FeCl_3(s)$$

The reaction is highly exothermic: $\Delta H_f^\circ = -405$ kJ/mole [1]. Under the conditions of this demonstration, the major product is $FeCl_3$. For iron(II) chloride, $FeCl_2$, the standard heat of formation is -341 kJ/mole [1]. Upon hydrolysis with water, $FeCl_2$ reacts violently to form hydrated iron(II) oxide and hydrochloric acid, while $FeCl_3$ yields hydrated iron(III) oxide and hydrochloric acid.

REFERENCE

1. "Selected Values of Chemical Thermodynamic Properties." *Natl. Bur. Stand. (U.S.)* **1952**; Circ. 500; p 263.

1.28

Reaction of Aluminum and Bromine

In a fume hood in a darkened room, small pieces of aluminum foil are added to liquid bromine. Flashes of light and small flames moving on the surface of the liquid are observed.

MATERIALS

10 ml bromine

aluminum foil, ca. 5 cm × 5 cm

10-ml graduated cylinder

gloves, plastic or rubber

600-ml beaker

12-cm watch glass

PROCEDURE

Wearing gloves and working in a fume hood, add 10 ml of bromine to the beaker. Cover the beaker with the watch glass. Tear the aluminum foil into small pieces (ca. 0.5 cm in diameter) and place the pieces on the watch glass.

Dim the room lights, lift the watch glass from the beaker, and add the aluminum foil to the bromine. Set the watch glass back on the beaker to reduce diffusion of the bromine into the room. In about a minute, flashes of light and small flames moving on the surface of the liquid are observed.

HAZARDS

Since bromine, a very strong oxidizing agent, vaporizes readily at room temperature to yield toxic fumes, adequate ventilation must be provided. Wear gloves while handling bromine containers. The reaction products, which include aluminum bromide and hydrogen bromide, are also toxic and cause burns.

DISPOSAL

The unreacted bromine should be allowed to evaporate. After the bromine has evaporated, the beaker should be rinsed to remove nonvolatile materials. Alternatively, bromine can be allowed to react with a solution of a mild reducing agent, such as sodium thiosulfate or sodium bisulfite, and then the aqueous mixture can be flushed down the drain with water.

DISCUSSION

The reaction between metallic aluminum and liquid bromine occurs exothermically, with accompanying flashes of light as the reaction proceeds. The reaction is

$$Al(s) \ + \ \frac{3}{2} \ Br_2(l) \longrightarrow AlBr_3(s)$$

The heat of formation of aluminum bromide is -526 kJ/mole [1]. The hygroscopic aluminum bromide fumes in air, producing a cloud which is not readily visible in the darkened room.

REFERENCE

1. "Selected Values of Chemical Thermodynamic Properties." *Natl. Bur. Stand. (U.S.)* **1952**; Circ. 500; p 321.

1.29

Reaction of White Phosphorus and Chlorine

When a small lump of white phosphorus is placed in a flask containing chlorine gas, sparks and a small flame erupt from the solid and a white coating forms on the inner surface of the flask.

MATERIALS

chlorine cylinder, with valve and rubber tubing

3–5 mm cube of white (yellow) phosphorus

2 corks to fit neck of 5-liter flask

combustion spoon

5-liter round-bottomed flask

cork ring to support flask

paper towel or sheet of filter paper

tongs, forceps, or spatula to handle phosphorus

PROCEDURE

Pierce 1 cork and insert the handle of the combustion spoon through the hole so that, when the cork is in the mouth of the flask, the bowl of the spoon will be in the center of the flask (see figure). Remove this assembly from the flask. In a hood, fill the flask with chlorine gas by upward displacement of air. Stopper the flask with the second cork.

To perform the demonstration, remove the phosphorus from the water in which it is stored and dry it with the filter paper. Use tongs or other suitable tool to handle the phosphorus. Work quickly, since white phosphorus ignites spontaneously in air.

Place the phosphorus in the combustion spoon, remove the cork from the flask, and place the spoon and its cork loosely in the neck of the flask. After 5–10 seconds, the phosphorus will emit sparks and then a small white flame. After about 1 minute, the sparks will die down, but the phosphorus will burn for 3–5 minutes.

HAZARDS

Chlorine gas irritates the eyes and mucous membranes and, if inhaled, can cause severe lung irritation and fatal pulmonary edema. In high concentrations, the gas irritates the skin.

Chlorine is a strong oxidizing agent, and combustible materials will burn in an atmosphere of Cl_2. Chlorine forms explosive mixtures with flammable vapors such as acetylene.

White phosphorus is spontaneously flammable in air. Combustion in air produces phosphorus pentoxide. Both white phosphorus and phosphorus pentoxide are very poisonous and can cause severe burns. Chronic effects can result from continued absorption of small amounts. Exposure of phosphorus to the air should be very brief, and all cutting should be done under water.

Phosphorus pentachloride is violently decomposed by water with formation of hydrochloric acid and phosphoric acid. Phosphorus pentachloride is an irritant which can burn the skin, eyes, and respiratory system.

DISPOSAL

In a hood, remove the cork and allow the smoke to be drawn off. Carefully rinse the inside walls of the flask with water and flush down the drain with water.

DISCUSSION

The exothermic reaction between phosphorus and chlorine yields phosphorus pentachloride:

$$P_4(s) + 10\ Cl_2(g) \longrightarrow 4\ PCl_5(s)$$

The standard heat of formation, ΔH_f°, for solid PCl_5 is -463 kJ/mole [1]. Phosphorus pentachloride is a solid which hydrolyzes upon reacting with water to yield phosphoric acid and hydrochloric acid. Above 300°C, PCl_5 dissociates to PCl_3 and Cl_2.

Phosphorus trichloride is a liquid at room temperature. Its standard heat of formation, ΔH_f°, is -311 kJ/mole [2]. It hydrolyzes upon reacting with water to yield phosphorous acid and hydrochloric acid.

REFERENCES

1. Mellor, J. W. "A Comprehensive Treatise on Inorganic and Theoretical Chemistry"; Longmans, Green and Co.: London, 1967; Vol. VIII, pp 475–77.
2. Mellor; p 466.

1.30

Reaction of Red Phosphorus and Bromine

The addition of powdered red phosphorus to liquid bromine produces a yellow flame and copious fumes.

MATERIALS

1 ml bromine

0.3 g red phosphorus

gloves, rubber or plastic

10-ml graduated cylinder

60-mm evaporating dish

spatula

PROCEDURE

Wearing gloves and working in a fume hood, add 1 ml of bromine to the evaporating dish. Using a spatula, add 0.3 g of red phosphorus to the bromine in the evaporating dish and step back quickly. A bright yellow flame and copious fumes will be produced instantly.

HAZARDS

Since bromine, a very strong oxidizing agent, vaporizes readily at room temperature to yield toxic fumes, adequate ventilation must be provided. Wear gloves while handling bromine containers. The reaction products include phosphorus tribromide, which hydrolyzes in moist air to yield phosphorous acid and hydrogen bromide. These products, which are expelled from the dish as a fuming smoke, are toxic and are skin irritants.

DISPOSAL

The unreacted bromine should be allowed to evaporate in a hood. Alternatively, bromine can be allowed to react with a solution of a mild reducing agent, such as sodium thiosulfate or sodium bisulfite, and then the aqueous mixture can be flushed down the drain with water.

DISCUSSION

The exothermic reaction between red phosphorus and bromine yields phosphorus tribromide, according to the reaction:

$$P_4(s) + 6\,Br_2(l) \longrightarrow 4\,PBr_3(l)$$

The standard heat of formation, ΔH_f°, for phosphorus tribromide is -199 kJ/mole [1].

Phosphorus tribromide is a clear, mobile, fuming liquid which hydrolyzes upon reacting with water to yield phosphorous acid and hydrogen bromide [2].

REFERENCES

1. "Selected Values of Chemical Thermodynamic Properties." *Natl. Bur. Stand. (U.S.)* **1952**; Circ. 500; p 79.
2. Mellor, J. W. "A Comprehensive Treatise on Inorganic and Theoretical Chemistry"; Longmans, Green and Co.: London, 1930; Vol. III, pp 1031–32.

1.31

Spontaneous Combustion of White Phosphorus

A solution of white phosphorus in carbon disulfide is poured onto a piece of filter paper placed over the top of a glass cylinder. Within a few minutes, the paper ignites and a sharp, bark-like sound is produced.

MATERIALS

2 g white (yellow) phosphorus

10 ml carbon disulfide, CS_2

gloves, rubber or plastic

120-mm evaporating dish

knife or spatula

forceps or tongs

50-ml beaker

filter paper, 1 piece for each cylinder, slightly larger in diameter than the mouth of each cylinder

several cylinders of various sizes, such as 2000 ml, 1000 ml, 500 ml, 250 ml

10-ml graduated cylinder

disposable pipette or dropper

small glass bottle, stoppered

PROCEDURE

Prepare the solution of white phosphorus in carbon disulfide prior to the demonstration. Wearing rubber or plastic gloves, place the white phosphorus under water in a large evaporating dish for cutting. (White phosphorus is usually stored under water.) With tongs or forceps, add the pieces to a tared beaker of water for weighing. Add 2 g of white phosphorus to 10 ml of carbon disulfide and swirl to dissolve. The solubility of white phosphorus in carbon disulfide is 1.25 g/ml [1].

Place a piece of filter paper on top of each cylinder. Remove all other combustible materials from the immediate area. Using the pipette, carefully add approximately 1 ml of the white phosphorus solution to the center of each piece of filter paper.

After an interval ranging from less than a minute to over 3 minutes depending on the amount of liquid and the temperature of the room, the paper suddenly catches fire.

The carbon disulfide in the cylinder burns rapidly, producing a sharp sound somewhat similar to a dog's bark. The burning filter paper flies into the air. If you want all the pieces of filter paper to ignite simultaneously, adjust, by trial and error, the volume of solution added to each piece.

HAZARDS

White phosphorus is spontaneously flammable in air. Combustion in air produces phosphorus pentoxide. Both white phosphorus and phosphorus pentoxide are very poisonous and can cause severe burns. Chronic effects can result from continued absorption of small amounts. Exposure of phosphorus to the air should be very brief, and all cutting should be done under water.

Carbon disulfide is extremely flammable and toxic. The explosive range is 1–50% (v/v) in air [2].

Wear gloves when making or using the solution of phosphorus and carbon disulfide. The glass stoppered bottle containing the solution should be kept tightly stoppered and away from heat or flames. A fluorocarbon lubricant provides a tight, stable seal, but the solution can be stored safely for only a few days.

DISPOSAL

The cooled, charred filter paper should be discarded in a waste container. The cylinders can be cleaned by scrubbing with soap and water.

Any remaining white phosphorus in carbon disulfide should be placed in a flat pan in a fume hood away from other combustible materials and allowed to evaporate and burn.

DISCUSSION

Carbon disulfide has a boiling point of 46.5°C and a vapor density of 2.67 compared to the air's vapor density of 1.00 [2]. As the denser carbon disulfide vapors diffuse downward into the cylinder, they form an explosive mixture with air. Mixtures of carbon disulfide vapor (1–50% by volume) and air are explosive [2]. The white phosphorus remaining on the filter paper combines with oxygen from the air in a highly exothermic reaction:

$$P_4(s) + 5\ O_2(g) \longrightarrow P_4O_{10}(s)$$

The heat of formation for P_4O_{10} is -3010 kJ/mole [3], which is the energy source for the ignition of the paper and for the rapid combustion of the carbon disulfide-oxygen mixture in the cylinder. This rapid combustion produces a bark-like sound [4]. The tone of the sound produced depends on the size and shape of the cylinder.

REFERENCES

1. Windholz, M., Ed. "The Merck Index," 9th ed.; Merck and Co.: Rahway, New Jersey, 1976; p 957.

2. Windholz; p 231.

3. "Selected Values of Chemical Thermodynamic Properties." *Natl. Bur. Stand. (U.S.)* **1952**; Circ. 500; p 73.

4. Alyea, H. N.; Dutton, F. B., Eds. "Tested Demonstrations in Chemistry"; Journal of Chemical Education: Easton, Pennsylvania, 1965; p 39.

1.32

Dehydration of Sugar by Sulfuric Acid

Concentrated sulfuric acid added to granulated sugar in a beaker produces a solid-liquid mixture that changes from white to yellow to brown to black. The mixture then expands out of the beaker accompanied by vapor and the smell of burned sugar.

MATERIALS

70 g granulated sugar, $C_{12}H_{22}O_{11}$

70 ml concentrated (18M) sulfuric acid, H_2SO_4

300-ml tall-form beaker

40-cm stirring rod

paper towels

gloves, plastic or rubber

100-ml graduated cylinder

PROCEDURE

Add the granulated sugar to the tall-form beaker and insert the stirring rod into its center. Place the beaker on a mat of paper towels. Add 70 ml of concentrated sulfuric acid to the sugar and stir briefly. As the column of black solid begins to grow, support it with the stirring rod. The column grows to twice the height of the beaker.

HAZARDS

Since concentrated sulfuric acid is a strong acid and a powerful dehydrating agent, it must be handled with great care. Spills should be neutralized with an appropriate agent, such as $NaHCO_3$, and then wiped up. The solid residue may contain unreacted acid. Prolonged contact with the steam produced in the reaction can cause burns.

DISPOSAL

Carbonaceous residue should be rinsed *thoroughly* with water and then discarded in a waste container.

DISCUSSION

Among the many important industrial properties of sulfuric acid is the ability of the concentrated acid to dehydrate materials. This property is related thermodynamically to the large energy change that occurs as the sulfuric acid becomes hydrated. The dehydration of sucrose, $C_{12}H_{22}O_{11}$, is an exothermic reaction (reaction 1). The heat of dehydration for sucrose is calculated from the standard heat of combustion of sucrose (reaction 2) and the heat of formation of carbon dioxide (reaction 3):

$$C_{12}H_{22}O_{11}(s) \longrightarrow 12\ C(graphite) + 11\ H_2O(l) \tag{1}$$

$$C_{12}H_{22}O_{11}(s) + 12\ O_2(g) \longrightarrow 12\ CO_2(g) + 11\ H_2O(l) \tag{2}$$

$$C(graphite) + O_2(g) \longrightarrow CO_2(g) \tag{3}$$

For reaction 2, the standard heat of combustion, ΔH°_{comb}, is -5640.9 kJ/mole; and for reaction 3, the standard heat of formation, ΔH°_f, of carbon dioxide is -393.5 kJ/mole [1]. To calculate ΔH°_{rxn} for reaction 1, multiply reaction 3 by 12, and subtract the product from reaction 2. Thus, ΔH°_{rxn} for reaction 1 is -918.9 kJ/mole.

In this demonstration, 70 g or 0.20 moles of sucrose are used. The heat evolved for this quantity of sugar is 180 kJ.

The water formed by the dehydration of the sugar dilutes the sulfuric acid and liberates heat:

$$H_2SO_4 \cdot nH_2O + m\ H_2O \longrightarrow H_2SO_4 \cdot n_1H_2O$$

where $n + m = n_1$.

The value of n, calculated from the density of 98% H_2SO_4, is 0.11 [2]. The reaction forms 11 times 0.20, or 2.2, moles of water; therefore $n_1 = 0.11 + 2.2 = 2.3$ moles of water.

Interpolation from a table of values of ΔH°_f for $H_2SO_4 \cdot nH_2O$ yields [3]

$$\Delta H^\circ_f(H_2SO_4 \cdot nH_2O) = -814.78\ \text{kJ/mole}\quad (n = 0.11)$$

$$\Delta H^\circ_f(H_2SO_4 \cdot nH_2O) = -855.36\ \text{kJ/mole}\quad (n = 2.3)$$

The heat of dilution for this reaction is

$$(-855.36\ \text{kJ/mole}) - (-814.78\ \text{kJ/mole}) = -40.58\ \text{kJ/mole}$$

Since 1.28 moles of $H_2SO_4 \cdot nH_2O$ are used in this demonstration, the dilution of the acid contributes $(-40.58\ \text{kJ/mole})(1.28\ \text{mole}) = -51.9$ kJ. The total heat evolved is 180 kJ + 51.9 kJ = 232 kJ.

REFERENCES

1. Weast, R. C., Ed. "CRC Handbook of Chemistry and Physics," 59th ed.; CRC Press: Cleveland, Ohio, 1978–79; pp D-69, D-325.
2. Weast; p F-7.
3. "Selected Values of Chemical Thermodynamic Properties." *Natl. Bur. Stand. (U.S.)* **1952**; Circ. 500; p 41.

1.33

Reaction of Potassium Chlorate and Sugar

A drop of concentrated sulfuric acid added to a pile of potassium chlorate and granulated sugar produces smoke and a purplish flame.

MATERIALS

6 g potassium chlorate, $KClO_3$

2 g granulated sugar

1 drop of concentrated (18M) sulfuric acid, H_2SO_4

spatula

ring stand with metal base

PROCEDURE

With a spatula, carefully mix the potassium chlorate and sugar. Pile the mixture in the middle of the base of the ring stand. With the spatula, make a small depression in the top of the pile. Add 1 drop of concentrated sulfuric acid to the depression and stand back.

The reaction starts slowly, evolving smoke after 1–2 seconds, and then the pile bursts into flame.

HAZARDS

Since potassium chlorate is a strong oxidizing agent, mixtures of $KClO_3$ with combustible materials can be flammable or explosive. Do not store the mixture of sugar and potassium chlorate.

Since concentrated sulfuric acid is a strong acid and a powerful dehydrating agent, it must be handled with great care. Spills should be neutralized with an appropriate agent, such as sodium bicarbonate ($NaHCO_3$), and then wiped up.

DISPOSAL

Very little residue remains after this reaction. After it cools, the solid on the base of the ring stand should be flushed down the drain with water.

DISCUSSION

The stoichiometry of the reaction between potassium chlorate, sugar, and sulfuric acid is not known. Cotton and Wilkinson [1] report that, at room temperature, mixtures of $KClO_3$, H_2SO_4, and a reducing agent such as oxalic acid produce the potentially explosive gas, chlorine dioxide (ClO_2). According to Mellor [2], chloric acid ($HClO_3$) is formed when $KClO_3$ and H_2SO_4 are mixed and "chloric acid decomposes organic substances very rapidly, often with inflammation." Thus, the exothermic nature of this reaction is readily observed, but the energy changes cannot be described quantitatively.

REFERENCES

1. Cotton, F. A.; Wilkinson, G. "Advanced Inorganic Chemistry: A Comprehensive Text"; Interscience Publishers, John Wiley and Sons: New York, 1966; p 566.
2. Mellor, J. W. "A Comprehensive Treatise on Inorganic and Theoretical Chemistry"; Longmans, Green and Co.: London, 1922; Vol. II, p 310.

1.34

Decomposition of Ammonium Dichromate

Ignition of a cone-shaped pile of ammonium dichromate crystals produces a voluminous green solid, sparks, and smoke. The changes resemble a volcanic eruption.

MATERIALS

15 g ammonium dichromate, $(NH_4)_2Cr_2O_7$

5 ml ethanol, C_2H_5OH, or saturated potassium nitrate solution, KNO_3

powder funnel

1-liter beaker

paper toweling, ca. 2.5 cm $\times$ 2.5 cm

tongs

wire gauze (without asbestos)

500-ml wide-mouth jar, with screw cap

PROCEDURE

In a well-ventilated area, pour 15 g of ammonium dichromate through the powder funnel into the center of the beaker to form a cone-shaped pile. Prepare a "fuse" by twisting a small piece of paper toweling and soaking it in ethanol. The fuse can also be prepared by soaking the twisted toweling in saturated KNO_3 and allowing it to dry overnight. Insert the fuse in the center of the cone. With tongs, hold a match to light the fuse. Cover the beaker with the wire gauze (see figure). Sparks and smoke emanate from the cone, and a fluffy green solid is produced.

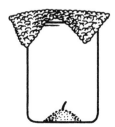

HAZARDS

Chromium salts can irritate the skin and cause skin lesions. Dust containing chromium(VI) species irritates the eyes and respiratory tract [1, 2]. Chromium compounds have been associated with cancer [1]. Do not inhale or touch either the chromium(VI) species or the green product. The toxicity of Cr_2O_3 has not been clearly established [3].

DISPOSAL

Before removing the wire gauze from the beaker, tap it gently with a spatula to dislodge any attached green product into the beaker. Remove the wire gauze and rinse with water. Cover the green chromium(III) oxide product with a small amount of water. Transfer the slurry to a jar, seal it, and discard it in a sanitary landfill.

DISCUSSION

Ammonium dichromate decomposes upon heating to produce nitrogen, water, and chromium(III) oxide [4, 5]:

$$(NH_4)_2Cr_2O_7(s) \longrightarrow N_2(g) + 4\,H_2O(g) + Cr_2O_3(s)$$

The standard heat of reaction, ΔH°_{rxn}, is -315 kJ/mole [6].

REFERENCES

1. Muir, G. D., Ed. "Hazards in the Chemical Laboratory"; The Chemical Society: London, 1977; p 199.
2. Windholz, M., Ed. "The Merck Index," 9th ed.; Merck and Co.: Rahway, New Jersey, 1976; pp 71, 289.
3. Ingerson, E. *Chem. Eng. News* **1980**, October 27, p 54.
4. Finholt, J. E. *J. Chem. Educ.* **1970**, *47*, 533.
5. Ondrus, M. G. *J. Chem. Educ.* **1976**, *53*, 228.
6. "Selected Values of Chemical Thermodynamic Properties." *Natl. Bur. Stand. (U.S.)* **1952**; Circ. 500; pp 9, 286, 291.

1.35

Reaction of Potassium Permanganate and Glycerine

The addition of about 1 ml of glycerine to a small pile of potassium permanganate crystals produces white smoke and a purple flame. The length of time before the reaction occurs varies with the state of subdivision of the solid.

MATERIALS

15 g crystalline potassium permanganate, $KMnO_4$

ca. 5 ml glycerine (glycerol), $C_3H_5(OH)_3$

2 90-mm evaporating dishes

mortar and pestle, ca. 100 ml

spatula

2 droppers

PROCEDURE

Pile 5 g of crystalline potassium permanganate in the middle of the evaporating dish. With a spatula, form a depression in the center of the pile to facilitate the permeation of the solid by glycerine. Using a dropper, carefully add approximately 1 ml of glycerine to the depression.

After several seconds a white puff of smoke is produced, followed by crackling, sparking, and a purple flame. Combustion continues until the glycerine is consumed. The product is a grayish solid with green regions.

To demonstrate the effect of surface area on the rate of reaction, prepare two piles of potassium permanganate instead of one. In one evaporating dish, place 5 g of the crystalline solid as commercially supplied. Using a mortar and pestle, grind another 5 g of potassium permanganate to a finely divided state and transfer it to the second evaporating dish. With a spatula, form a depression in the center of each pile. Using two droppers, simultaneously add glycerine to each pile. Glycerine reacts faster with finely divided solid $KMnO_4$ than with the larger crystals.

HAZARDS

This reaction yields both flame and sparks. In addition, pieces of solid $KMnO_4$ are frequently expelled several centimeters from the reaction site. Handle $KMnO_4$ with great care, since explosions can occur if it is brought into contact with organic or other readily oxidizable substances, either in solution or in the dry state [1, 2].

DISPOSAL

Unreacted $KMnO_4$ and the product (K_2MnO_4) should be flushed down the drain with water. The insoluble residue should be discarded in a waste container. After rinsing soluble materials with water, any insoluble residue of manganese oxide in the evaporating dish should be removed by brushing with a paste of ferrous ammonium sulfate and dilute sulfuric acid.

DISCUSSION

This demonstration illustrates spontaneous combustion. The relatively slow initial oxidation of glycerine by $KMnO_4$ speeds up as the system generates heat and eventually catches fire. This demonstration also illustrates the effect of an increase in the surface area on the rate of a heterogeneous reaction.

The reactions involved are varied and not completely identified. They include [3]

$$14\ KMnO_4(s)\ +\ 4\ C_3H_5(OH)_3\ (l) \longrightarrow$$
$$7\ K_2CO_3(s)\ +\ 7\ Mn_2O_3(s)\ +\ 5\ CO_2(g)\ +\ 16\ H_2O(g)$$

Manganese(III) oxide (Mn_2O_3) is black and potassium carbonate (K_2CO_3) is white. Inspection of the solid residue, which includes a greenish crystalline material, indicates that other products are formed. Addition of water to this residue yields a dark green solution and an insoluble solid. We presume that the green color is due to potassium manganate (K_2MnO_4) and that the dark insoluble solid contains Mn_2O_3 and/or MnO_2.

REFERENCES

1. Windholz, M., Ed. "The Merck Index," 9th ed.; Merck and Co.: Rahway, New Jersey, 1976; p 993.
2. Haight, G. P.; Phillipson, D. *J. Chem. Educ.* **1980**, *57*, 325.
3. Scheer, R. *J. Chem. Educ.* **1959**, *36*, A219.

1.36

Thermite Reaction

Addition of a small amount of glycerine to a pile of iron oxide, aluminum powder, and potassium permanganate produces a flame, sparks, and molten iron.

MATERIALS FOR PROCEDURE A

50–55 g iron(III) oxide powder, Fe_2O_3

15 g aluminum powder, 325 mesh or finer

20–25 g crystalline potassium permanganate, $KMnO_4$

5–6 ml glycerine (glycerol), $C_3H_5(OH)_3$

spatula

2 clay flower pots, ca. 2½ in. (ca. 6.5 cm) inside top diameter with 1-cm hole in bottom

filter paper or paper towel to fit the bottom of one pot

iron ring with inside diameter of 6–7 cm

1-m ring stand

heat-resistant pad, ca. 1 m × 1 m

large metal bucket, ca. 20 cm in diameter and 20 cm deep, half full of sand

transparent safety shield

10-ml beaker

tongs

heat-protective gloves

MATERIALS FOR PROCEDURE B

3 g iron(III) oxide powder, Fe_2O_3

1 g aluminum powder

1 g potassium chlorate, $KClO_3$

1 g granulated sugar, $C_{12}H_{22}O_{11}$

1 drop of concentrated (18M) sulfuric acid, H_2SO_4

heat-resistant pad, ca. 30 cm × 30 cm

2 spatulas

Optional Materials for Procedure B

400 ml sand

brass tubing, ca. 2.5 cm in diameter, 3 cm long, 1 mm wall thickness

aluminum foil, ca. 7 cm × 7 cm

heat-protective gloves

tongs

PROCEDURE A

If you have not performed this demonstration, we recommend that you try it outdoors first.

This demonstration should be performed only in a large well-ventilated room or outdoors. The reaction produces a large quantity of smoke. Sparks may be thrown 2 m vertically and 5 m horizontally.

Keep flammable material away from the experiment area.

Place a scrap of paper towel or filter paper over the bottom opening of one clay pot and place that pot inside the other clay pot. Mix the iron oxide and aluminum powder. Scoop the mixture into the pot. Form a small cone-shaped indentation in the center of the mixture approximately 2 cm deep and 1–2 cm wide. Fill this indentation with 20–25 g of potassium permanganate crystals. Form another small cone-shaped indentation in the $KMnO_4$ crystals. Place the clay pots inside a metal ring clamped to the top of a ring stand. Put the ring stand in the middle of the large heat-resistant pad. To catch the molten iron, place a sand bath about 1 m below the reaction vessel (see figure). The bucket should be one-half to two-thirds full of sand. Use the safety shield for protection from sparks.

Place 5–6 ml of glycerine in a small beaker. Quickly pour the glycerine onto the depression in the $KMnO_4$ crystals. Step back immediately. Ignition of the mixture is achieved in 15–60 seconds. If the reaction fails to ignite, wait 1–2 minutes after the initial flare up that results from the glycerine-$KMnO_4$ reaction, then add more $KMnO_4$ and glycerine (see Demonstration 1.35). Flame, flying sparks, smoke, and dust are produced. Molten iron runs through the hole in the pot into the sand bath. When the reaction is over, use tongs to pick up the red-hot iron. Allow the iron to cool before touching.

PROCEDURE B

Mix the iron oxide and aluminum powder and pile in the middle of the heat-resistant pad. Flatten the pile with a spatula to form a circular top with a diameter of about 2.5 cm. With a spatula, mix the sugar and potassium chlorate (do not grind!) and pour the mixture on the flattened top of the iron oxide–aluminum pile. Form a slight cavity in the top of the sugar–potassium chlorate mixture.

Add 1 drop of the 18M sulfuric acid to the cavity in the pile and step back. The reaction is quite violent, so keep the hands and body away from the mixture. Within seconds, a flame is produced, followed by sparks and smoke. Afterward, solid iron mixed with aluminum oxide remains on the pad.

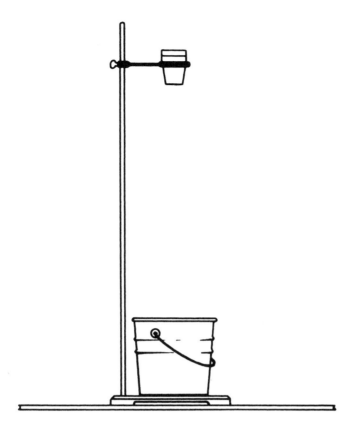

Alternatively [*1*], pile 400 ml of sand on the heat-resistant pad. Cover one end of the brass tubing with aluminum foil to form a cup. Fill the tubing to within 0.5 cm of the top with the iron oxide and aluminum mixture. Add 2 g of the sugar and potassium chlorate mixture to the tubing. Then form a small cone-shaped indentation in this mixture. Set the filled tubing on top of the sand layer. Add 1 drop of concentrated H_2SO_4 and step back. Flame, smoke, and sparks will emerge from the brass tubing, and an iron button will be formed in the tubing. Use tongs to handle the hot materials.

HAZARDS

This demonstration produces intense heat and molten metal. A fire extinguisher should be readily available at all times. Water should *not* be used to extinguish the reaction, since addition of water to hot iron produces potentially explosive hydrogen gas. Since fires resulting from thermite reactions can be difficult to control, the chemicals should not be used in larger amounts than suggested. A safety shield should be used for protection from sparks.

Care should be taken to avoid exposure to molten iron. Heat-protective gloves should be worn, and the hot products should be handled only with tongs.

Since potassium chlorate is a strong oxidizing agent, mixtures of $KClO_3$ with combustible materials can be flammable or explosive. Do not store the mixture of sugar and potassium chlorate.

Since concentrated sulfuric acid is a strong acid and a powerful dehydrating agent, it must be handled with great care. Spills should be neutralized with an appropriate agent, such as sodium bicarbonate ($NaHCO_3$), and then wiped up.

Handle $KMnO_4$ with great care, since explosions can occur if it is brought into contact with organic or other readily oxidizable substances, either in solution or in the dry state.

Because of the smoke and dust released, the demonstration should be performed in a well-ventilated room at the end of the class period.

DISPOSAL

Allow the solids produced to cool to room temperature. All solids can be scraped off into a waste container. The inner flower pot invariably cracks and should not be re-used.

DISCUSSION

The reaction of iron oxide and aluminum can be represented by the equation:

$$Fe_2O_3(s) + 2\ Al(s) \longrightarrow Al_2O_3(s) + 2\ Fe(s)$$

This reaction is one of a class of reactions known as the Goldschmidt or "thermite" process, which has been used industrially for welding, the preparation of metals from their oxides, and the production of incendiary devices. The process is initiated by heat but then becomes self-sustaining. The reaction can proceed in the absence of air since the mixture provides its own oxygen.

The thermite reaction is initiated by the heat released from the mixture of potassium chlorate, sugar, and sulfuric acid or from the mixture of potassium permanganate and glycerine. Other means of initiation have been suggested [2], but the ones described here are less complicated and more reliable than others, such as the magnesium strip and metal peroxide method [3, 4].

The changes in enthalpy, entropy, and free-energy for this reaction can be calculated from thermodynamic data [5]. For Fe_2O_3, ΔH_f° is -824 kJ/mole and S° is $+87.40$ J/mole·K; for Al_2O_3, ΔH_f° is -1673 kJ/mole and S° is $+50.92$ J/mole·K. Using these data, an enthalpy change (ΔH_{298}°) of -849 kJ/mole is obtained. The entropy change (ΔS°) is -36.48 J/mole·K; and the free-energy change at 298 K, $\Delta G = \Delta H - T\Delta S$, is -838 kJ/mole.

In this reaction, an unfavorable entropy change is offset by a large negative enthalpy. The large amount of heat produced is sufficient to raise the temperature of the products to the melting point of iron (1530°C) and to melt it ($\Delta H_{fusion} = 14.9$ kJ/mole) [6]. The iron produced is known to remain red hot even under water [4]; welding procedures have been developed as modifications of this reaction.

Roebuck [7] describes other mixtures of metal oxides and aluminum, including Cr_2O_3, MnO_2, Mn_3O_4, CuO, NiO, CoO, and Fe_3O_4. We have not used any oxide other than Fe_2O_3. We urge that other oxides not be used, since CuO/Al and Mn_3O_4/Al mixtures have been reported to react explosively [8].

REFERENCES

1. Kindler, L. I. *J. Chem. Educ.* **1965**, *42*, A607.
2. Bozzelli, J. W.; Barat, R. B. *J. Chem. Educ.* **1979**, *56*, 675.

3. Alyea, H. N.; Dutton, F. B., Eds. "Tested Demonstrations in Chemistry," 6th ed.; Journal of Chemical Education: Easton, Pennsylvania, 1965; p 17.

4. Arthur, P. "Lecture Demonstrations in General Chemistry"; McGraw-Hill: New York, 1939; p 362.

5. "Selected Values of Chemical Thermodynamic Properties." *Natl. Bur. Stand. (U.S.)* **1968**, **1971**; Technical Notes 270-3 and 270-4; p 208 and p 75.

6. Weast, R. C., Ed. "CRC Handbook of Chemistry and Physics," 54th ed.; CRC Press: Cleveland, Ohio, 1973; p B-244.

7. Roebuck, P. J. *Educ. in Chem.* (Britain) **1979**, *16*, 178.

8. Crellin, J. R. *Educ. in Chem.* (Britain) **1980**, *17*, 93.

1.37

Combustion of Magnesium in Carbon Dioxide

Ignition of magnesium and an oxidizing agent in a block of dry ice results in a brilliant flare of light and a black and white residue [1, 2].

MATERIALS

block of dry ice, solid CO_2, ca. 10 cm $\times$ 10 cm $\times$ 4 cm

2 pieces 6-cm magnesium metal, ribbon

3 g magnesium metal, powder

2 g magnesium metal, turnings

1 g potassium chlorate, $KClO_3$

gloves suitable for handling dry ice

600-ml beaker

wooden splint

tongs

burner

tool to form hole in dry ice block, e.g., scoopula or heated stainless steel rod

heat-resistant pad, ca. 0.5 m $\times$ 0.5 m

50-ml beaker

PROCEDURE

Do not look directly at burning magnesium.

Place a small chunk of dry ice inside the 600-ml beaker and wait for enough dry ice to sublime to fill the beaker. Ignite a wooden splint and slowly lower the flame into the beaker. Observe that combustion ceases.

Holding a 6-cm strip of magnesium ribbon with tongs, ignite it in a burner flame. Hold the burning magnesium ribbon inside the beaker and note that magnesium continues to burn.

Form a hole 2 cm in diameter and 2 cm deep in one of the large faces of the block of dry ice. Place the block on a heat-resistant pad about 0.25 m². In a small beaker, mix about 2 g powdered magnesium and 2 g magnesium turnings. Pour the mixture into the hole in the dry ice. Insert a 6-cm strip of magnesium ribbon in the mixture as a fuse without letting the ribbon touch the dry ice. In the small beaker, cautiously mix

1 g of powdered magnesium and 1 g of potassium chlorate. Pour this mixture around the base of the fuse (see figure). With a burner, ignite the end of the magnesium fuse and step back quickly. Do not look directly at the flare.

When the residue is cool, break it open and examine the black material.

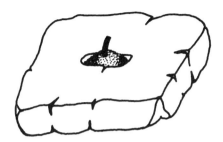

HAZARDS

Inhalation of the dust from magnesium powder is irritating to the respiratory system. Magnesium powder dispersed in air is a serious explosion hazard. Magnesium metal is very reactive, and its contact with any one of many chemical substances (such as carbonates, chlorinated hydrocarbons, oxidizing agents, iodine, phosphates, etc.) may result in ignition or explosion. The mixture of potassium chlorate and magnesium should not be stored.

The burning of magnesium in air produces intense heat, which can cause burns and initiate combustion in flammable materials. Since a carbon dioxide fire extinguisher will not extinguish burning magnesium, a dry-powder fire extinguisher must be used.

Potassium chlorate is a strong oxidizing agent and may be explosive when mixed with combustible material. It is harmful if ingested. Avoid contact with skin, eyes, or clothing.

Burning magnesium produces a light of sufficient intensity to yield temporary loss of sight. Avoid looking directly at the flare.

The reaction ejects burning pieces of magnesium up to a distance of 15 cm, white powder up to 50 cm, and a large amount of white smoke.

DISPOSAL

The block of dry ice should be allowed to evaporate in an open area. The residues should be discarded in a waste container.

DISCUSSION

The reaction between magnesium and carbon dioxide yields magnesium oxide and carbon:

$$2 \text{ Mg(s)} + CO_2\text{(g)} \longrightarrow 2 \text{ MgO(s)} + \text{C(s)}$$

The standard heat of reaction, ΔH°_{rxn}, is -809 kJ [3]. The black residue is a mixture of carbon and magnesium nitride, Mg_3N_2.

Magnesium is so powerful a reducing agent that it burns in carbon dioxide. This demonstration shows why carbon dioxide fire extinguishers cannot be used to put out magnesium fires.

REFERENCES

1. Arthur, P. "Lecture Demonstrations in General Chemistry"; McGraw-Hill: New York, 1939; p 296.
2. Driscoll, J. A. *J. Chem. Educ.* **1978,** *55,* 450.
3. Stull, D. R.; Prophet, H. "JANAF Thermochemical Tables," 2nd ed.; *Natl. Bur. Stand. (U.S.)* **1971**; unpaginated.

1.38

Pyrophoric Lead

Finely divided lead shaken from a tube spontaneously ignites in a shower of bright sparks accompanied by smoke.

MATERIALS

20 g tartaric acid, $HO_2CCH(OH)CH(OH)CO_2H$

1 liter distilled water

670 ml 0.2M lead nitrate, $Pb(NO_3)_2$

15 ml concentrated (16M) nitric acid, HNO_3

1 liter 0.5M sodium hydroxide, NaOH

2-liter beaker

magnetic stirrer and stirring bar

1-liter separatory funnel

ring stand with ring, to support separatory funnel

stirring rod

dropper

pH paper

Büchner funnel

filter paper

filter flask

aspirator

drying oven

several test tubes, 16 mm $\times$ 150 mm (one per sample)

1-hole cork with about 15 cm of 6–8 mm glass tubing inserted

tube furnace or Meker burner

heat-protective gloves

rubber stoppers to fit test tubes

metal spatula

PROCEDURE

Dissolve the tartaric acid in approximately 400 ml of distilled water in a 2-liter beaker. Put a stirring bar in the beaker and place the beaker on the magnetic stirrer.

While stirring the solution of tartaric acid, slowly add the lead nitrate solution from a separatory funnel. When the mixture begins to appear cloudy, stop adding lead nitrate. Using the dropper, slowly add 16M nitric acid until the mixture clears. Continue the alternate addition of lead nitrate and nitric acid until all the lead nitrate is added. The final solution should be clear, but do not add a large excess of nitric acid. After rinsing the separatory funnel with distilled water, fill it with 0.5M sodium hydroxide. While stirring the mixture, add the sodium hydroxide solution dropwise. With a stirring rod and pH paper, measure the pH of the solution as the neutralization proceeds. Stop adding sodium hydroxide at a pH of 8. Using a Büchner funnel, filter the precipitate and wash thoroughly with distilled water. Dry the precipitate of lead tartrate overnight in a drying oven (~60°C) or air dry.

To prepare pyrophoric lead, place about 7 g of lead tartrate in a test tube and stopper with a 1-hole cork and glass tubing. In a hood, heat the test tube in a tube furnace or with a Meker burner at about 450°C. Heat for about 10 minutes or until water and yellowish fumes are no longer released through the glass tubing. Wearing heat-protective gloves, remove the test tube from the heat source, remove the cork and glass tubing, and tightly stopper the test tube with a rubber stopper. Allow the tube to cool to room temperature.

Perform the demonstration in a well-ventilated area. Wearing gloves, unstopper the tube and quickly stir the contents with a metal spatula. From a height of about 1 m above a fireproof surface, sprinkle the contents from the tube. In the air, the black lead powder rapidly oxidizes, producing a stream of fire accompanied by smoke.

To produce a dramatic cascade of fire and smoke, carefully pour the lead from one or more tubes onto a heat-resistant pad. Quickly stir with a spatula. Then, from a height of about 1 m above a fireproof surface, tip the pad to spill the powder. We do not recommend putting the powder in your hand and then allowing it to slide off [1].

If the pyrophoric lead must be stored for several days or longer before use, the test tubes can be sealed by using a glassblower's torch before they cool. To use, scratch the tip with a file and break it off.

HAZARDS

Lead and its compounds are toxic. Lead poisoning, either acute or chronic, can result from exposure to dust or fumes. The demonstration must be performed in a well-ventilated area. Contact with combustible materials must be avoided.

DISPOSAL

Sealed tubes containing pyrophoric lead can be stored in a fireproof container until needed. The opened tubes should be cleaned by washing with dilute nitric acid. The resulting solution should then be flushed down the drain with water.

DISCUSSION

An important aspect of pyrophoric behavior is the size of the metallic particle. The size must be small enough to yield a high ratio of surface area to mass so that the heat generated by oxidation will be great enough to raise the temperature to the ignition point [2]. Lead tartrate, $PbC_4H_4O_6$, can be heated to decomposition, producing a finely divided, pyrophoric powder. Other finely divided metals, such as cobalt, zinc, and iron, may ignite spontaneously upon exposure to air.

REFERENCES

1. Alyea, H. N.; Dutton, F. B., Eds. "Tested Demonstrations in Chemistry," 6th ed.; Journal of Chemical Education: Easton, Pennsylvania, 1965; pp 7, 8.
2. Beddall, K. A. *School Sci. Review* **1966**, *47*, 498.

1.39

Explosive Decomposition of Nitrogen Triiodide

When the brown solid, nitrogen triiodide, is gently touched with a feather mounted on a long pole, a sharp explosion and a puff of violet smoke occur.

MATERIALS FOR PROCEDURE A

2–3 g iodine

15 ml concentrated (15M) aqueous ammonia, NH_3

50-ml beaker

stirring rod

spatula

8 sheets filter paper (9 cm or larger)

transparent tape

feather, mounted on a long pole (at least 2 m)

ear plugs or other hearing protection

ADDITIONAL MATERIALS FOR PROCEDURE B

ring stand and 3 iron rings

3 sheets filter paper (somewhat larger than the rings)

Caution! When dry, nitrogen triiodide (NI_3) is extremely unstable and can detonate unexpectedly. A slight touch or even movement of air can set off the explosion. In contrast, wet nitrogen triiodide is relatively safe to handle.

PROCEDURE A

To prepare nitrogen triiodide, add 2–3 g of iodine to 15 ml of concentrated aqueous ammonia in a 50-ml beaker, stir, and let stand for at least 5 minutes [1].

Complete the steps in this paragraph within 5 minutes. Retaining the solid in the beaker, decant the supernatant liquid into a sink and flush with water. With a spatula, scrape the brown residue of nitrogen triiodide onto a stack of four sheets of filter paper, which absorbs most of the remaining liquid. Divide the damp solid into four equal parts. Transfer each part to a separate, dry sheet of filter paper. Arrange the sheets of filter paper in the lecture area at least 0.5 m apart. Tape the papers in position to prevent accidental movement.

Allow the solid to dry undisturbed for at least 30 minutes. To detonate, touch the solid with a feather mounted on a long pole.

PROCEDURE B

To prepare nitrogen triiodide, add 2–3 g of iodine to 15 ml of concentrated aqueous ammonia in a 50-ml beaker, stir, and let stand for at least 5 minutes.

Complete the steps in this paragraph within 5 minutes. Retaining the solid in the beaker, decant the supernatant liquid into a sink and flush with water. With a spatula, scrape the brown residue of nitrogen triiodide onto a stack of four sheets of filter paper, which absorbs most of the remaining liquid. Divide the damp solid into three equal parts. Transfer each part to a separate, dry sheet of filter paper and spread uniformly. Tape each paper on top of an iron ring. Arrange the three rings on the stand, one near the bottom, one in the middle, and one at the top (see figure). Place the ring stand assembly where it can be seen and where it will not be disturbed. Let it dry for at least 30 minutes.

To detonate, touch the bottom circle with a feather mounted on a long pole. The explosion will detonate the other two portions.

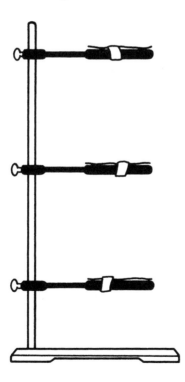

HAZARDS

Nitrogen triiodide is extremely sensitive to touch and is a very powerful explosive. Larger amounts of this substance should not be prepared, and great care should be taken when handling even the modest amounts used in this demonstration. Do not store nitrogen triiodide.

Concentrated aqueous ammonia solution causes burns and is irritating to the skin, eyes, and respiratory system.

Iodine causes burns. Iodine vapor is harmful to the eyes and respiratory system.

In small, enclosed areas, the noise from the explosion can cause a ringing in the ears. Even in large areas, the demonstrator should wear ear plugs or other hearing protection and the students should be cautioned to protect their ears.

DISPOSAL

The detonation will completely destroy all traces of the dry nitrogen triiodide. If a sample does not explode, rub it with the pole to encourage detonation or allow it more time to dry. Otherwise, carefully pour water on the lecture table so that it slowly flows into the sample. When totally wet, the sample should be flushed down the drain with water.

Any nitrogen triiodide that remains in the preparation beaker or on the spatula should be decomposed by rinsing with ethanol. Let stand overnight and then flush the solution down the drain with water.

DISCUSSION

The discovery and characterization of nitrogen iodide, a touch-sensitive compound, stems from the work of Courtois in 1813. "A chocolate-coloured amorphous powder which was made by B. Courtois about 1813 by the action of an aqueous solution of ammonia on solid iodine was thought . . . to be nitrogen iodide, NI_3, analogous to nitrogen chloride, NCl_3. . . . " [2].

Subsequent work has shown that the preparation of nitrogen triiodide involves several reactions, including [2]:

$$NH_3 + H_2O + I_2 \rightleftharpoons NH_4I + HOI$$

$$2\ NH_3 + 3\ HOI \rightleftharpoons NH_3 \cdot NI_3 + 3\ H_2O$$

$$NH_3 + 2\ HOI \rightleftharpoons NHI_2 + 2\ H_2O$$

$$NH_3 + 3\ HOI \rightleftharpoons NI_3 + 3\ H_2O$$

The decomposition of nitrogen triiodide has been characterized by this reaction [2]:

$$8\ NH_3 \cdot NI_3 \rightleftharpoons 5\ N_2 + 6\ NH_4I + 9\ I_2$$

Lack of thermodynamic data for the original species prevents calculation of the energy changes.

REFERENCES

1. Alyea, H. N.; Dutton, F. B., Eds. "Tested Demonstrations in Chemistry," 6th ed.; Journal of Chemical Education: Easton, Pennsylvania, 1965; p 37.
2. Mellor, J. W. "A Comprehensive Treatise on Inorganic and Theoretical Chemistry"; Longmans, Green and Co.: London, 1928; Vol. VIII, pp 605 ff.

1.40

Explosive Reactions
of the Allotropes of Phosphorus
and Potassium Chlorate

A mixture of either red phosphorus and potassium chlorate or white phosphorus and potassium chlorate is placed on the base of a ring stand. When a 1-kg weight is dropped on the mixture, a loud noise is heard and sparks are emitted [1].

MATERIALS FOR PROCEDURE A

red phosphorus (about the size of a match head)

potassium chlorate, $KClO_3$ (about the size of a match head)

ring stand with metal base and clamp

1-kg weight

6 m heavy string

2 microspatulas

ear plugs or other hearing protection

MATERIALS FOR PROCEDURE B

1 g potassium chlorate, $KClO_3$

white (yellow) phosphorus in carbon disulfide (2 g P_4/10 ml CS_2) in glass stoppered bottle (See Demonstration 1.31 for preparation.)

ring stand with metal base and clamp

1-kg weight

6 m heavy string

dropper

ear plugs or other hearing protection

PROCEDURE A

Place the clamp about 0.5 m high on the ring stand. Tie one end of the string to the neck of the weight. Pass the string over the clamp so that the weight can be raised or lowered by pulling the string from a distance of at least 5 m. Pull the string to raise

99

the weight about 25–30 cm above the base of the ring stand. Tie the other end of the string to hold the weight in position.

With a microspatula, place the red phosphorus on the base of the ring stand directly beneath the brass weight. Using a second microspatula, cover the red phosphorus with the potassium chlorate (see figure).

Standing at least 5 m away from the ring stand, release the weight so that it falls and strikes the mixture. A very sharp report is heard and sparks are produced.

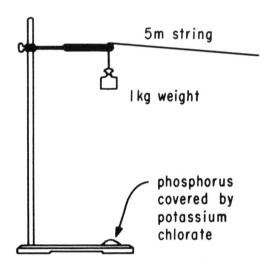

PROCEDURE B

Position the weight above the ring stand as described in Procedure A.

Place 1 g of potassium chlorate on the base of the ring stand directly beneath the weight. Using a dropper, add approximately 1 ml of the solution of white phosphorus in carbon disulfide to the pile of potassium chlorate. Allow the mixture to stand for about 5 minutes.
Caution: the dry mixture can explode or burn spontaneously.

From a distance of at least 5 m, release the weight. A very loud noise is heard and sparks are produced.

HAZARDS

Potassium chlorate is a strong oxidizing agent. Mixtures of potassium chlorate with combustible materials can be flammable or explosive. The mixtures of $KClO_3$ with phosphorus must not be stored.

White phosphorus is spontaneously flammable in air. Both red and white phosphorus are highly flammable and are explosive when mixed with oxidizing agents. Combustion in air produces phosphorus pentoxide. White phosphorus and phosphorus pentoxide are poisonous and cause severe burns. Chronic effects can result from continued absorption of small amounts.

Carbon disulfide is extremely flammable and toxic. The explosive range is 1–50% (v/v) in air [2].

Gloves should be worn when making up or using the solution of white phosphorus in carbon disulfide. The glass stoppered bottle containing the solution should be kept tightly stoppered and away from heat or flames. A fluorocarbon lubricant provides a tight, stable seal, but the solution cannot be safely stored for more than a few days.

The mixture of potassium chlorate with either form of phosphorus can be set off by friction or static electricity. The mixture with white phosphorus may detonate spontaneously when dry.

Larger than specified amounts of the chemicals should not be used. Since the noise from the explosion can lead to a ringing in the ears, persons close to the demonstration should use ear plugs or other hearing protection.

DISPOSAL

The completeness of the reaction leaves little residue. The base of the ring stand and the surrounding area should be wiped with a damp cloth.

DISCUSSION

The mixing of strong oxidizing agents such as chlorates with combustible or oxidizable substances yields mixtures which "are very prone to decompose with explosive violence when struck with a hammer or heated" [3].

The presumed equation for the reaction of potassium chlorate and red phosphorus is

$$10 \; KClO_3(s) \; + \; 3 \; P_4(s) \longrightarrow 3 \; P_4O_{10}(s) \; + \; 10 \; KCl(s)$$

The heat of reaction for this equation is -9425 kJ [4]. For the reaction with white phosphorus, the heat of reaction is -9479 kJ. The major contribution to both large energy values is the heat of formation of P_4O_{10}, since the potassium salts have similar values for their heats of formation:

	ΔH_f° (kJ/mole)
$KClO_3(s)$	-391
$P_4(red)(s)$	-18
$P_4(white)(s)$	0
$P_4O_{10}(s)$	-3010
$KCl(s)$	-436

The rapid release of energy expands the air in the confined space between the weight and the base of the ring stand, which results in an explosion.

In Procedure B, a solution of white phosphorus in carbon disulfide is added to the potassium chlorate. The slow evaporation of the carbon disulfide exposes the white phosphorus to the air while forming a coating on the potassium chlorate crystals. As white phosphorus reacts with air, heat is produced which may initiate a spontaneous reaction between the phosphorus and potassium chlorate before the weight is dropped.

REFERENCES

1. Alyea, H. N.; Dutton, F. B., Eds. "Tested Demonstrations in Chemistry," 6th ed.; Journal of Chemical Education: Easton, Pennsylvania, 1965; p 39.
2. Windholz, M., Ed. "The Merck Index," 9th ed.; Merck and Co.: Rahway, New Jersey, 1976; p 231.
3. Mellor, J. W. "A Comprehensive Treatise on Inorganic and Theoretical Chemistry"; Longmans, Green and Co.: London, 1922; Vol. II, p 310.
4. "Selected Values of Chemical Thermodynamic Properties." *Natl. Bur. Stand. (U.S.)* **1952**; Circ. 500; pp 72, 73, 487.

1.41

Explosions of Lycopodium and Other Powders

When lycopodium powder is dispersed and then ignited in a metal can, an explosion blows off the lid of the can [1]. In a second procedure, balloons filled with either oxygen or compressed air and one of several powders explode when ignited.

MATERIALS FOR PROCEDURE A

0.5 g to 1.0 g lycopodium powder

glass-bending torch

long-stem glass funnel, 7 cm in diameter at top

1-hole rubber stopper

2-gallon metal can, ca. 35 cm high and 19 cm in diameter

1 m rubber tubing, to fit funnel stem

rubber bulb

candle

spatula

tongs

rubber dam, 25 cm × 25 cm

MATERIALS FOR PROCEDURE B

0.5 g to 1.0 g lycopodium powder

2 g corn starch

2 g flour

compressed air

oxygen cylinder, with valve and tubing

powder funnel

6 balloons (2 sets of 3 colors)

ring stand and ring

candle, mounted on a meter stick

PROCEDURE A

First, assemble the explosion can. Using a torch, bend the stem of the funnel at a right angle about 2 cm below the cone of the funnel. Insert the funnel stem in the rubber stopper with the large end of the stopper toward the cone. Near the bottom of the metal can, make a round hole in the side wall of the can to fit the small diameter of the rubber stopper. Place the stopper-funnel assembly tightly in the hole from inside the can so that the funnel is vertical. Attach the rubber tubing to the funnel stem and the rubber bulb to the other end of the tubing. Squeeze the bulb several times to make sure that all connections are tight and that air blows through the funnel. Place the candle in the can, so that the wick is approximately level with the funnel top (see figure).

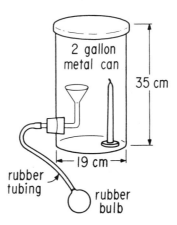

Lycopodium powder is highly combustible when dispersed. With a spatula, add the lycopodium powder to the funnel. Using tongs to hold the match, light the candle. Place the rubber dam over the top of the can and firmly press the lid on, using the rubber dam to make the seal uniformly tight. So that the candle does not burn the dam, quickly pick up the bulb, move away from the can as far as practical, and squeeze the bulb firmly. An explosion, accompanied by flame, will blow the lid high in the air. A slight, pleasant odor is produced. If preferred, leave the lid off and show how the powder burns when dispersed over the candle flame.

PROCEDURE B

Place about 0.5 g (about 1 teaspoon) of lycopodium powder in each of two like-colored balloons, about 1 g (about 1 teaspoon) of corn starch in each of two more balloons, and about 1 g (about 1 teaspoon) of flour in the last two balloons. Inflate one balloon in each pair with compressed air and the other with oxygen. In sequence, place each balloon on a ring attached to a stand and ignite with a candle mounted on a meter stick. Observe each explosion and note the differences in sound intensity and flame size. Note the effect of combustion in the oxygen-filled balloons compared to the air-filled balloons.

Although Reinstein and Shaver [2] suggest using additional solids, such as powdered milk, powdered sugar, and coal (ground with methanol and allowed to dry), we have not had success with these substances.

HAZARDS

Since lycopodium powder is flammable when dispersed in air, it should be handled carefully. The powder will not burn unless it is air-borne as dust. The flame accompanying the explosion can start a fire if combustible materials are in the immediate vicinity. Inhaling the powder or its dust should be avoided, since its effect is not known.

DISPOSAL

The combustion of dispersed lycopodium powder leaves no residue. Any spilled powder should be wiped up with a damp cloth and discarded in a waste container.

DISCUSSION

The undetermined chemical nature of lycopodium powder precludes the writing of specific combustion reactions. Derived from club-moss spores, the yellow powder is odorless and tasteless. It is easily dispersed, and upon ignition its dust explodes violently.

This demonstration illustrates that a combustible material, when finely divided and dispersed in air, will explode upon ignition. The exothermic reaction releases energy in the forms of heat and sound. The safety concern for dust explosions in coal mines, flour mills, and grain elevators is based on this effect.

REFERENCES

1. Alyea, H. N.; Dutton, F. B., Eds. "Tested Demonstrations in Chemistry"; Journal of Chemical Education: Easton, Pennsylvania, 1965; p 8.
2. Reinstein, J.; Shaver, R. "Chemical Demonstrations Proceedings"; Western Illinois University and Quincy-Keokuk Section of the American Chemical Society, 1978; p 33.

1.42

Explosive Reaction of Hydrogen and Oxygen

When four balloons filled with different gases are ignited, the balloons emit differing intensities of sound. The loudest sound occurs when a hydrogen-oxygen mixture is ignited. Hydrogen-oxygen mixtures can also be ignited in a Coca-Cola bottle, soap bubbles, or a special cannon [1].

MATERIALS FOR PROCEDURE A

hydrogen cylinder, with valve

oxygen cylinder, with valve

4 balloons, minimum inflated diameter ca. 25 cm

thread, ca. 4 m

candle, mounted on a meter stick

ear plugs or other hearing protection

MATERIALS FOR PROCEDURE B

hydrogen cylinder, with valve and rubber tubing 1 m long

oxygen cylinder, with valve and rubber tubing 1 m long

Coca-Cola bottle (glass, 16-oz, and thick-walled)

transparent tape

3-liter beaker, half full of water

rubber stopper to fit Coca-Cola bottle

burner

ear plugs or other hearing protection

MATERIALS FOR PROCEDURE C

hydrogen cylinder, with valve

oxygen cylinder, with valve

gas mixing apparatus, consisting of two 500-ml Erlenmeyer flasks fitted with rubber stoppers, glass-tubing bends, a glass tubing "Y," and rubber tubing (Figure 1)

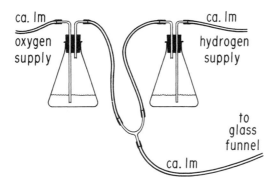

Figure 1.

candle, mounted on meter stick, or special burner (see Procedure C)

glass funnel, 35 mm in diameter at top

150-ml beaker

100 ml soap solution (sold in toy stores for blowing bubbles)

ear plugs or other hearing protection

MATERIALS FOR PROCEDURE D

hydrogen cylinder, with valve and rubber tubing

oxygen cylinder, with valve and rubber tubing

special "cannon" (see Procedure D)

400-ml beaker, half full of water

waxed cork to fit muzzle of cannon

Tesla coil leak-tester

ear plugs or other hearing protection

PROCEDURE A

Fill the four balloons to the same size as follows. Fill one with compressed air or by blowing up by mouth. Fill another with hydrogen gas by fitting the balloon over the gas nozzle and carefully opening the cylinder. Fill the third with oxygen gas the same way. Fill the fourth with about 2 volumes of hydrogen gas and about 1 volume of oxygen gas. Tie a thread of approximately 2 m to the hydrogen balloon and a similar thread to the hydrogen-oxygen balloon. Tie the threads so that the hydrogen and hydrogen-oxygen balloons are about 1.5 m above the lecture table. The air and oxygen balloons can rest on the edge of the desk or can be held aloft using a thread and ring stand.

Light the candle on the meter stick and ignite the balloons in this sequence: air, oxygen, hydrogen, and hydrogen-oxygen. You may wish to repeat the hydrogen and hydrogen-oxygen explosions in a darkened room to observe the color and size of each flame.

PROCEDURE B

Attach the pieces of rubber tubing to the cylinders of gas. After wrapping the Coca-Cola bottle with transparent tape, fill it with water and invert it in the beaker of water. Adjust the valve on the oxygen cylinder to deliver oxygen slowly. Insert the tubing in the inverted bottle under water until approximately one third of the water in the bottle is displaced. Remove the tubing from the bottle and turn off the valve on the oxygen tank.

Adjust the valve on the hydrogen cylinder to deliver hydrogen slowly, and insert the tubing in the inverted bottle under water. Displace the remaining water with hydrogen and remove the tubing from the bottle. Turn off the valve on the hydrogen cylinder. Insert the rubber stopper in the inverted bottle and remove it from the beaker.

To demonstrate the reaction, light a burner, hold the bottle firmly, pointed away from anyone, and remove the stopper. Immediately bring the mouth of the bottle to the burner. A flash and a resounding explosion will occur.

PROCEDURE C

To produce a quiet, controllable flame at a safe distance from the demonstrator, we have constructed a special burner (Figure 2). A rigid tube of stainless steel with

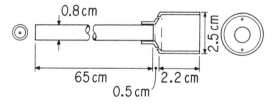

Figure 2.

walls about 0.1 cm thick is soldered to a burner mouth turned from brass. Two holes (about 0.1 cm in diameter) are drilled through the oblique surface of the mouth to provide an air supply. The burner is connected to a gas-cock with a 3-m piece of rubber tubing, thus allowing mobility for the demonstrator. Depending upon gas pressure at the cock, you may want to restrict gas flow with a brass plug, machine-turned to fit the inside diameter of the stainless steel tube and drilled with a suitable hole; ours has a hole about 0.2 cm in diameter. Gas flow can also be regulated at the gas-cock.

You will need to train two assistants to help with this procedure.

Place 100 ml of soap solution in the 150-ml beaker. Connect the gas-mixing apparatus to the hydrogen and oxygen cylinders and to the funnel stem. Ignite the candle or special burner. The first assistant opens the hydrogen cylinder valve, and the second holds the funnel stem. After about 15 seconds of hydrogen gas flow (allowing air to be swept from the system), assistant 2 momentarily dips the funnel into the soap solution to form a film across its mouth. Assistant 1 adjusts the hydrogen gas flow to blow a soap bubble about every 3 seconds. By shaking the funnel sharply, assistant 2 releases bubbles from the funnel when they are 5–10 cm in diameter. The soap solution on the funnel must be replenished occasionally.

Taking care not to bring the flame near the funnel, ignite the bubbles as they rise, commenting on the rate at which the bubbles rise, the nature of the flame, and so on. Then, ask assistant 1 to begin adding oxygen to the gas forming the bubbles. Assistant 1 opens the oxygen cylinder valve, regulating the flow by observing the bubbling in the Erlenmeyer flask and taking care to increase the oxygen flow slowly and not to exceed a flow of about half the hydrogen flow. Assistant 2 continues to form bubbles. Continue to ignite the bubbles, commenting on their changing rate of ascent and the nature of the explosions as each bubble is ignited. When finished, shut off all gas supplies.

PROCEDURE D

The barrel of the special "cannon" is a 20-cm length of 1-in (2.54-cm) black iron pipe with a standard external pipe thread at one end. An adaptor is made from brass or aluminum, about 4 cm long, with a matching internal pipe thread at one end and a thread to fit a standard automobile spark plug at the other (Figure 3). Assemble the

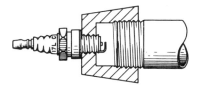

Figure 3.

pipe, adaptor, and spark plug tightly together. Mount the cannon barrel in a clamp and ring stand arrangement or in a special holder (Figure 4). A similar cannon can be constructed from Plexiglas [2].

Place the delivery tubes from the hydrogen and oxygen cylinders beneath the surface of the water in a 400-ml beaker. Open the valves and adjust the flow of gases so that hydrogen bubbles are streaming from the tube at about twice the rate of the oxygen bubbles. Brace the cannon so it stands muzzle down. Insert both tubes, with

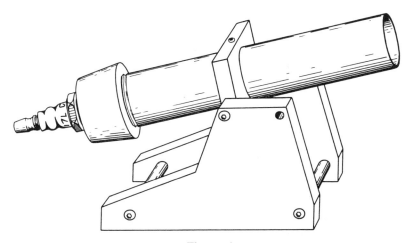

Figure 4.

gas flowing through them, into the muzzle of the cannon and allow the flow to continue for 3 minutes. This displaces air from the barrel and replaces it with a mixture of hydrogen and oxygen gases. Remove the tubes and immediately cork the muzzle tightly.

Aim the cannon to avoid hitting anyone or anything breakable. The flying cork can cause serious injury, because the vigorous explosion propels it at least 20 m. To set off the explosion, plug in the Tesla coil leak-tester and touch it to the spark plug in the breech of the cannon.

HAZARDS

Hydrogen gas is very flammable and yields explosive mixtures with air and oxygen. The explosion of the mixture of hydrogen and oxygen is quite loud. Tests should be made prior to the demonstration to adjust the total volume in the balloons so that the sound of the explosion is tolerable in the room.

Since oxygen gas is an excellent oxidizing agent, materials not readily burned in air become quite flammable in pure oxygen.

Although we know of no instance in which the glass Coca-Cola bottle has shattered during the explosion, it should be wrapped with tape as a precaution.

DISCUSSION

When ignited, hydrogen gas combines with oxygen gas explosively in proportions ranging from 4.1–71.5% hydrogen [3]. The ignition temperature for the reaction of the pure mixture of hydrogen and oxygen gases is the same as that for hydrogen and air: 580°C–590°C [4].

The gaseous reaction between hydrogen and oxygen to form water is

$$2 \text{ H}_2(g) + \text{O}_2(g) \longrightarrow 2 \text{ H}_2\text{O}(g)$$

This exothermic reaction yields 232 kJ/mole of water formed [5].

The rapid release of a considerable amount of energy causes the surrounding air to expand suddenly, resulting in a sharp explosion. When pure hydrogen is ignited, the reaction with the surrounding air is less rapid, the sound is less loud, and a significantly larger flame is produced. The comparatively mild sounds emitted when the air and oxygen balloons are ignited are caused by the rupturing of the balloons and the escape of each gas. The explosions are caused by a "sudden pressure effect through the action of heat on produced or adjacent gases" [6].

According to Atkins [7] explosions occur for two basic reasons.

The basic reason for a thermal explosion is the exponential dependence of the reaction rate on temperature. When the energy of an exothermic reaction cannot escape, the temperature of the reaction system increases and the reaction accelerates. This process results in heat production at an even greater rate, the heat cannot escape, and so the reaction goes even faster.

The other type of explosion depends on a chain reaction in which free radicals are produced. When there are steps in a chain reaction that increase the number of free radicals in the system, the reaction rate will cascade into an explosion. A chain mechanism involving free radicals (such as H', HO', etc.) is proposed for the H_2-O_2 reaction. Some of the steps are initiation: $H_2 + O_2 \longrightarrow HO_2^{\cdot} + H^{\cdot}$

propagation: $H_2 + HO_2^{\cdot} \longrightarrow HO^{\cdot} + H_2O$

$H_2 + HO^{\cdot} \longrightarrow H^{\cdot} + H_2O$

$H^{\cdot} + O_2 \longrightarrow HO^{\cdot} + O^{\cdot}$ (branching)

$O^{\cdot} + H_2 \longrightarrow HO^{\cdot} + H^{\cdot}$ (branching)

The last two steps are called branching steps because they increase the number of free radicals, which increases the reaction rate very rapidly, which in turn leads to the characteristic explosion when H_2 and O_2 are sparked together.

The occurrence of an explosion depends on the temperature and pressure of the reacting system. The regions of explosion for hydrogen and oxygen are shown in Figure 5. At very low pressure, the system is outside the explosion limits, and

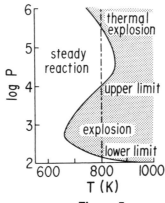

Figure 5.
Source: Figure 26.9 [7]

therefore when the mixture is sparked reaction occurs without explosion. At these very low pressures, the radicals produced in the branching reactions can reach the walls of the container, where they can combine and give up their excess energy. (The efficiency of this exchange depends on the composition of the walls.) Raising the pressure takes the system through the *first explosion limit* if the temperature exceeds about 400°C. Now when the mixture is sparked, it explodes because the radicals react before reaching the walls and the branching reactions are explosively efficient. If the pressure is increased still further, the system passes through the *second explosion limit* into a region where the reaction proceeds without explosion. The pressure is now large enough for the radicals produced in a branching reaction to combine in the gas, the molecules of the reaction mixture being able to soak up the excess energy. If the pressure is raised still further, the system passes through the *third explosion limit,* and the reaction reverts to an explosion, apparently a thermal explosion.

In this demonstration we use soap bubbles as carriers for exploding hydrogen and oxygen gases. For detailed descriptions of other experiments with soap bubbles, see C. V. Boys [8].

REFERENCES

1. Alyea, H. N.; Dutton, F. B., Eds. "Tested Demonstrations in Chemistry"; Journal of Chemical Education: Easton, Pennsylvania, 1965; pp 5, 9, 10, 58.

2. Olene, M. E. *J. Chem. Educ.* **1962**, *39*, A796.
3. Thorpe, E. "Dictionary of Applied Chemistry"; Longmans, Green and Co.: London, 1921; Vol. II, p 701.
4. Thorpe; p 699.
5. "Selected Values of Chemical Thermodynamic Properties." *Natl. Bur. Stand. (U.S.)* **1952;** Circ. 500; p 9.
6. Standen, A. "Kirk-Othmer Encyclopedia of Chemical Technology," 2nd ed.; Interscience Publishers, John Wiley and Sons: New York, 1965; Vol. VIII, p. 581.
7. Atkins, P. W. "Physical Chemistry"; W. H. Freeman and Company: San Francisco, 1978; pp 878–80.
8. Boys, C. V. "Soap Bubbles and the Forces Which Mould Them"; Educational Services Incorporated, A Doubleday Anchor Book: New York, 1959.

1.43

Combustion of Methane

Mixtures of methane and oxygen gases are ignited either under continuous gas flow or static conditions. Controlled gas flow in a burner results in a smooth continuous flame. Combustion of methane in a balloon or in soap bubbles [1] results in an explosion.

MATERIALS FOR PROCEDURE A

> burner
>
> source of natural gas

MATERIALS FOR PROCEDURE B

> methane cylinder, with valve and rubber tubing
>
> oxygen cylinder, with valve and rubber tubing
>
> 2 balloons, minimum inflated diameter ca. 25 cm
>
> thread, ca. 2 m long
>
> ring stand and iron ring with clamp
>
> special burner (see Procedure C of Demonstration 1.42) or candle mounted on a meter stick

MATERIALS FOR PROCEDURE C

> glass funnel, 35 mm in diameter at top
>
> rubber tubing to fit stem of funnel, ca. 1 m long
>
> 100 ml soap solution (sold in toy stores for blowing bubbles)
>
> 150-ml beaker
>
> special burner (see Procedure C of Demonstration 1.42) or candle mounted on a meter stick

PROCEDURE A

Close off the air-supply opening on the burner. Attach the burner tubing to a natural gas outlet and open the valve at the outlet. Light the burner and note the large yellow flame. Slowly open the air-supply inlet on the barrel to allow air to mix with the gas. The flame will become blue. When the correct gas-air mixture is obtained, the flame will have three cone-shaped regions. The bright blue cone will be surrounded by a deeper blue cone. If this is not observed when the air inlet is fully opened, the gas-air ratio is too high and must be reduced by slightly closing the gas outlet valve or by adjusting the needle valve at the base of the burner. If too much air is allowed to enter, the flame will rise off the barrel and may blow out.

Occasionally, a burner will flash back. This means that the gas is burning inside the barrel at its base instead of its top. The barrel becomes very hot. The burner should be turned off and allowed to cool.

PROCEDURE B

Fill one balloon with methane from a cylinder. Fill the other balloon with about 2 volumes of methane and no more than 1 volume of oxygen. Tie the methane-filled balloon with thread so that it floats about 1.5 m above the lecture table. Place the other balloon on a ring clamped high on a ring stand.

Use the special burner or a candle mounted on a meter stick to ignite each balloon separately. Observe the differences between each combustion reaction. You may wish to repeat each explosion in a darkened room to observe the color and size of each flame.

PROCEDURE C

You will need to train one assistant to help with this procedure.

Attach the rubber tubing to the funnel stem and to a gas outlet. Pour the soap solution into the beaker. Light the candle or special burner, keeping the flame at a safe distance from the funnel.

The assistant momentarily dips the funnel into the soap solution to form a film across the funnel mouth. The assistant then adjusts the gas flow to a moderate rate, so that a bubble of approximately 10–15 cm in diameter is formed. With a slight, quick movement of the funnel, the assistant releases the bubble. The first bubbles may contain mostly air. Repeat the bubble-forming process until a bubble rises. Taking care not to bring the flame near the funnel, ignite each bubble before it rises out of reach.

HAZARDS

Since methane is highly flammable, the gas source must be isolated from possible accidental ignition. The balloons and bubbles should be ignited from a distance because the flames produced spread well beyond the size of the balloons and bubbles.

In Procedure B keep the methane-oxygen ratio high to avoid loud explosions that could injure ear drums.

DISCUSSION

The combustion of methane under controlled-flow conditions results in a smooth, highly energetic flame, while the ignition of methane gas in a balloon or soap bubble results in an explosion. A flame can be defined as a gas rendered luminous by the liberation of chemical energy [2]. Flames are characterized by the emission of light and rapid temperature rise and are the result of highly exothermic reactions between gases.

The equation for the combustion reaction of methane to carbon dioxide and water is

$$CH_4(g) + 2 O_2(g) \longrightarrow CO_2(g) + 2 H_2O(g)$$

The heat of this combustion reaction is -802.3 kJ/mole of methane burned [3]. The standard heat of formation values, ΔH_f°, for $CH_4(g)$, $CO_2(g)$, and $H_2O(g)$ are -74.85 kJ/mole, -393.5 kJ/mole, and -241.8 kJ/mole, respectively [3].

Prior to Bunsen's invention of the familiar laboratory burner around 1855, simple flames of the diffusion type were used. These produced smoke and soot, and their temperatures were low. The Bunsen burner produces clear flames with much more intense combustion, higher temperatures, and no soot problems. This is accomplished by premixing gas and air and passing the mixture through a burner barrel at a speed which is sufficient to prevent the flame from striking back down the barrel. Thus, the mixture burns at the top of the burner, and the combustion is aided by the air surrounding the flame. The geometry of the burner, the velocity of the gas jet, and the air flow through the burner determine the size and power of the flame. For detailed discussions of burners and flames, see Gaydon and Wolfhard [4] and references therein.

The structure of a burner flame is shown in Figure 1. According to Dean [2], the unburned gas mixture emerges from region A and passes into preheating region B,

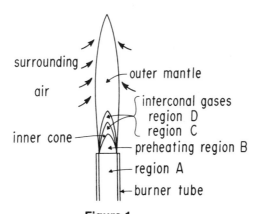

Figure 1.
Source: Figure 3–1 [2]

which is about 1 mm in thickness. There, the mixture is heated by conduction and radiation from region C and by diffusion of radicals which initiate oxidation. At atmospheric pressure, region C is 0.02 mm to 0.2 mm in thickness depending on the fuel. Gases emerging from region C consist mainly of CO, H_2, CO_2, H_2O, and N_2, with lesser amounts of $H^\bullet$, $O^\bullet$, and $HO^\bullet$. Region D is about 1 mm in thickness. Regions C and D constitute a brightly luminous layer, which is conical in shape, consists of interconal gases, and is separated from the preheating zone and the outer mantle. The outer mantle contains the burned gas mixture. Emission of radiation in flame photometry takes place in the outer mantle above region D. In a methane-air flame, the blue color results from emissions around 400 nm due to CH species and around 470 nm due to C_2 species. The yellow color results from emissions around 510 nm by C_2

species [5]. The temperature distribution within a flame of natural gas and air is shown in Figure 2.

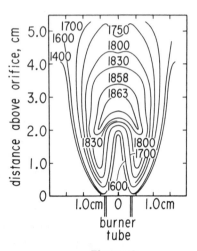

Figure 2.
Source: Figure 32 [6]

The pressure of natural gas in distribution pipelines is controlled to prevent accidents in home furnaces, stoves, and burners. The pressure is so low that a balloon cannot be filled with natural gas from the familiar laboratory gas outlet.

Two interesting characteristics of the methane soap bubbles are their tendency to rise and their flammability. The calculated difference in density between methane (0.714 g/liter at STP) and air (1.29 g/liter at STP) accounts for the ability of methane to provide "lift" to soap bubbles. A bubble must be sufficiently large to obtain net lift. This lift results from the difference in weight between the air displaced and the methane in the bubble. Soap bubbles filled with a mixture of methane and oxygen will rise only if the CH_4/O_2 ratio is high. For detailed descriptions of other experiments with soap bubbles, see C. V. Boys [7].

For an excellent discussion of the chemistry of flames, see W. C. Gardiner, Jr. [8].

REFERENCES

1. Snipp, R.; Meltson, B.; Hardy, W. "Spectacular Gas Density Demonstration Using Methane Bubbles"; Creighton University, Omaha, Nebraska, 1979; unpublished.
2. Dean, J. A. "Flame Photometry"; McGraw-Hill: New York, 1960; pp 15–16.
3. "Selected Values of Chemical Thermodynamic Properties." *Natl. Bur. Stand. (U.S.)* **1952**; Circ. 500; pp 9, 99, 100.
4. Gaydon, A. G.; Wolfhard, H. G. "Flames: Their Structure, Radiation and Temperatures"; Chapman & Hall Ltd.: London, 1960.
5. Pearse, R. W. B.; Gaydon, A. G. "Identification of Molecular Spectra"; Interscience Publishers, John Wiley and Sons: New York, 1963; Plate 9.
6. Lewis, B.; Van Elbe, G. *J. Chem. Phys.*, **1943**, *11*, 75.
7. Boys, C. V. "Soap Bubbles and the Forces Which Mould Them"; Educational Services Incorporated, A Doubleday Anchor Book: New York, 1959.
8. Gardiner, W. C., Jr. *Sci. Am.*, **1982**, *246*, 110.

1.44

Explosive Reaction of Nitric Oxide and Carbon Disulfide

When a mixture of nitric oxide gas and carbon disulfide vapor is ignited at the top of a large glass tube, the flame travels down the tube, producing a bright blue light and a loud noise.

MATERIALS

Cylinder of nitric oxide gas, NO (with valve and rubber tubing)

   or

   35 g light copper turnings

   100 ml concentrated (16M) nitric acid, HNO_3

   thistle tube or funnel with long stem

   2-hole rubber stopper to fit mouth of flask

   500-ml Florence or Erlenmeyer flask

   2 right angle bends of 8-mm glass tubing (ca. 5 cm each arm)

   50 cm rubber tubing to fit glass bends

2.5 ml carbon disulfide, CS_2

glass tube, 58 mm in diameter and 122 cm long

2 rubber stoppers to fit glass tube

ring stand and iron ring

pneumatic trough

syringe, 3-ml or larger, with needle

PROCEDURE

Stopper one end of the glass tube and fill the tube with water. Stopper the other end. In a hood, place one end of the tube in a pneumatic trough, and remove the stopper from that end. Support the tube vertically so that it will not fall over.

If your source of nitric oxide is a cylinder, fill the tube with nitric oxide gas by water displacement. The large amount of water displaced must be removed to prevent the trough from overflowing. Stopper the tube underwater. Remove the tube from the trough and place it upright through the iron ring on the ring stand. The ring stand and tube should be placed on the floor.

If you do not have a cylinder of nitric oxide, it may be prepared as described in the next four paragraphs.

Place the copper turnings in the flask. Insert the thistle tube in the rubber stopper so that it extends to the bottom of the flask (see figure). Insert one glass bend in each end of the rubber tubing. Insert one of the glass bends in the other hole of the rubber stopper and place the stopper in the mouth of the flask.

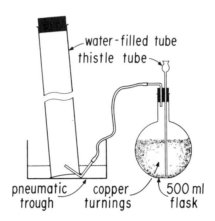

Prepare the nitric oxide in a hood. Through the thistle tube, slowly add about 2–3 ml concentrated nitric acid to the copper. The gas in the flask will turn reddish brown due to the formation of nitrogen dioxide from nitric oxide and oxygen trapped in the flask. After about 1 minute, place the delivery tube under the large glass tube in the pneumatic trough so that the large tube fills with colorless NO gas. Continue to add small portions of nitric acid slowly to the reaction flask until the tube is filled. Adding small portions will keep the rate of gas evolution controllable. Agitate the flask periodically to allow fresh copper to contact the nitric acid. The large amount of water displaced must be removed to prevent the trough from overflowing. When the tube is full, stopper the tube underwater.

Quench the reaction by filling the flask with water. Flush the liquid down the drain. Rinse the copper several times with water and store for future use.

Remove the tube from the trough and place it upright through the iron ring on the ring stand. The ring stand and tube should be placed on the floor.

Under a hood, draw 2.5 ml carbon disulfide into a syringe. Gently loosen but do not remove the upper stopper of the tube. Insert the needle of the syringe between the stopper and the wall of the tube. Quickly inject all the CS_2 into the tube and remove the syringe. With one hand keep the loosened stopper in place but do not replace it tightly. The CS_2 should be added carefully to avoid introducing air into the tube (colorless NO is rapidly oxidized to brown NO_2).

Keeping one hand on the loosened stopper, remove the tube from the ring stand and invert it several times to insure that the CS_2 is vaporized and well mixed with the NO gas. During the mixing, pressure will build in the tube. Release the pressure by slightly opening and quickly shutting the loosened stopper. Continue inverting the tube and releasing the pressure until you no longer hear the gas escaping from the tube. Tighten the stopper and place the tube through the ring on the ring stand. The tube is now ready for the demonstration. Perform the demonstration as soon as possible.

Keep your hands and face away from the top of the tube, since the reaction is quite violent. To ignite the mixture, carefully remove the upper stopper and drop a lighted match into the tube. This operation should be performed carefully and quickly to avoid introducing air into the tube. As the flame travels down through the mixture of gases, a roaring sound is heard. The blue flame is best observed in a darkened room.

HAZARDS

Nitric oxide is rapidly oxidized in air to nitrogen dioxide, an extremely toxic gas. Nitrogen dioxide is irritating to the respiratory system; inhaling it may result in severe pulmonary effects which are not apparent until several hours after exposure.

Carbon disulfide is extremely flammable and toxic. The explosive range is 1–50% (v/v) in air [1]. The vapor is irritating to the eyes and malodorous.

In addition to their individual hazards, these gases form an explosive mixture as in this demonstration.

DISPOSAL

After the reaction the tube may contain unreacted gases as well as gaseous products. Therefore, it should be stoppered and then opened under a hood. After the gases have been vented, the solid remaining on the walls of the tube can be removed by scrubbing with a long-handled brush and a detergent solution. The residue is easier to remove immediately after the demonstration.

DISCUSSION

Nitric oxide boils at $-157°C$, is colorless, and has a density of 1.343 g/liter at $0°C$. Carbon disulfide boils at $46°C$, is colorless, and has a density of 1.26 g/ml at $20°C$. It is quite volatile (its vapor pressure at $28°C$ is 400 mm of Hg).

The combustion of carbon disulfide in nitric oxide produces nitrogen, carbon monoxide, carbon dioxide, sulfur dioxide, and sulfur [2]. Nitrous oxide and carbon oxysulfide (COS) can be found in trace amounts depending on the reaction conditions [3]. The stoichiometry is fairly complex and depends on initial reactant concentrations.

The NO/CS_2 combustion reaction has been studied in closed vessels and found to follow a nonchain mechanism [3]. The reaction is third order overall: second order in NO and first order in CS_2. The overall rate constant is 7.7×10^{-9} mm^{-2} sec^{-1} and the activation energy is 290 kJ/mole. The activation energy for this reaction corresponds very closely to the energy required to remove a sulfur atom from CS_2 (290–330 kJ/mole).

The bright blue emission accompanying the reaction corresponds to a spectroscopic continuum from 490 nm to 310 nm. The emission has been ascribed to a triplet-singlet transition in SO_2 [3]. Generation of triplet SO_2 may follow this reaction path:

$$SO(^3\Sigma) + N_2O(^1\Sigma) \longrightarrow SO_2(^3\Sigma) + N_2(^1\Sigma)$$

The formation of N_2O probably involves the abstraction of O by CS_2 or by CS from a collision dimer (N_2O_2).

A much less intense emission of COS ($^3\Pi$) can be observed in the continuum of 550–700 nm. Several even weaker emissions can be observed for SO and S_2.

The best visible emission is observed when the mixture of gases is approximately 30% CS_2 by volume [4, 5]. The procedure outlined here produces a mixture that is 31% CS_2. When tubes of different size are used, sounds of different pitch are heard. Addition of 0.75–0.80 ml of CS_2 per liter capacity of the tube will produce a mixture of the proper proportions. The amount of CS_2, however, should not vary greatly from that described because larger amounts of CS_2 tend to suppress the flame and smaller amounts tend to produce more violent explosions.

REFERENCES

1. Windholz, M., Ed. "The Merck Index," 9th ed.; Merck and Co.: Rahway, New Jersey, 1976; p 231.
2. van Liempt, J. A. M. *Chem. Weekblad* **1934**, *31*, 706.
3. Roth, W.; Rautenberg, T. H. *J. Phys. Chem.* **1956**, *60*, 379–81.
4. Mellor, J. W. "A Comprehensive Treatise on Inorganic and Theoretical Chemistry"; Longmans, Green and Co: London, 1967; Vol. VIII, pp 232–33.
5. van Liempt, J. A. M.; DeVriend, J. A. *Recueil des Travaux Chimiques des Pays-Bas* **1933**, *52*, 160, 549; **1934**, *53*, 760.

1.45

Photochemical Reaction
of Hydrogen and Chlorine

A cork-stoppered test tube filled with a mixture of hydrogen and chlorine gases is clamped to a ring stand. Light from a slide projector is aimed at the test tube. A loud report is heard and the stopper is ejected from the test tube.

MATERIALS

chlorine cylinder with valve and tubing

hydrogen cylinder with valve and tubing

1-liter beaker

thick-walled test tube, 20 cm long and 2.5 cm in diameter, wrapped with one thickness of transparent tape

waxed cork to fit test tube

ring stand with clamp

slide projector, with at least a 300-watt bulb

ear plugs or other hearing protection

PROCEDURE

Nearly fill the beaker with water and place it in a hood. Wrap the test tube with one thickness of transparent tape. Fill the test tube with water. Without admitting air, place the test tube in the beaker in an inverted position. Purge the delivery tubes on the hydrogen and chlorine cylinders. By water displacement, half fill the test tube with chlorine gas. Complete the filling with hydrogen gas and stopper the test tube with the cork. Perform the demonstration within 5 minutes of preparing the test tube. Clamp the test tube in the ring stand at a 45° angle.

To demonstrate the reaction, place the ring stand with the clamped test tube in a well-ventilated area. Aim the test tube away from anyone or anything breakable. Put on ear plugs or other hearing protection. From a distance of 1 m or less, aim the light beam from a slide projector at the bottom of the test tube (see figure). Within a second or two, a loud noise is heard and the cork is ejected vigorously.

121

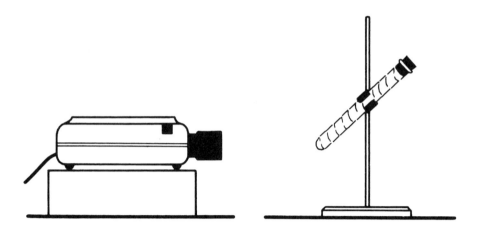

HAZARDS

Chlorine gas irritates the eyes and mucous membranes and, if inhaled, can cause severe lung irritation and fatal pulmonary edema. In high concentrations, the gas irritates the skin.

Chlorine is a strong oxidizing agent, and combustible materials will burn in an atmosphere of Cl_2.

Hydrogen gas is very flammable and yields explosive mixtures with air and chlorine. The explosion of the mixture of hydrogen and chlorine is quite loud.

Hydrogen chloride is a toxic and corrosive gas, highly irritating to the eyes and respiratory tract.

In small, enclosed areas the noise from the explosion can cause a ringing in the ears. Even in large areas, the demonstrator should wear ear plugs or other hearing protection, and the students should be cautioned to protect their ears.

Although we know of no instance in which the test tube has shattered during the explosion, it should be wrapped with tape as a precaution.

DISPOSAL

The test tube should be rinsed thoroughly with water.

DISCUSSION

This demonstration is a modification of the procedure described by Alyea and Hornbeck [1]. Other procedures [2, 3], in which balloons are filled with H_2/Cl_2 mixtures, are not as convenient or safe.

The standard heat of formation, ΔH_f°, of gaseous HCl is -537 kJ/mole [4]. The photochemical reaction of hydrogen and chlorine is a chain reaction [4]:

$$Cl_2 + h\nu \longrightarrow 2\ Cl^\cdot \tag{1}$$

$$Cl^\cdot + H_2 \longrightarrow HCl + H^\cdot \tag{2}$$

$$H^\cdot + Cl_2 \longrightarrow HCl + Cl^\cdot \tag{3}$$

$$H^{\bullet} + O_2 \longrightarrow HO_2 \tag{4}$$

$$Cl^{\bullet} + O_2 \longrightarrow ClO_2 \tag{5}$$

Reaction 1, the chain-initiation step, involves the absorption of light by a chlorine molecule and generates 2 chlorine atoms. Reactions 2 and 3 are chain-propagation steps. Reactions 4 and 5 illustrate how oxygen gas acts as an inhibitor. In the presence of 1% O_2, the chain length is reduced from approximately 10^6 to approximately 10^3. Chain-termination steps are complex and include collisions of radicals with the test tube wall.

REFERENCES

1. Alyea, H. N.; Hornbeck, L. G. *J. Chem. Educ.* **1967**, *44*, A84.
2. Bowman, L. H., Arkansas Tech University, personal communication, 1980.
3. Ramette, R. W., Carleton College, personal communication, 1981.
4. Gimblett, F. G. R. "Introduction to the Kinetics of Chemical Chain Reactions"; McGraw-Hill: London, 1970; pp 13–19.

Dangerous Demonstrations

We are aware of the following systems and we judge them unsuitable as demonstrations. In all cases the systems are extremely hazardous.

Manganese Heptoxide Oxidation of Ethanol
     Alfthan, V. *J. Chem. Educ.* **1967**, *44*, A465.
     Haight, G. P.; Phillipson, D. *J. Chem. Educ.* **1980**, *57*, 325.
Permanganate Volcano
     Haight, G. P.; Phillipson, D. *J. Chem. Educ.* **1980**, *57*, 325.
     Haight, G. P.; Phillipson, D. *Chem. Eng. News* **1980**, March 31, p 3.
     Ingerson, E. *Chem. Eng. News* **1980**, June 9, p 108.
     Mann, L. T., Jr. *Chem. Eng. News* **1980**, May 5, p 2.
     Schaffrath, R. E. *Chem. Eng. News* **1980**, June 9, p 4.
     Weand, B. L. *Chem. Eng. News* **1980**, June 9, p 4.
Explosive Decomposition of Lead Picrate
     Alyea, H. N..; Dutton, F. B., Eds. "Tested Demonstrations in Chemistry," 6th ed.;
          Journal of Chemical Education: Easton, Pennsylvania, 1965; p 158.
Phosphine Fire Flask
     Dillard, C. R. *J. Chem. Educ.* **1956**, *33*, 137.
Spontaneous Combustion of Diethylzinc
     Taliaferro, O. A., University of Wisconsin-Madison, personal communication, 1971.
Explosive Decomposition of Nitroglycerine
     Taliaferro, O. A., University of Wisconsin-Madison, personal communication, 1971.
Oxidation of Methanol by Chromium(VI) Oxide
     Taliaferro, O. A., University of Wisconsin-Madison, personal communication, 1971.
Reaction of Potassium and Bromine
     Taliaferro, O. A., University of Wisconsin-Madison, personal communication, 1971.
Endothermic Reactions of Hydrated Metal Chlorides with Thionyl Chloride
     Matthews, G. W. J. *J. Chem. Educ.* **1966**, *43*, 476.
     Conrad, J., University of Wisconsin-River Falls, personal communication, 1980.

2

Chemiluminescence

Rodney Schreiner, Mary Ellen Testen, Bassam Z. Shakhashiri, Glen E. Dirreen, and Lloyd G. Williams

Many combustion reactions produce light by thermal means (see Chapter 1). These reactions release sufficient thermal energy to heat the reaction mixture to incandescence. A famous example of light production through chemically induced incandescence is the bright white light emitted by calcium oxide pellets when heated in a hydrogen-oxygen flame, the original "lime light." In contrast, some reactions release light energy at room temperature. The nonthermal production of visible light by a chemical reaction leads to the term "cool light" and the process is called *chemiluminescence*. Although chemiluminescent reactions are not rare, the production of "cool light" holds such fascination for both chemists and nonchemists that demonstrations of chemiluminescent reactions are always well-received. A wide range of sources provides general information regarding chemiluminescence [1–5] as well as related topics [6–8].

Bioluminescence is the production of chemiluminescence by living organisms. Perhaps the most familiar example of such an organism is the firefly. In a biochemical reaction within the body of this insect, light is produced by the action of an enzyme on its substrate, luciferin. In addition to a number of insect species, other organisms that produce bioluminescence are certain bacteria, algae, coelenterates such as the sea pansy, and crustaceans such as cypridina.

Chemiluminescence occurs when an energy-releasing reaction produces a molecule in an electronically excited state and that molecule, as it returns to the ground state, releases its energy as a photon of light. "Lightning" exemplifies gas-phase chemiluminescence. When an electrical discharge occurs in the atmosphere, gas molecules, such as N_2, O_2, etc., are excited from the ground state to higher energy levels. In addition, N, O, and other atoms are produced. The recombination of these atoms into molecules, as well as the return of excited-state molecules to the ground state, releases energy in the form of visible light. Chemiluminescence also occurs in the liquid and solid phases, as illustrated by the demonstrations in this chapter. The chemiluminescent process, then, is distinctly different from the photoluminescent processes, *fluorescence* and *phosphorescence*, both of which occur after the excited state of a molecule is produced from its ground state by the absorption of light energy. Thus, fluorescent species, whose excited states have lifetimes of only 10^{-9} to 10^{-6} seconds, emit light only while being irradiated. Phosphorescent species, whose excited states have lifetimes of 10^{-3} seconds to several minutes, may appear to emit light without being irradiated, but this emitted energy had to have been absorbed at an earlier time. Chemiluminescent reactions, however, produce light without any prior absorption of radiant energy. (A subsequent volume in this series will include demonstrations of fluorescence and phosphorescence.)

The wavelength of the light emitted by a molecule is related to the energy of its excited state by the equation

$$E = hc/\lambda \tag{1}$$

where h is Planck's constant (6.6×10^{-34} J s),

λ is the wavelength of the emitted light in meters,

c is the speed of light (3.0×10^8 m/s), and

E is the energy difference in joules between the emissive excited state and the ground state of the molecule.

Thus, a photon having a wavelength of 650 nm (650×10^{-9} m) is produced by a molecule whose excited state is 3.0×10^{-19} J above its ground state. The amount of energy released in the emission of one mole of photons of a given wavelength is equivalent to the energy of 6.023×10^{23} photons. The magnitude of this energy is called an einstein. Thus, an einstein of 650-nm photons is equivalent to

$$E = (6.023 \times 10^{23} \text{ photons}) (3.0 \times 10^{-19} \text{ J/photon}) \tag{2}$$

$$= 1.8 \times 10^5 \text{ J} = 180 \text{ kJ}$$

The table summarizes the energies corresponding to a range of wavelengths of visible light.

Selected Wavelengths and Energies of Visible Radiation

Wavelength (nm)	Energy of one photon ($\times 10^{-19}$ J)	Energy of one einstein (kJ)
400	5.0	300
450	4.4	260
500	4.0	240
550	3.6	220
600	3.3	200
650	3.0	180
700	2.8	170

The language of quantum theory best describes the nature of the electronic states of molecules. Many fine textbooks discuss quantum theory, several of which are listed in the references [9]. All of these books discuss one application of quantum theory, molecular orbital (MO) theory, which is of particular value in discussing luminescence. According to MO theory, the energy of electrons in a molecule can be described in terms of molecular orbitals (MOs), which are combinations of the atomic orbitals of the atoms in the molecule. The molecular orbitals that determine the chemical properties of compounds are the highest-energy occupied molecular orbitals (HOMOs) and the lowest-energy unoccupied molecular orbitals (LUMOs).

The energies of LUMOs and HOMOs are frequently represented with molecular orbital diagrams, such as those in Figure 1. Figure 1a represents the ground state for most molecules, particularly luminescent organic molecules; that is, the HOMOs are

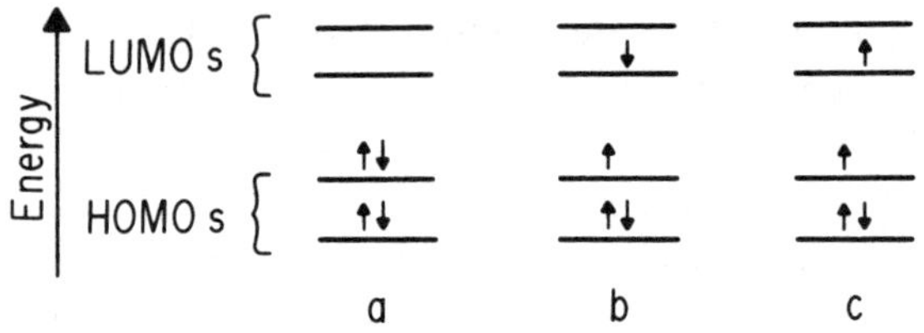

Figure 1.

occupied by paired electrons, while the LUMOs are, of course, empty. In most molecules, the ground state consists of completely filled orbitals, so that all electron spins are paired. A state having all electron spins paired is called a *singlet state*. When a molecule is in an excited state, at least one of its electrons is in a higher-energy orbital, leaving half filled an orbital that is normally filled in the molecule's ground state. Since the electron in the higher-energy orbital and the one in the lower-energy orbital are alone in their orbitals, either may have α or β spin. If the spins of these electrons are opposite, all electron spins in the molecule are paired, and the molecule is in a singlet state (see Figure 1b). If the spins of the two electrons in different orbitals are the same, the molecule has two unpaired spins. Such a state is called a *triplet state* (Figure 1c). Thus, the arrangement of two electrons in a molecule can produce either of two states, a singlet or a triplet.

The singlet and triplet states that arise from a particular orbital electron configuration do not possess the same energy, the triplet state always being lower in energy. State diagrams, such as Figure 2, are frequently used to describe the energy relationships between the states of a molecule. In these diagrams, horizontal lines (S_0, S_1, S_2, T_1, T_2) represent the energy levels of the different states. The vertical distance between these lines corresponds to the energy difference between the states. In Figure 2, the singlet states are shown on the left and the triplet states on the right, each accompanied by a small MO diagram indicating the electron configuration that produces that state. (These small MO diagrams are usually not included in state diagrams.) The singlet ground state is labelled S_0, the states S_1 and T_1 correspond to the singlet and triplet states of the first excited electronic configuration, and states S_2 and T_2 correspond to those of the second excited configuration.

The production of chemiluminescence depends on a number of factors [10]. First, the reaction must provide sufficient energy to create an electronically excited state. Second, the reaction must also produce a species that is capable of forming an elec-

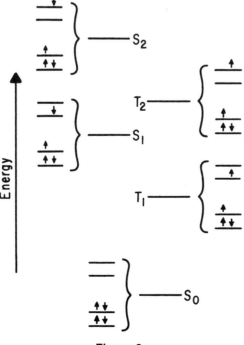

Figure 2.

tronically excited state. Third, the reaction must follow a mechanism that favors the production of this excited state over direct formation of the ground state. Fourth, the reaction mixture must contain a molecule that deactivates from an electronically excited state by the emission of a photon. Last, for the luminescence to be visible, the rate of production of the excited-state species must be sufficiently greater than the rate of quenching. The following paragraphs describe these factors in greater detail.

For a chemical reaction to produce light, it must provide sufficient energy. To produce blue light with a wavelength of 450 nm, a minimum energy of 260 kJ/mole must be provided. For the emission of red light, with a wavelength of 650 nm, 180 kJ/mole is required. Because of these energy requirements, only chemical reactions that are highly energy-releasing can produce chemiluminescence.

For a species to produce visible light, it must receive the excitation energy produced by the reaction. The only way for chemical energy to be transformed into electronic energy is through a species directly involved in the reaction mechanism, that is, through a species that is produced as an intermediate in the reaction. This species must be created in the excited state, and therefore it must possess an electronically excited state with an energy very close to that released by the step that forms it. Since this step of the mechanism converts chemical energy into electronic energy, it is called the *chemielectronic step*.

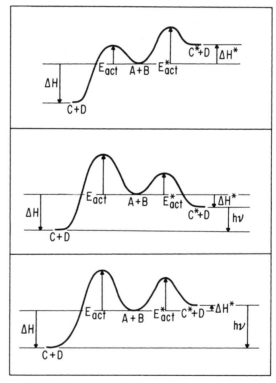

Figure 3.

The mechanistic and energetic relationships required for a chemical excitation process are illustrated by the reaction coordinate diagrams of Figure 3. These diagrams refer to a generalized reaction in which A and B react to form the products C and D in the ground state, or to form C* and D, where C* is a product in an excited state. In the upper curve, the direct production of C in the excited state is strongly endothermic. This situation occurs if C has no electronically excited states below the energy of the reaction, ΔH. In this case, the reaction has no pathway other than the direct formation of the ground state. In the central curve, either the ground-state product or the excited-state product is exothermic. This reaction may be chemiluminescent, depending on the relative activation energies for the production of the ground-state and the excited-state products. The situation depicted by the central curve favors the production of the excited-state product because of its lower activation energy. In the bottom curve of Figure 3, the formation of the excited state is slightly endothermic. In this case, chemiluminescence may be observed if the activation energy for the formation of the excited state is comparable to the activation energy for the formation of the ground-state product and if the former is small enough to be provided by room-temperature thermal activation. The observed luminescence energy in this case can actually be somewhat greater than the thermodynamic difference between the ground-state products and reactants.

The energy possessed by the excited-state species created in the chemielectronic step of the reaction mechanism may be dissipated in any of three general processes: chemical reaction, radiationless conversions, and emission. In chemical reactions, the electronically excited product is simply an unstable intermediate in the reaction mechanism.

In the second case, a radiationless conversion transforms one electronic state to another state without the emission (or absorption) of electromagnetic radiation. All radiationless processes involve some sort of energy transfer from the excited molecule to its environment. The energy released in such a transition may be converted to excess vibrational energy and dispersed as heat by collisions with surrounding molecules. This process is represented in equation 3, where A* signifies the excited species produced by the chemielectronic step, and A is this species in its ground state:

$$A^* \longrightarrow A + heat \tag{3}$$

Additionally, the energy may be transferred to another molecule as electronic energy, producing the excited state of this second molecule. This situation is illustrated by equation 4, where B and B* are the ground and excited states of the second molecule:

$$A^* + B \longrightarrow A + B^* \tag{4}$$

In terms of what happens to A*, this process is termed the *quenching of A**, and B is called the *quencher*.

The third process in which excess electronic energy may be released is through emission, which is the process that occurs in a chemiluminescent reaction. This process is represented by equation 5:

$$A^* \longrightarrow A + h\nu \tag{5}$$

The rate of emission from an excited state depends upon several factors, one of which is the spin of the excited state relative to that of the ground state. Transitions between states of like spin are quantum-mechanically allowed; therefore, the higher-energy state has a short lifetime, from 10^{-9} to 10^{-6} seconds. Since transitions between states of differing spins are quantum-mechanically forbidden, the higher-energy state has a relatively long lifetime, from 10^{-3} to 10 seconds. However, in condensed phases (solids and liquids), vibrational and electronic energy dissipates from excited states at a rate that is very rapid compared to the rate of either type of emission. In spite of the short lifetime of the excited state produced in a chemiluminescent reaction, the emission may persist for quite a long time, since the excited state is being generated at a rate comparable to its rate of decay.

For an excited state to produce visible radiation, the emissive pathway for the deactivation of the excited state must be kinetically competitive with other deactivation routes. This requirement is fulfilled by molecules with widely spaced vibrational energy levels. Such molecules have no ready internal mechanism for converting the excess electronic energy into vibrational (heat) energy. These molecules are either very small molecules with few vibrational degrees of freedom or molecules with rigid aromatic ring systems. When the emissive excited-state molecules are produced directly by the reaction mechanism, the observed luminescence is called *direct chemiluminescence*. The luminescence may also arise from acceptor molecules that receive excitation energy from the directly produced excited-state species (a process represented by equation 4) and subsequently release it as photons of a particular wavelength. Such luminescence is termed *sensitized chemiluminescence,* and the acceptor molecule is called the *sensitizer* of the emission.

For the luminescence produced by a reaction to be observable, the kinetics of the reaction must be sufficiently rapid to produce a photon flux great enough to be detected. Furthermore, the rate at which photons are emitted depends on other factors in addition to the reaction rate. The combined effect of these factors is expressed in the *chemiluminescence quantum efficiency* of the reaction, ϕ_c. This may be defined as the ratio of the number of photons emitted by a reaction to the number of molecules of reactant consumed. For the overall reaction, $A + B \longrightarrow C + h\nu$, the chemiluminescence quantum yield is given by equation 6:

$$\phi_c = \frac{\text{moles of photons produced}}{\text{moles of A reacted}} \tag{6}$$

The chemiluminescence quantum efficiency is a product of three separate quantum efficiencies: one for the production of the emitting species, ϕ_p; the second for the production of its excited state, ϕ_e; and the third for radiation from that excited state, ϕ_r. This product is expressed in equation 7:

$$\phi_c = \phi_p \cdot \phi_e \cdot \phi_r \tag{7}$$

The first two terms of this product are characteristics of the reaction mechanism, while the last is identical to the normal emission quantum efficiency of the emitter. Thus, a low chemiluminescence efficiency can result from a low efficiency in the production of the excited state or from a low emission efficiency for this state.

In studying the mechanism of a chemiluminescent reaction, researchers attempt to discover a number of its characteristics. A primary concern is to identify the luminescent species, that is, the molecule that emits the visible radiation. This identification is generally made by comparing the chemiluminescent spectrum to photoluminescent spectra of species that may be present in the reaction system. When the chemiluminescent spectrum is nearly identical to the fluorescent or phosphorescent spectrum of some species that can be present in the reaction mixture, then this species may be identified as the emitting molecule in the reaction. This method of identification is by no means infallible, since it requires both that the excited state responsible for the chemiluminescence be accessible by photoexcitation and that there be a unique emitting molecule in the chemiluminescent system. These requirements are generally met, but not always. For example, in the peroxide oxidation of lucigenin (Demonstration 2.7), two emitting species have been identified.

Another interesting feature of chemiluminescent reactions is the nature of the chemielectronic step, in which chemical energy is converted to the energy of an electronically excited state of a molecule. The identification of the chemielectronic step is based on a variety of kinetic data, by which the entire mechanism may be elucidated. The chemielectronic step may produce the emitting excited-state molecule directly, or it may produce an excited-state intermediate which transfers its energy to the emitting molecule. In the latter case, the emitting molecule acts as a quencher of the initially produced excited-state molecule. The effect of other quenchers on the chemiluminescence may aid in identifying this primary excited-state molecule by determining its energy. If a quencher of known energy shuts off the chemiluminescence, then its energy is similar to or lower than that of the chemically produced excited state.

To interpret fully the effects of various experimental parameters such as solvent, concentration, quenchers, etc., on a chemiluminescent reaction, the entire mechanism of the reaction must be determined. Once the mechanism is determined, it may be possible to choose conditions that improve the efficiency of the chemiluminescent reaction or change its spectrum. The articles in reference 5 illustrate the experimental and theoretical considerations involved in determining mechanisms for a number of chemiluminescent reactions.

Although chemiluminescence can be produced in the liquid phase by electrical means, the intensity of the emission is too weak for lecture demonstration applications. Electro-generated chemiluminescence occurs in a variety of solvents in which different species can be oxidized or reduced to radical ions at the electrodes. Chemiluminescence can result from several mechanisms including reduction or oxidation of the radicals at an electrode surface and the reaction of an electrically produced cation and anion to form an excited singlet state [3, 10–13].

The phenomenon of triboluminescence, the emission of light when a mechanical stress is applied to a crystal, is exhibited by many inorganic and organic substances [14–17]. For example, when table sugar or a piece of candy is ground with a spoon in a totally darkened room, flashes of light are observed. The emission is from absorbed or adsorbed N_2 gas, which is released in an excited state when mechanical stress is applied to sugar crystals [15]. Since emission cannot be observed at distances greater than an arm's length, triboluminescence emission is too weak for lecture demonstration applications.

Near the end of the chapter, we list titles of demonstrations whose use we do not recommend. Sources and structures of dyes are listed at the end of the chapter.

REFERENCES

1. Gunderman, K. D. *Angew. Chem. Intl. Ed. Engl.* **1965**, *4*, 566.
2. McCapra, F. *Quart. Rev.* (London) **1966**, 485.
3. Haas, J. W., Jr. *J. Chem. Educ.* **1967**, *44*, 396.
4. Seybold, P. G. *Chemistry* **1973**, *46*, 6.
5. Cormier, M. J.; Hercules, D. M.; Lee, J., Eds. "Chemiluminescence and Bioluminescence"; Plenum Press: New York, 1973.
6. Hercules, D. M., Ed. "Fluorescence and Phosphorescence Analysis"; Interscience Publishers, John Wiley and Sons: New York, 1966.
7. Bowen, E. J., Ed. "Luminescence in Chemistry"; Van Nostrand: New York, 1968.
8. Legg, K. D. *J. Chem. Educ.* **1973**, *50*, 848.
9. Murrell, J. N.; Kettle, S. F. A.; Tedder, J. M. "Valence Theory"; Interscience Publishers, John Wiley and Sons: New York, 1965; Hanna, M. W. "Quantum Mechanics in Chemistry," 2nd ed.; Benjamin/Cummings: Menlo Park, California, 1969; Karplus, M.; Porter, R. N. "Atoms and Molecules"; Benjamin/Cummings: Menlo Park, California, 1970.

10. Hercules, D. M. *Accts. Chem. Res.* **1969,** *2,* 301.

11. Hercules, D. M. *Science* **1964,** *145,* 808.

12. Legg, K. D.; Hercules, D. M. *J. Am. Chem. Soc.* **1969,** *91,* 1902.

13. Tokel, N. E.; Bard, A. J. *J. Am. Chem. Soc.* **1972,** *94,* 2862.

14. Erikson, J. *J. Chem. Educ.* **1972,** *49,* 688.

15. Angelos, R.; Zink, J. I.; Hardy, G. E. *J. Chem. Educ.* **1979,** *56,* 413.

16. Chandra, B. P.; Zink, J. I. *Inorg. Chem.* **1980,** *19,* 3098.

17. Chandra, B. P.; Zink, J. I. *J. Phys. Chem.* **1980,** *73,* 5933.

2.1

Singlet Molecular Oxygen

The red chemiluminescence of singlet molecular oxygen can be observed when aqueous solutions of Cl_2 (or hypochlorite ion) and hydrogen peroxide are mixed [1–3]. The bright emission results from excited gaseous O_2 which is trapped in bubbles that form in the reaction mixture. Several procedural variations, some involving added sensitizers to enhance and prolong emission, are described. The color of emission varies with the sensitizers.

MATERIALS FOR PROCEDURE A

100 ml 30% hydrogen peroxide, H_2O_2

125 ml 6M sodium hydroxide, NaOH

chlorine cylinder, with valve

ca. 0.005 g violanthrone† (If violanthrone is unavailable, other sensitizers [dyes] can be used. Sensitizers that produce red chemiluminescence include iso-violanthrone, rubrene, phenathrenequinone, 9,10-diphenylanthracene, rhodamine 6G, and cresyl violet. Other sensitizers include 13,13′-dibenzanthronyl, which initially produces a red emission that after a few seconds changes to orange; luminol [3-aminophthalhydrazide], which produces a bluish lavender glow; and lucigenin [bis-N-methylacridinium nitrate], which produces a bluish green glow.)

30 ml dichloromethane, CH_2Cl_2

gloves, plastic or rubber

2 500-ml gas washing bottles

1 m Tygon or rubber tubing (to fit gas washing bottles and cylinder valve)

MATERIALS FOR PROCEDURE B

500 ml 30% hydrogen peroxide, H_2O_2

500 ml commercial bleach (5% NaOCl solution)

100 ml 0.1% violanthrone in dichloromethane

2 1-liter separatory funnels

250-ml separatory funnel

2 glass Y-shaped connecting tubes (to fit tubing with outside diameter of 12 mm)

50 cm Tygon tubing, with outside diameter of 12 mm

†See pp 202–4 for chemical structures and sources.

2-liter Erlenmeyer flask or 2-liter round-bottomed flask on cork ring

ring stand with 3 rings to support separatory funnels

gloves, plastic or rubber

MATERIALS FOR PROCEDURE C

50 ml 30% hydrogen peroxide, H_2O_2

0.005 g violanthrone in 30 ml of dichloromethane

100 ml commercial bleach (5% NaOCl solution)

gloves, plastic or rubber

250-ml beaker

50-ml beaker

150-ml beaker

hypodermic syringe, 10–30 ml capacity

hypodermic needle (for example, 20 gauge, 5 cm long)

PROCEDURE A

Caution: Chlorine gas is used in this procedure. A gas trap is required to absorb the excess chlorine gas. Perform this demonstration under a hood or in a well-ventilated area.

Wearing gloves, add 100 ml of 30% hydrogen peroxide and 25 ml of 6M sodium hydroxide to one of the 500-ml gas washing bottles. Close the gas washing bottle. Connect the Tygon or rubber tubing to the inlet tube of the other gas washing bottle. Add 100 ml of 6M sodium hydroxide to the second gas washing bottle and close the bottle. Using the Tygon or rubber rubing, connect the chlorine gas cylinder to the inlet of the first gas washing bottle. Connect the gas trap to the outlet of the gas washing bottle. Make sure that the chlorine gas will be introduced *below* the liquid surface in both vessels (Figure 1).

After dimming the room lights, adjust the chlorine gas flow to obtain a gentle flow of chlorine gas into the first gas washing bottle. Glowing red bubbles of singlet molecular oxygen will appear where the chlorine gas reacts with the alkaline hydrogen peroxide solution. The temperature of the mixture will rise quickly because of the exothermic nature of both the Cl_2/H_2O_2 reaction, which produces O_2 and hydrochloric acid, and the neutralization reaction with sodium hydroxide.

To enhance the red glow, follow these steps carefully. In 30 ml of dichloromethane, suspend approximately 0.005 g of violanthrone (or any one of isoviolanthrone, 13,13'-dibenzanthronyl, 9,10-diphenylanthracene, and phenanthrenequinone; rubrene, rhodamine 6G, and cresyl violet are water soluble and may be used as solids). Do not add too much sensitizer or you will quench the excitation reaction. Close the chlorine cylinder valve and make sure the temperature of the mixture in the first gas washing bottle is close to room temperature (dichloromethane boils at 40°C). Lift the top of this bottle and slowly add the violanthrone-dichloromethane suspension (or other sensitizer). The aqueous and organic liquids are immiscible, and the aqueous phase will be

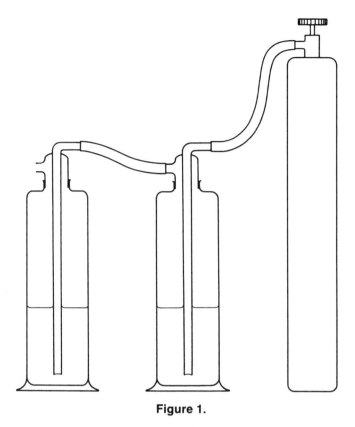

Figure 1.

on top of the denser organic phase. Close the gas washing bottle. Adjust the flow of chlorine gas from the cylinder to obtain a gentle flow of chlorine gas. Spectacular reddish purple bubbles will appear in the reaction vessel. Close the chlorine cylinder valve.

To produce a bluish lavender glow, open the first gas washing bottle and quickly add about 0.005 g of luminol. Close the gas washing bottle and swirl the mixture to dissolve the luminol. Adjust the chlorine gas flow to obtain a gentle flow of chlorine gas into the gas washing bottle. The intense red of the violanthrone emission will still appear but will be accompanied by the bluish lavender luminol emission in the aqueous layer. A bluish green glow may be produced by substituting lucigenin for luminol in this procedure.

The bluish lavender luminol emission can also be generated by adding 0.005 g of luminol directly into the first gas washing bottle after the red glow of singlet molecular oxygen is observed. If the violanthrone-dichloromethane solution is then added, the red solution sinks to the bottom of the gas washing bottle. When gas flow is resumed, bright reddish purple bubbles will appear in a glowing blue background.

PROCEDURE B†

Arrange the glassware as shown in Figure 2. The tube from the lower Y-tube to the receiving flask should be as long as possible. Wearing gloves, pour 500 ml of 30% hydrogen peroxide into one of the large separatory funnels and pour 500 ml of bleach solution (5% NaOCl) into the other. Add 100 ml of 0.1% violanthrone in dichloromethane solution to the 250-ml separatory funnel.

†We are grateful to Professor John G. Burr of the University of Oklahoma for suggesting this procedure.

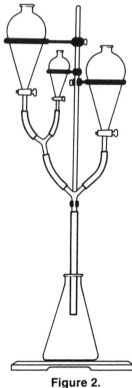

Figure 2.

Darken the room as much as possible. Open the stopcocks on both large separatory funnels and allow the liquids to flow into the 2-liter flask. Observe the red glow where the solutions mix. While the two solutions are flowing, open the stopcock on the 250-ml separatory funnel, which adds the violanthrone solution to the mixture. Observe the brighter red glow. Regulate the liquid flow with the stopcocks to minimize heating and foaming as the solutions mix.

PROCEDURE C

Wearing gloves, pour 50 ml of 30% H_2O_2 into the 250-ml beaker. In the 50-ml beaker, add 0.005 g of violanthrone to 30 ml of dichloromethane. Pour this purple-dye suspension into the 250-ml beaker. Pour 100 ml of bleach into the 150-ml beaker. Fill the syringe with bleach and place the needle inside the large beaker near the bottom. The tip of the needle should be near the side of the beaker. Dim the room lights, *rapidly* inject the bleach solution, and observe the red chemiluminescence.

Red emission appears when either Cl_2 or Br_2 is dissolved in dichloromethane and is then injected rapidly with a syringe into an alkaline H_2O_2 solution. Table 1 summarizes the results of several qualitative experiments in which a syringe was used to mix the reagents as described.

HAZARDS

Chlorine gas irritates the eyes and mucous membranes and, if inhaled, can cause severe lung irritation and fatal pulmonary edema. In high concentrations, the gas irritates the skin.

Table 1. Summary of Experimental Observations in Procedure C

Reagent in syringe	Reagent[a] in beaker	Initial pH of H_2O_2 solution	Observation[b]
bleach	H_2O_2	3.6	moderate glow
bleach	H_2O_2	9.5	faint glow
bleach	H_2O_2	1.0	faint glow
H_2O_2	bleach	3.6	no glow
HOCl[c]	H_2O_2	3.6	faint glow
HOCl[c]	H_2O_2	9.6	bright glow
Cl_2 in CH_2Cl_2	H_2O_2	3.6	no glow
Cl_2 in CH_2Cl_2	H_2O_2	9.5	bright glow
Br_2 in CH_2Cl_2	H_2O_2	3.6	no glow
Br_2 in CH_2Cl_2	H_2O_2	9.5	moderate glow
I_2 in CH_2Cl_2	H_2O_2	3.6	no glow
I_2 in CH_2Cl_2	H_2O_2	9.5	no glow

[a]Approximately 80 ml of 30% H_2O_2 was made acidic or basic by the addition of about 5 ml of 6M HCl or 6M NaOH.
[b]The intensity of the observed red emission was classified qualitatively as "faint," "moderate," or "bright."
[c]Chloride-free HOCl [1].

Chlorine is a strong oxidizing agent, and combustible materials will burn in a chlorine atmosphere. Chlorine forms explosive mixtures with vapors such as hydrocarbons or other flammable organic compounds.

Since 30% hydrogen peroxide is a strong oxidizing agent, contact with skin and eyes must be avoided. In case of contact, immediately flush with water for at least 15 minutes; get immediate attention if the eyes are affected.

Avoid contact between 30% hydrogen peroxide and combustible materials. Avoid contamination from any source, since any contaminant, including dust, will cause rapid decomposition and the generation of large quantities of oxygen gas. Store 30% hydrogen peroxide in its original closed container, making sure that the container vent works properly.

The toxic and carcinogenic properties of violanthrone, isoviolanthrone, rubrene, phenanthrenequinone, 9,10-diphenylanthracene, 13,13′-dibenzanthronyl, cresyl violet, lucigenin, and luminol are not known, although most should be assumed to be poisonous. Rhodamine 6G has been shown to be carcinogenic in animal tests.

Dichloromethane vapor is narcotic in high concentrations and is irritating to the eyes.

DISPOSAL

When the demonstration is completed, close the chlorine gas cylinder, disconnect the reaction vessel and gas trap and place them in the fume hood. All the solutions *except* the dichloromethane should be washed down the drain with large volumes of cold water. The dichloromethane solution should be discarded in a solvent waste container.

Expel all the chlorine gas from the gas washing bottle and other glassware by filling each completely with cold water and then emptying it. Repeat the filling and emptying several times. This operation should be performed in the fume hood.

DISCUSSION

The red chemiluminescence resulting from singlet molecular oxygen appears when aqueous solutions of Cl_2 (or hypochlorite ion) and hydrogen peroxide are mixed:

$$Cl_2(aq) + H_2O_2(aq) \longrightarrow O_2(g) + 2\,H^+(aq) + 2\,Cl^-(aq) \tag{1}$$

$$OCl^-(aq) + HO_2^-(aq) \longrightarrow O_2(g) + Cl^-(aq) + OH^-(aq) \tag{2}$$

In 1927 Mallet [4] first reported the red glow accompanying the OCl^-/H_2O_2 reaction. Kasha and co-workers [5] studied this reaction in some detail and have attributed the chemiluminescence to the simultaneous transition of two excited oxygen molecules $[O_2\,(^1\Delta_g)]$ to the ground state in a two-molecule one-photon process:

$$2\,O_2(^1\Delta_g) \longrightarrow 2\,O_2(^3\Sigma_g^-) + h\nu \ (\sim 630 \text{ nm}) \tag{3}$$

The red emission results from excited gaseous O_2 which is trapped in bubbles that form in the reaction mixture [6]. Brabham and Kasha [7] point out that the intensity of the emission is highly sensitive to the rate of $O_2\,(^1\Delta_g)$ formation for two reasons: (a) the emission is produced by a simultaneous transition in two excited oxygen molecules, and therefore the intensity depends on the square of the singlet oxygen concentration; and (b) the singlet oxygen is rapidly quenched in solution. For both reactions 1 and 2, the apparatus design and reaction conditions strongly influence the rate of singlet oxygen production.

Bubbling Cl_2 into hydrogen peroxide solution in a standard gas washing apparatus results in rapid evolution of O_2 accompanied by vigorous frothing (Procedure A). The rate of bubbling should be controlled so that no frothing occurs in the region of injection; otherwise, the red glow will not be observed. Frothing probably assists the quenching of singlet oxygen, thereby interfering with the emission. The intensity of emission can be enhanced by making the hydrogen peroxide solution alkaline. Mechanistic studies of the chlorine oxidation of hydrogen peroxide suggest [8, 9] that the species Cl_2, $HOCl$, and OCl^- oxidize hydrogen peroxide (hydroperoxide ion) to produce $O_2(^1\Delta_g)$. In alkaline solution, the rate law is $d[O_2]/dt = k_3[H_2O_2]\,[OCl^-]$. At 25°C, the value of k_3 is $\sim 10^7 M^{-1}sec^{-1}$. Gaseous singlet oxygen ($^1\Delta_g$) is produced initially in yields up to 40–50% of the total oxygen generated. The hydrochloric acid formed during the reaction (equation 1) eventually causes a decrease in alkalinity which effectively stops singlet oxygen evolution. The lifetime of $O_2(^1\Delta_g)$ in aqueous solution is about 3 microseconds [10] while its lifetime in the gas phase is about 45 minutes [5].

Singlet oxygen chemiluminescence is observed when household bleach (5% NaOCl solution) and 30% hydrogen peroxide are mixed rapidly (Procedures B and C). The bleach solution is sufficiently alkaline to produce maximum glow. Injecting 30% H_2O_2 into household bleach does *not* produce emission, probably because of the large difference in concentration between the reactants [7] (30% H_2O_2 is approximately 10M, whereas 5% NaOCl is approximately 0.7M). When the relatively dilute NaOCl solution is injected into peroxide, all OCl^- is consumed at the point of injection and chemiluminescence is observed. However, when peroxide is injected into household bleach, all OCl^- near the point of injection is consumed immediately and the excess peroxide must diffuse into the bulk of the solution before reaction can take place. This diffusion process results in a slow production of singlet oxygen, and chemiluminescence is not observed. When the reactants' concentrations are comparable, the order of injection is no longer important and the red glow is observed either way.

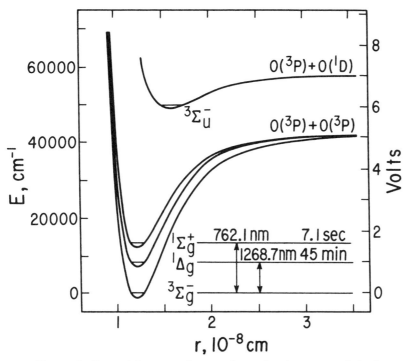

Figure 3. Potential energy diagram for molecular oxygen [13, 5].

Other sources of singlet oxygen can also be employed [11], but the chlorine-peroxide reaction is probably the most convenient to use. Singlet oxygen ($^1\Delta_g$) is of considerable significance in organic and biochemical systems [12].

Spectroscopic Considerations

A potential energy diagram for some of the lower states of molecular oxygen is shown in Figure 3. The lowest energy configuration is split by electron repulsion into three distinct states: $^1\Sigma_g^+$, $^1\Delta_g$, and $^3\Sigma_g^-$. See Table 2 for the physical constants of these states. The term symbols used to designate diatomic molecular states are analogous to atomic term symbols. The electronic states are designated Σ, Π, Δ, etc., just as s, p, d, etc., denote atoms. The left superscript indicates the multiplicity. The $+$ or $-$ right superscript for sigma (Σ) states indicates whether the electronic wave function remains the same or changes sign upon reflection through a plane passing through both nuclei. If the two nuclei in a diatomic molecule have the same charge, the molecule has a

Table 2. Physical Constants of the Three Lowest Electronic States of Molecular Oxygen [13]

State	Mean lifetime (sec)	Equilibrium internuclear distance (Angstroms)	ν (cm^{-1})	λ nm	E (kJ/mole)
$_g^+$	12[a]	1.227	13,120.91[c]	7,621.4	156.9
$_g$	2,700[b]	1.216	7,882.39[d]	12,687	94.26
$_g^-$	—	1.207	—	—	—

[a]Wallace and Hunten [14].   [c]Babcock and Herzberg [16].
[b]Badger, Wright, and Whitlock [15].   [d]Herzberg and Herzberg [17].

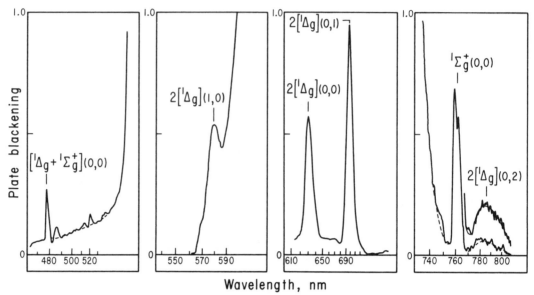

Figure 4. Four regions of the chemiluminescence spectrum from the chlorine–hydrogen peroxide and hypochlorite ion–hydrogen peroxide reactions [5].

center of symmetry, and the electronic wave function may be even or odd with respect to inversion. This even or odd property is designated by a subscript g or u, respectively (from the German *gerade* or *ungerade*).

On the basis of Hund's Rule, which holds for molecules as well as atoms, the ground state of O_2 is $^3\Sigma_g^-$. The transition between $^3\Sigma_g^-$ and $^3\Sigma_u^-$ states is allowed for electric dipole radiation and is readily observed as the Schuman-Runge band system in the ultraviolet region of the spectrum [13]. The electric dipole transitions between the triplet ground state and the $^1\Sigma_g^+$ and $^1\Delta_g$ states are forbidden on the bases of spin and symmetry. In addition, the transition between the ground state and $^1\Delta_g$ is orbitally forbidden. The low transition probabilities are reflected in the long radiative lifetimes of the two states: 12 seconds for $^1\Sigma_g^+$ and 45 minutes for $^1\Delta_g$ [14, 15]. These lifetimes are in the limit of zero pressure and may be considerably modified by perturbations from surrounding gases. In fact, the transitions have been observed as the atmospheric oxygen bands ($^1\Sigma_g^+ \longleftrightarrow {}^3\Sigma_g^-$) at 762.1 nm and the infrared atmospheric oxygen bands ($^1\Delta_g \longleftrightarrow {}^3\Sigma_g^-$) at 1,268.7 nm [13].

The spectrum of the red chemiluminescence produced in the Cl_2/H_2O_2 and OCl^-/H_2O_2 reactions has been studied by several researchers. Khan and Kasha [5] observed two prominent bands at 633.4 nm and 703.2 nm (Figure 4) as well as weaker bands at 578.0 nm and 786.0 nm. They have assigned these bands to vibrational components of the double-molecule transition (two-molecule, one-photon):

$$O_2(^1\Delta_g) + O_2(^1\Delta_g) \longrightarrow O_2(^3\Sigma_g^-) + O_2(^3\Sigma_g^-) \tag{4}$$

Khan and Kasha also observed a band at 762.1 nm which they assigned as the 0,0 band of the $^1\Sigma_g^+ \longrightarrow {}^3\Sigma_g^-$ transition.

Simultaneous transitions of the type in equation 4 have also been observed [5] in the high pressure (150 atm) absorption spectrum of oxygen (Figure 5). The ordinarily spin-forbidden transition becomes allowed for the simultaneous transition because in the double-molecule $O_2(^3\Sigma_g^-)O_2(^3\Sigma_g^-)$ the coupling of the spins produces a singlet component. The transition can be represented as

$$^1[(^1\Delta_g)\,(^1\Delta_g)] \longleftrightarrow^{1,\,3,\,5} [(^3\Sigma_g^-)\,(^3\Sigma_g^-)] \tag{5}$$

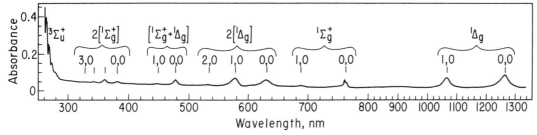

Figure 5. Absorption spectrum of molecular oxygen at high pressure [5].

Since absorption requires a three-body process (two O_2 molecules and a photon) which has a very low probability of occurring, high pressures must be used to observe it. Emission, on the other hand, requires only a two-body process and can be observed at atmospheric pressure, even with relatively low partial pressures of singlet oxygen. The energy of the photon emitted or absorbed in the simultaneous transition (633.4 nm) is double the energy for the single-molecule process (1,268.7 nm).

Bonding (Molecular Orbital Theory)

A molecular orbital analysis is useful to elucidate the mechanism of singlet oxygen production in the peroxide-hypochlorite system. The chloroperoxy ion ($OOCl^-$) has been postulated as an intermediate in the Cl_2 and OCl^- oxidation of H_2O_2 [5]. The ground state of this ion is a singlet, and ionic fission of the species to produce O_2 and Cl^- (^1S) must leave the oxygen in a singlet state.

Khan and Kasha [5] have used the spin-state correlation diagram (Figure 6) to deduce that $^1\Delta_g$ is the only product of the fission of $OOCl^-$. Since the $^1\Delta_g$ state of O_2

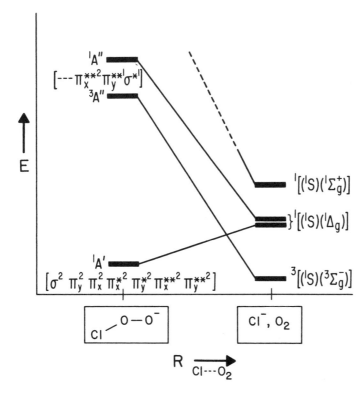

Figure 6. Spin-state correlation diagram for chloroperoxy ion and its products, chloride ion and molecular oxygen [5].

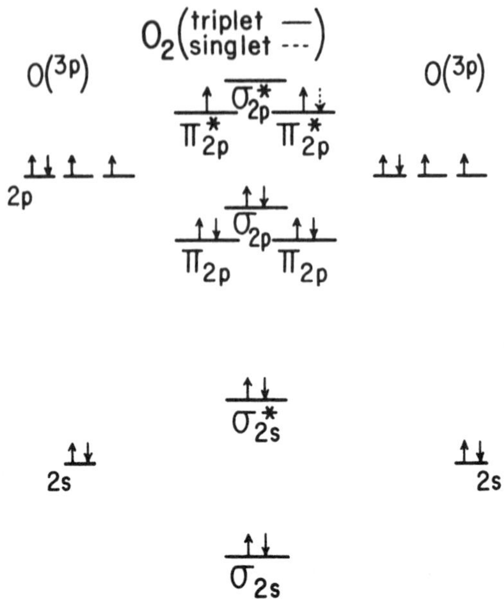

Figure 7. Molecular orbital diagrams for atomic and molecular oxygen.

is doubly degenerate (Figure 7), even higher excited states of $OOCl^-$ must be used to correlate with $^1\Sigma_g^+$. The spin-state correlation excludes the direct production of $O_2(^1\Sigma_g^+)$ in the breakdown of chloroperoxy ion. Since emission from this state is observed in the spectrum of the chemiluminescence, the most likely mechanism of production is

$$^1\Delta_g + {}^1\Delta_g \longrightarrow {}^1\Sigma_g^+ + {}^3\Sigma_g^-$$

The $^1\Sigma_g^+$ is present in very small amounts with the ratio $[^1\Sigma_g^+]/[^1\Delta_g]$ less than 10^{-6}.

Energy Transfer from Singlet Oxygen to Acceptors

Although a molecule of gaseous oxygen has no states in the visible region of the spectrum, visible chemiluminescence is readily observed from acceptor molecules in a solution containing chlorine species and alkaline hydrogen peroxide. The various poly-aromatic sensitizers used in this demonstration are excited by the singlet oxygen ($^1\Delta_g$) produced in the reactions of chlorine species with hydrogen peroxide. With violan-throne, the red glow of singlet oxygen is replaced by a very bright red emission which is 2–4 orders of magnitude brighter than the original oxygen emission. The oxygen emission is of low quantum efficiency because of quenching, but the violanthrone emission is only slightly quenched and is efficiently excited by the singlet oxygen before the latter species can be otherwise quenched. The emission spectra of violan-throne and other dyes are shown in Figures 8a–e. The suggested mechanism for this energy transfer [18, 19] involves the lowest singlet state as the emitting species:

$$^1A_0 + O_2(^1\Delta_g) \longrightarrow {}^3A_1 + O_2(^3\Sigma_g^-)$$
$$^3A_1 + O_2(^1\Delta_g) \longrightarrow {}^1A_1 + O_2(^3\Sigma_g^-)$$
$$^1A_1 \longrightarrow {}^1A_0 + h\nu$$

where $^1A_0$ = ground singlet state of violanthrone,
$^3A_1$ = triplet state, and
$^1A_1$ = excited singlet state.

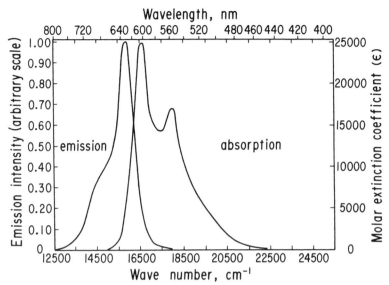

Figure 8a. Emission and absorption spectra of violanthrone.

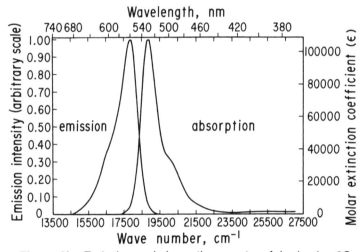

Figure 8b. Emission and absorption spectra of rhodamine 6G.

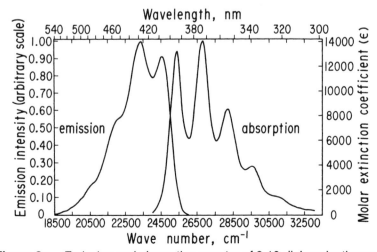

Figure 8c. Emission and absorption spectra of 9,10-diphenylanthracene.

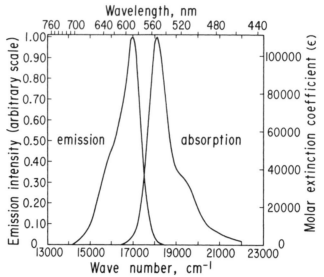

Figure 8d. Emission and absorption spectra of rhodamine B.

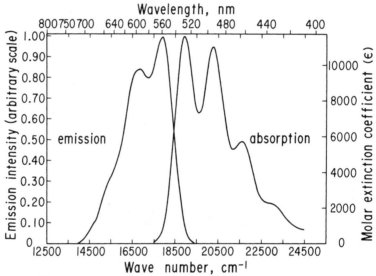

Figure 8e. Emission and absorption spectra of rubrene.

In all cases of polyaromatic hydrocarbons, the sensitizer emission energy is almost equal to or lower than that of 2 $O_2(^1\Delta_g)$ (see Figure 9). One sensitizer, 1,1-dibenzanthronyl, is reported to emit a green glow [20]. The mechanism for this green emission is not understood, but it is unlikely to result from excitation of $O_2(^1\Sigma_g^+)$ since the lifetime of this species in solution is probably much smaller than the 3 microsecond lifetime of $O_2(^1\Delta_g)$.

In the case of luminol, the mechanism by which the bluish lavender emission is produced is not yet resolved. It may be caused by oxidation of luminol by chlorine species [21, 22].

This demonstration can be used along with another demonstration that displays the paramagnetism and blue color of liquid oxygen [23]. The blue absorption transition of liquid oxygen is the reverse of the red chemiluminescence emission. Both the paramagnetism and color are intrinsic properties of oxygen and are not bulk properties of the liquid phase.

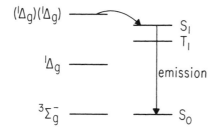

Figure 9. Energy levels of oxygen species, *left,* and emission sensitizer, *right.*

REFERENCES

1. Shakhashiri, B. Z.; Williams, L. G. *J. Chem. Educ.* **1976**, *53*, 358.
2. Bailey, P. S.; Bailey, C. A.; Koski, P. G.; Andersen, J.; Techsteiner, C. *J. Chem. Educ.* **1975**, *52*, 524.
3. Sharpe, S. "The Alchemist's Cookbook: 80 Demonstrations"; Shell Canada Centre for Science Teachers, McMaster University: Hamilton, Ontario, undated; p 25.
4. Mallet, L. *C. R. Acad. Sci. Paris* **1927**, *185*, 352.
5. Khan, A. U.; Kasha, M. *J. Am. Chem. Soc.* **1970**, *92*, 3293 and references therein.
6. Khan, A. U.; Kasha, M. *Nature* **1964**, *204*, 241.
7. Brabham, D. E.; Kasha, M. personal communication, 1974.
8. Held, A. M.; Halko, D. J.; Hurst, J. L. *J. Am. Chem. Soc.* **1978**, *100*, 5733.
9. Hurst, J. K.; Carr, P. A. G.; Hovis, F. E.; Richardson, R. J. *Inorg. Chem.* **1981**, *20*, 2435.
10. Matheson, I. B. C.; Massoudi, R. *J. Am. Chem. Soc.* **1980**, *102*, 1942.
11. Peters, J. W.; Bekowies, P. J.; Winer, A. M.; Pitts, J. N., Jr. *J. Am. Chem. Soc.* **1975**, *97*, 3299 and references therein.
12. Gorman, A. A.; Rodgers, M. A. *J. Chem. Soc. Rev.* **1981**, *10*, 205.
13. Herzberg, G. "Spectra of Diatomic Molecules"; Van Nostrand: New York, 1950.
14. Wallace, L.; Hunten, D. M. *J. Geophys. Res.* **1968**, *73*, 4813.
15. Badger, R. M.; Wright, A. C.; Whitlock, R. F. *J. Phys. Chem.* **1965**, *43*, 4345.
16. Babcock, H. D.; Herzberg, L. *Astrophys. J.* **1948**, *108*, 167.
17. Herzberg, L.; Herzberg, G. H. *Astrophys. J.* **1947**, *105*, 353.
18. Abbott, S. R.; Ness, S.; Hercules, D. M. *J. Am. Chem. Soc.* **1970**, *92*, 1128.
19. Ogryzlo, E. A.; Pearson, A. E. *J. Phys. Chem.* **1968**, *72*, 2913.
20. Slabaugh, W. H. *J. Chem. Educ.* **1970**, *47*, 522.
21. Fuchsman, W. H.; Young, W. G. *J. Chem. Educ.* **1976**, *53*, 548.
22. Eriksen, T. E.; Lind, J.; Mereny, G. *Trans. Faraday Soc. 1* **1981**, *77*, 2125.
23. Shakhashiri, B. Z.; Dirreen, G. E.; Williams, L. G. *J. Chem. Educ.* **1980**, *57*, 373.

2.2

Lightsticks

A commercially available chemiluminescent system produces a yellowish green cold light which is easily visible in a darkened room [1]. The light, initially bright enough for reading, gradually fades over a period of about 12 hours.

MATERIALS

3 "CYALUME" lightsticks [2]

500 ml ice

400 ml hot water

2 500-ml beakers

thermometer, $-10°C$ to $+110°C$

light meter (optional)

PROCEDURE

To observe the chemiluminescence, follow the instructions on the package (remove the wrapper foil, bend the plastic tube slightly to break the thin vial inside, and shake). Since the light is generated without heat or flame, the "cool-light" source may be passed around the room to enable students to inspect it more closely.

To demonstrate the effect of temperature on the intensity of light emission, immerse the second activated lightstick in an ice bath. As the contents of the lightstick cool, the intensity of the emitted light decreases. Place the third lightstick in a beaker of hot water and observe the increase in light intensity. To avoid melting the plastic tube, the temperature must not exceed 70°C, and the beaker should not be on a hot plate or above a burner. The plastic container for the lightstick slows the cooling or heating of its contents. Note the temperature of each bath and the duration of immersion of each lightstick. Observe the effect of temperature on the intensity of light emission by removing the lightsticks from both baths and holding them on opposite sides of the first lightstick, which has been kept at room temperature.

In a totally darkened room, a light meter can be used to measure the glow intensity. To make light-intensity measurements as a function of temperature, place the lightstick for a fixed period of time, 2–3 minutes, in baths of different temperature. The activation energy for the reaction can be obtained from an Arrhenius plot. We have obtained a value of 56.4 kJ/mole in the $-5°C$ to 50°C range.

HAZARDS

According to the manufacturer, the mixture of the two components is nontoxic either to the skin or through ingestion and the mixture is not irritating to the eyes. In the event of rupture of the plastic container, wash thoroughly with water. The solution in the lightstick will stain clothing and soften or mar paint and varnish.

DISPOSAL

The lightsticks can be given to students to keep. When the glow is no longer visible, the lightsticks can be thrown away in a waste can.

DISCUSSION

This simple demonstration has been well received by a wide variety of age groups. Elementary school children, college students, parents, and others of different scientific background are able to infer on the basis of observing the temperature effect on the glow intensity that raising the temperature increases the rate of reaction. Most observers accurately predict that the glow intensity of the lightstick immersed in hot water will fade before that of the one kept at room temperature, which in turn will fade before the one immersed in the ice bath. Some realize that such predictions rely on the assumption that all lightsticks have the same concentrations of reagents.

We encourage students to place a glowing lightstick in a freezer to see how long the lightstick can retain its glow upon warming to room temperature. One of the authors has observed faint glow from an activated lightstick kept in a home freezer for over 6 months.

According to the manufacturer, the lightstick is wrapped in an airtight, foil wrapper because of sensitivity to humidity. If the wrapper is punctured and the lightstick not used within a few days, the light emission will be reduced. If the lightstick is left in a damaged wrapper for a longer period of time, it may become totally deactivated. The manufacturer claims that if the foil wrapper has not become damaged, the lightstick has a shelf life of up to 4 years.

The CYALUME lightstick contains dilute hydrogen peroxide in a phthalic ester solvent contained in a thin glass ampule, which is surrounded by a solution containing a phenyl oxalate ester and the fluorescent dye 9,10-bis(phenylethynyl)anthracene. The concentration of H_2O_2 in the ampule was found to be less than 0.5%.† When the ampule is broken, the H_2O_2 and oxalate ester react:

During the course of the reaction, an intermediate is produced which transfers energy to a dye molecule. Visible light is emitted when the excited dye molecule returns to the ground state. Although details of a plausible mechanism for the reaction are described

†We thank Lawrence D. David for performing the analysis.

in the next section, the following equations summarize the essential features of the chemiluminescence:

$$\text{(phenyl)}-O-\underset{\underset{O}{\|}}{C}-\underset{\underset{O}{\|}}{C}-O-\text{(phenyl)} + H_2O_2 \longrightarrow \text{(phenyl)}-O-\underset{\underset{O}{\|}}{C}-\underset{\underset{O}{\|}}{C}-O-OH + \text{(phenyl)}-OH$$

$$\text{(phenyl)}-O-\underset{\underset{O}{\|}}{C}-\underset{\underset{O}{\|}}{C}-O-OH \longrightarrow \underset{\underset{O}{\diagup}}{\overset{O-O}{\underset{}{C-C}}}\underset{O}{\diagdown} + \text{(phenyl)}-OH$$

$$\underset{\underset{O}{\diagup}}{\overset{O-O}{\underset{}{C-C}}}\underset{O}{\diagdown} + \text{Dye} \longrightarrow \text{Dye*} + 2\,CO_2$$

$$\text{Dye*} \longrightarrow \text{Dye} + h\nu$$

The chemicals used in this demonstration can be synthesized [3], and it is possible to produce blue emission by using the dye 9,10-diphenylanthracene. The chemicals can also show the effect of a catalyst on the reaction rate [3].

Mechanistic Details

A chemiluminescent reaction usually involves three steps [4]:
(a) preliminary reactions producing a key intermediate;
(b) an excitation step converting the intermediate's chemical energy into electronic energy in a sensitizer; and
(c) emission from the excited reaction product of the excitation step.

Peroxyoxalate chemiluminescence differs from most chemiluminescent reactions since the fluorescence spectrum of an added compound determines the chemiluminescence spectrum, independent of the reactants. Through the use of a variety of sensitizers, emissions spanning most of the visible spectrum have been obtained [5].

The overall reaction for general oxalate ester chemiluminescent reactions is

$$\text{RO-}\underset{\underset{O}{\|}}{C}\text{-}\underset{\underset{O}{\|}}{C}\text{-OR} + H_2O_2 + S \longrightarrow 2\,ROH + 2\,CO_2 + S \qquad (1)$$

where S = sensitizer.

A tentative mechanism, proposed by Rauhut [4], is

$$\text{RO}\underset{\underset{O}{\|}}{C}\text{-C}\underset{\underset{O}{\|}}{}\text{OR} + H_2O_2 \underset{\text{base}}{\rightleftarrows} \text{RO}\underset{\underset{O}{\|}}{C}\text{-}\underset{\underset{OOH}{\overset{OH}{|}}}{C}\text{—OR} \qquad (2)$$

I

$$ROC\text{-}C\underset{\overset{|}{OOH}}{\overset{\overset{O\;\;OH}{\|\;\;|}}{}}OR \longrightarrow$$

I

$$\overset{O\;\;O}{\overset{\|\;\;\|}{ROC\text{-}COOH}} + ROH \quad (3a)$$

II

$$\longrightarrow \text{decomposition} \quad (3b)$$

$$\overset{O\;\;O}{\overset{\|\;\;\|}{ROC\text{-}COOH}} \longrightarrow$$

II

$$\overset{O\;\;O}{\overset{\|\;\;\|}{\underset{\underset{O\text{-}O}{|\;\;|}}{C\text{-}C}}} + ROH \quad (4a)$$

III

$$\underset{H_2O_2}{\longrightarrow} \overset{O\;\;O}{\overset{\|\;\;\|}{HOOC\text{-}COOH}} + ROH \quad (4b)$$

IV

$$\overset{O\;\;O}{\overset{\|\;\;\|}{HOOC\text{-}COOH}} \rightleftharpoons \overset{O\;\;O}{\overset{\|\;\;\|}{\underset{\underset{O\text{-}O}{|\;\;|}}{C\text{-}C}}} + H_2O_2 \quad (5)$$

IV   **III**

$$\overset{O\;\;O}{\overset{\|\;\;\|}{\underset{\underset{O\text{-}O}{|\;\;|}}{C\text{-}C}}} \longrightarrow$$

III

$$\overset{S}{\longrightarrow} 2\,CO_2 + S^* \quad (6a)$$

$$\longrightarrow \text{decomposition} \quad (6b)$$

$$S^* \longrightarrow S + h\nu \quad (7)$$

where S* is an excited state of the sensitizer, S.

In benzene, the sensitizer 9,10-bis(phenylethynyl)anthracene has a fluorescent short-wavelength band at 486 nm. (See discussion of the sensitizer on page 151.)

Reaction 2, base-catalyzed nucleophilic addition of H_2O_2 to an ester carbonyl, is known to be rapid with electronegatively substituted aryl acetates [4]. Intermediate **I** is expected to have transient existence by analogy to ester hydrolysis mechanisms [4]. Intermediate **I** could decompose (reaction 3b) or undergo elimination of ROH, producing **II**, a peroxyacid (reaction 3a). The activity of the leaving group in reaction 3a helps determine the relative rates of reactions 3a and 3b [4]. As the acidity of ROH increases, reaction 3a becomes faster, and the quantum yield is increased since more of **I** is converted to **II**.

Intermediate **II** could undergo two possible reactions, both of which can lead to chemiluminescence [4]: reaction 4a, an intramolecular nucleophilic displacement producing 1,2-dioxetanedione, **III**; and reaction 4b, further reaction with H_2O_2, yielding diperoxyoxalic acid, **IV**. Infrared analysis shows that the stoichiometry of the overall reaction is 1:1 H_2O_2/oxalate [4]; and, since varying the H_2O_2/oxalate ratio does not affect the chemiluminescent quantum yield, the chemiluminescent reaction is also 1:1 H_2O_2/oxalate. Since **IV** requires 2:1 stoichiometry, it is not an essential intermediate. However, formation of **IV** is expected at high H_2O_2 ratios, and it could provide the key chemiluminescent intermediate, through reaction 5. Intermediate **III**, produced by reaction 4a, could lead directly to chemiluminescence. Evidence for the existence of **III** is mostly indirect [6], but a mass spectral peak for C_2O_4 has been observed in a similar system [7]. This peak decayed at the same rate as the chemiluminescence.

Since the chemiluminescent reaction rate is independent of the sensitizer concentration in the presence of base [4], reaction 6 is fast relative to prior reactions. Rauhut [4] proposes a dark reaction, 6b, which is competitive with an excitation reaction, 6a, because the quantum yield is increased by increasing the sensitizer concentration.

In an attempt to elucidate the chemielectronic step (in which chemical energy is converted into electronic energy) for the H_2O_2/oxalate ester reaction, Lechtken and Turro [8] concluded:

(a) the oxalate ester system is capable of generating excited states whose excitation energies approach 439 kJ/mole;

(b) in the presence of energy acceptors, the formation of "free" excited CO_2 singlet states is improbable;

(c) the energy transfer to the dye (sensitizer) probably occurs during the chemielectronic step; and

(d) the oxalate system does not involve triplet states.

Other experiments [4] indicate that a metastable intermediate is formed in the absence of a sensitizer and that the sensitizer catalyzes a reaction of the intermediate, leading to chemiluminescence. These experiments indicate that **II** could be the intermediate; but **II** does not appear to be responsible for sensitizer excitation, for the following reasons:

(a) the decomposition of **II** should produce equal amounts of CO and CO_2, but CO has been excluded as a product of the chemiluminescent reaction;

(b) high quantum yields require H_2O_2 (replacing H_2O_2 with t-butyl hydroperoxide or peroxybenzoic acid results in weak chemiluminescence);

(c) reaction of an active oxalate with H_2O_2 in an evacuated system with a pad moistened with sensitizer above the solution results in a nonluminescent solution but a brightly lit pad, indicating formation of a volatile chemiluminescent intermediate. Since **II** is not believed to be volatile, **III** is proposed as the intermediate causing sensitizer excitation.

Some experiments [4] require that the sensitizer, acting as an energy acceptor, function as a catalyst for the decomposition of the chemiluminescent intermediate. If the Woodward-Hoffman symmetry rules are obeyed, the concerted decomposition of **III** would produce an excited product. Concerted decomposition in conjunction with a sensitizer requiring 293 kJ/mole or less for excitation would be favored. Decomposition of a charge-transfer complex between **III** and a sensitizer could provide a short-lived mixed eximer of carbon dioxide and fluorescer, which is capable of dissociation to ground-state CO_2 and an excited fluorescer.

McCapra [9] hypothesizes that dioxetanedione, which is highly strained and can be extremely energetic, is preserved long enough by orbital-symmetry prohibitions that it reacts by a lower-energy pathway involving formation of the easily accessible first singlet excited state of the sensitizer. He mentions that an interesting possibility for the excitation step might be the electron transfer luminescence:

$$\underset{\substack{| \ | \\ O\text{-}O}}{\overset{\substack{O \ O \\ \| \ \| }}{C\text{-}C}} + DPA \longrightarrow \underset{\substack{| \ | \\ O\text{-}O}}{\overset{\substack{O \ O \\ \| \ \| }}{C\text{-}C}}{}^{\cdot -}DPA^{\cdot +} \longrightarrow DPA^{\cdot +}CO_2^{\cdot -} + CO_2$$

$$DPA^{\cdot +}CO_2^{\cdot -} \longrightarrow DPA^* + CO_2$$

where DPA = diphenylanthracene.

Rauhut [10] suggests that the excitation step involves initial formation of a charge-transfer complex between the intermediate and the sensitizer, with the sensitizer acting as electron donor. This proposal is supported experimentally since the rate of excitation increases as the ionization potential of the sensitizer decreases.

The sensitizer 9,10-bis(phenylethynyl)anthracene is bright yellowish green [11]. The maximum of its short-wavelength fluorescence band in benzene occurs at 486 nm, with an absolute fluorescence quantum yield, ϕ_F, of 0.96 einstein/mole. The maximum of the long-wavelength absorption band is at 455 nm, with log ϵ = 4.52.

The curves for absorption and fluorescence for 9,10-bis(phenylethynyl)anthracene are reproduced in the figure. The fluorescence spectrum represents the relative photon flux per unit wave number increment, and the absorption curve represents the molar extinction coefficient (l/mole·cm) versus wave number [12].

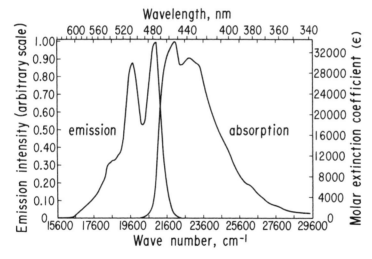

Emission and absorption spectra of 9,10-bis(phenylethynyl)anthracene.

REFERENCES

1. Shakhashiri, B. Z.; Williams, L. G.; Dirreen, G. E.; Francis, A. *J. Chem. Educ.* **1981**, *58*, 70.
2. CYALUME is a trademark of the American Cyanamid Company. Lightsticks can be obtained from the Ventron Corporation (Alfa Products), the Aldrich Chemical Company, or Markson Science, Inc. Lightsticks are also available in some hardware stores, sporting goods stores, and gasoline service stations. Since both regular and high intensity lightsticks are available, make sure all three are of the same intensity. Amusement parks and carnivals often have CYALUME in the shapes of bracelets and necklaces.
3. Mohan, A. G.; Turro, N. J. *J. Chem. Educ.* **1974**, *51*, 528.

4. Rauhut, M. M. *Accts. Chem. Res.* **1969**, *2*, 80 and references therein (footnotes 13 and 32).

5. Rauhut, M. M.; Roberts, G. B.; Maulding, D. R.; Bergmark, W.; Coleman, R. *J. Org. Chem.* **1975**, *40*, 330.

6. McCapra, F. *Pure Appl. Chem.* **1970**, *24*, 611.

7. Cordes, H. F.; Richter, H. P.; Heller, C. A. *J. Am. Chem. Soc.* **1969**, *91*, 7209.

8. Lechtken, P.; Turro, N. J. *Mol. Photochem.* **1974**, *6*, 95.

9. McCapra, F. *Prog. Org. Chem.* **1971**, *8*, 231.

10. Rauhut, M. M.; Bollyky, L. J.; Roberts, G. B.; Loy, M.; Whitman, R. H.; Iannotta, A. V.; Semsel, A. M.; Clarke, R. A. *J. Am. Chem. Soc.* **1967**, *89*, 6515.

11. Maulding, D. R.; Roberts, B. G. *J. Org. Chem.* **1969,** *34*, 1734.

12. Berlman, I. B. "Handbook of Fluorescence Spectra of Aromatic Molecules," 2nd ed.; Academic Press: New York, 1971; p 369.

2.3

Sensitized Oxalyl Chloride Chemiluminescence

In a darkened room, light is emitted when a clear liquid is added to flasks containing solutions of various colors. The intensity, color, and duration of emission vary with the sensitizer used [1, 2].

MATERIALS FOR SOLUTION 1 (for each sensitizer used)

25 ml dichloromethane, CH_2Cl_2

4 ml 3% hydrogen peroxide, H_2O_2

0.005 g violanthrone† (Other sensitizers [dyes] which can be used include isoviol-anthrone, rhodamine 6G, fluorescein disodium salt, rubrene, 9,10-diphenyl-anthracene, 13,13'-dibenzanthronyl.)

125-ml Erlenmeyer flask with cork stopper

10-ml graduated cylinder

MATERIALS FOR SOLUTION 2

2 ml oxalyl chloride, $Cl_2C_2O_2$ (**See Hazards section before handling oxalyl chloride.**)

50 ml dichloromethane, CH_2Cl_2

gloves, plastic or rubber

125-ml Erlenmeyer flask with cork stopper

10-ml graduated cylinder

PROCEDURE

To prepare solution 1, pour 25 ml of dichloromethane into a 125-ml Erlenmeyer flask. Add 4 ml of 3% hydrogen peroxide and 0.005 g of sensitizer. Do not add too much sensitizer or you will quench the excitation reaction. Swirl the flask to dissolve the sensitizer. Dichloromethane and the peroxide solution are immiscible.

Prepare solution 2 under a fume hood. Wear gloves. Add 2 ml of oxalyl chloride to 50 ml of dichloromethane in a clean, *dry* 125-ml Erlenmeyer flask. Swirl to mix. This solution is stable indefinitely if kept stoppered and in the dark [1].

†See pp 202–4 for chemical structures and sources.   **153**

To perform the demonstration, add 2 ml of solution 2 to solution 1. The color, intensity, and duration of the chemiluminescence vary according to the sensitizer used (see table). To maintain optimum brightness, constant swirling is required. Use of 4 ml of 30% hydrogen peroxide instead of 3% gives luminescence of greater intensity but shorter duration.

Summary of Emission Characteristics

Sensitizer	Color	Intensity	Approximate duration
violanthrone	red	moderate	30 seconds
isoviolanthrone	orange	moderate	30 seconds
rhodamine 6G	orange	very bright	30 seconds
fluorescein	yellow	dim	5 seconds
rubrene	yellow	very bright	2 minutes
9,10-diphenylanthracene	blue	very bright	3 minutes
tetracene	green	moderate	30 seconds
13,13'-dibenzanthronyl[a]	green to yellow to red	bright	40 seconds
mixture of rubrene and 9,10-diphenylanthracene[a]	yellow to blue	bright	3 minutes overall

[a]These two systems are the most "exocharmic."

HAZARDS

Oxalyl chloride is poisonous, toxic by inhalation, and severely irritating to skin, eyes, and respiratory tract. Wear plastic or rubber gloves and dispense only under a fume hood.

A 3% solution of hydrogen peroxide is a topical antiseptic and cleaning agent, but it decomposes rapidly with evolution of oxygen upon exposure to light, heat, or foreign materials.

Dichloromethane vapor is narcotic in high concentrations and is irritating to the eyes.

Rhodamine 6G has been shown to be carcinogenic in animal tests. Toxicity and carcinogenicity of the other dyes used in this demonstration are not known, although most may be assumed to be poisonous.

DISPOSAL

The solutions used in this demonstration should be discarded in a waste solvent container.

DISCUSSION

We have not found it necessary to use a slurry of alumina or sodium peroxycarbonate ($Na_2CO_3 \cdot H_2O_2$) in solution 1 [1]. The glow is bright enough without enhancement by a heterogeneous surface.

   This demonstration is another example of sensitized chemiluminescence in which molecular oxygen is not involved as the energy donor. Although the mechanistic details have not been fully elucidated, the behavior of the hydrogen peroxide–oxalyl chloride reaction system is similar to that of the hydrogen peroxide–phenyl oxalate ester system (see Demonstration 2.2). The reaction scheme is believed to involve the formation of a monoperoxyoxalic acid intermediate as the first step [3, 4]:

$$Cl\text{-}\overset{\overset{O}{\|}}{C}\text{-}\overset{\overset{O}{\|}}{C}\text{-}Cl + H_2O_2 + H_2O \longrightarrow HO\text{-}O\text{-}\overset{\overset{O}{\|}}{C}\text{-}\overset{\overset{O}{\|}}{C}\text{-}OH + 2\,HCl \qquad (1)$$

$$HO\text{-}O\text{-}\overset{\overset{O}{\|}}{C}\text{-}\overset{\overset{O}{\|}}{C}\text{-}OH + S \longrightarrow H_2O + 2\,CO_2 + S^* \qquad (2)$$

$$S^* \longrightarrow S + h\nu \qquad (3)$$

where S = sensitizer.

   The second step involves the simultaneous decomposition of the intermediate mono-peroxyoxalic acid and the intermolecular energy transfer to a sensitizer molecule which then releases energy in the visible region of the spectrum. The color of emission seems to vary only with the sensitizer (see listing in table). The close correspondence between the chemiluminescent and fluorescent spectra indicates that the emitting species is the singlet state of the sensitizer [5]. See Figure 8 of Demonstration 2.1 for the emission spectra of selected dyes. The quantum yield, depending on the sensitizer, can be as high as 0.05, making the hydrogen peroxide–oxalyl chloride reaction one of the most efficient chemiluminescent systems [3].

REFERENCES

1. Bramwell, F. B.; Goodman, S.; Chandross, E. A.; Kaplan, M. *J. Chem. Educ.* **1979**, *56*, 111.
2. Chandross, E. A. *Tetrahedron Letters* **1963**, *12*, 761.
3. McCapra, F. *Prog. Org. Chem.* **1971**, *8*, 231.
4. Rauhut, M. M.; Roberts, B. G.; Semsel, A. M. *J. Am. Chem. Soc.* **1966**, *88*, 3604.
5. Haas, J. W., Jr. *J. Chem. Educ.* **1967**, *44*, 396.

2.4

Oxidations of Luminol

In a darkened room, a colorless liquid and a blue liquid are poured simultaneously into a large glass funnel with a spiral delivery tube emptying into a large glass container. The mixed liquids emit a blue chemiluminescent glow as they flow through the tube and into the container. The glow persists for more than 2 minutes. When the room lights are turned on, the mixture in the receiving vessel is olive green. Other mixtures and procedures for observing luminol chemiluminescence are described.

MATERIALS FOR PROCEDURE A

4.0 g sodium carbonate (anhydrous), Na_2CO_3

3 liters distilled water

0.2 g luminol (3-aminophthalhydrazide), $C_8H_7O_2N_3$

24.0 g sodium bicarbonate, $NaHCO_3$

0.5 g ammonium carbonate monohydrate, $(NH_4)_2CO_3 \cdot H_2O$

0.4 g copper(II) sulfate pentahydrate, $CuSO_4 \cdot 5H_2O$, or 0.25 g copper(II) chloride dihydrate, $CuCl_2 \cdot 2H_2O$

50 ml 3% hydrogen peroxide, H_2O_2

cotton (optional)

2 1-liter Erlenmeyer flasks with rubber stoppers or 2 1-liter round-bottomed flasks with rubber stoppers and cork support rings

glass funnel (15–20 cm in diameter with long stem to fit tubing)

ring stand (1 m or taller) and ring with clamp holder to support funnel

2 m of transparent Tygon tubing (0.5 cm in diameter) or glass spiral assembly (See footnote in Procedure A.)

metal wire to fasten tubing to funnel

6–8 extension clamps or small extension clamps and clamp holders

3-liter Erlenmeyer flask or 3-liter round-bottomed flask with cork support ring

gloves, plastic or rubber (optional)

MATERIALS FOR PROCEDURE B

2 liters distilled water

4.0 g sodium carbonate (anhydrous), Na_2CO_3

24.0 g sodium bicarbonate, $NaHCO_3$

50 ml 3% hydrogen peroxide, H_2O_2

0.1 g luminol (3-aminophthalhydrazide), $C_8H_7O_2N_3$

0.1 g bovine hemoglobin

0.005 g fluorescein disodium salt, $Na_2C_{20}H_{10}O_5$ (optional)

2-liter beaker

spatula

magnetic stirrer

large stirring bar (2–3 cm)

MATERIALS FOR PROCEDURE C

0.02 g luminol (3-aminophthalhydrazide), $C_8H_7O_2N_3$

0.4 g bovine hemoglobin

0.4 g sodium perborate tetrahydrate, $NaBO_3 \cdot 4H_2O$

3.0 g sodium orthophosphate dodecahydrate, $Na_3PO_4 \cdot 12H_2O$

3.0 g sucrose, $C_{12}H_{22}O_{11}$ (Table sugar may be used.)

200 ml distilled water

mortar and pestle

250-ml Erlenmeyer flask with rubber stopper

Optional Materials for Procedure C

0.010 g fluorescein disodium salt, $Na_2C_{20}H_{10}O_5$

0.010 g rhodamine B, $C_{28}H_{31}O_3N_2$

0.8 g sodium carbonate (anhydrous), Na_2CO_3

4.8 g sodium bicarbonate, $NaHCO_3$

0.04 g luminol (3-aminophthalhydrazide), $C_8H_7O_2N_3$

0.08 g copper(II) sulfate pentahydrate, $CuSO_4 \cdot 5H_2O$, or 0.04 g copper(II) chloride dihydrate, $CuCl_2 \cdot 2H_2O$

0.10 g ammonium carbonate monohydrate, $(NH_4)_2CO_3 \cdot H_2O$

1.0 g sodium perborate tetrahydrate, $NaBO_3 \cdot 4H_2O$

500 ml distilled water

600-ml beaker

magnetic stirrer and stirring bar

scoopula

MATERIALS FOR PROCEDURE D

100 ml dimethylsulfoxide (DMSO), $(CH_3)_2SO$ (**See Hazards section before handling dimethylsulfoxide.**)

10 ml 3M potassium hydroxide, KOH

0.05 g luminol (3-aminophthalhydrazide), $C_8H_7O_2N_3$

0.01 g fluorescein disodium salt, $Na_2C_{20}H_{10}O_5$ (optional)

250-ml Erlenmeyer flask with rubber stopper

rubber gloves

PROCEDURE A

Solution A-1. In a 1-liter flask, dissolve 4.0 g of sodium carbonate in 500 ml of distilled water. Add 0.2 g of luminol and stir to dissolve. Add 24.0 g of sodium bicarbonate, 0.5 g of ammonium carbonate monohydrate, and 0.4 g of copper(II) sulfate pentahydrate and stir until all the solid dissolves. Dilute to a final volume of 1 liter with distilled water. The pH of the solution will be approximately 9.

Solution A-2. In the other 1-liter flask, dilute 50 ml of 3% hydrogen peroxide to 1 liter with distilled water.

Place the funnel in the ring and clamp it as high as possible on the ring stand. Slip the Tygon tubing over the funnel stem. If the fit is loose, tighten it by placing a metal wire around it. Place 6–8 clamps at about equal distance from each other along the ring stand. Form a coil around the ring stand with the Tygon tubing and hold the tubing in place with the clamps.† Place the lower end of the tube in the 3-liter receiving flask (see Figure 1).

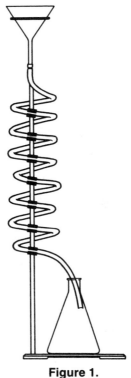

Figure 1.

†At the University of Wisconsin-Madison, we use a glass spiral assembly made of 18-mm glass tubing. The spiral has 4½ turns with a separation of about 7 cm between adjacent turns. The diameter of the coil is 14 cm and its height is 50 cm. One end of the coil is 6 cm long from the turn while the other is 9 cm long. The shorter end of the coil is connected to a 200-mm glass funnel (200-mm top diameter, 18-mm stem diameter). The longer end of the coil is placed directly into the receiving flask.

Tygon tubing can be used to construct a variety of shapes. Numerals and letters can be produced by bending the tubing and spraying appropriate parts with black paint. Clearly, the only requirement is to avoid restricting the liquid flow from the funnel to the receiving flask.

To perform the demonstration, dim the room lights but allow sufficient light for the demonstrator to see the flask and coil assembly. Slowly pour solutions A-1 and A-2 simultaneously into the funnel. A strong blue glow will be observed. Continue pouring until both flasks are empty. The glowing liquid runs through the spiral and collects in the receiving vessel. The solution will continue to glow for approximately 2 minutes after it first exits the spiral. Before disconnecting the coil assembly, flush it with distilled water.

In an alternative procedure, wear gloves and soak a piece of cotton in about 100 ml of solution A-1. In the dark, immerse the cotton in about 100 ml of solution A-2 and wring out the cotton [1]. The cotton will glow and "drip fire."

PROCEDURE B

Solution B. Pour 1.5 liters of distilled water into the 2-liter beaker. Add 4.0 g of sodium carbonate and 24.0 g of sodium bicarbonate. Stir to dissolve. Add 50 ml of 3% H_2O_2 and dilute to 1800 ml with distilled water.

Powder B-1. Mix 0.1 g of luminol and 0.1 g of bovine hemoglobin.

Powder B-2 (optional). Combine 0.005 g of fluorescein with 0.05 g of powder B-1.

To perform the demonstration, place the beaker containing solution B on the magnetic stirrer. Adjust the stirring speed to form a slight vortex (2–3 cm in depth). Dim the lights. Sprinkle about 0.04 g of powder B-1 in the vortex and 0.04 g of the same powder near the walls of the beaker. A blue glow will appear in the center of the beaker. The same amounts of powder B-2 will produce a yellow glow. The powder sprinkled near the walls of the beaker will gradually be drawn into the vortex.

To observe a glowing yellow "tornado" effect amid a glowing blue solution, first sprinkle 0.04 g of powder B-1 into the vortex and then sprinkle about the same amount of powder B-2 near the walls of the beaker.

PROCEDURE C

Powder C-1. Mix 0.02 g of luminol, 0.4 g of bovine hemoglobin, 0.4 g of sodium perborate tetrahydrate, 3.0 g of sodium orthophosphate dodecahydrate, and 3.0 g of sucrose in a mortar. Grind until the powder is uniformly mixed. Since the powder is hygroscopic, it should be stored in a tightly sealed container until used [1].

Powder C-2 (optional). Mix 1.0 g of powder C-1 with 0.005 g of fluorescein disodium salt.

Powder C-3 (optional). Mix 1.0 g of powder C-1 with 0.005 g of rhodamine B.

Powder C-4 (optional). Mix 0.8 g of sodium carbonate, 4.8 g of sodium bicarbonate, 0.04 g of luminol, 0.08 g of copper(II) sulfate pentahydrate, 0.10 g of ammonium carbonate monohydrate, and 1.0 g of sodium perborate tetrahydrate in a mortar. Grind until the powder is uniformly mixed.

Powder C-5 (optional). Combine 1.0 g of powder C-4 with 0.005 g of fluorescein disodium salt.

Powder C-6 (optional). Combine 1.0 g of powder C-4 with 0.005 g of rhodamine B.

To perform the demonstration, measure 200 ml of distilled water into a 250-ml Erlenmeyer flask. Dim the room lights. Add 1.0 g of powder C-1. Stopper the flask and swirl or shake to mix. The solution will glow blue for about 2 minutes. A solution of powder C-4 will also glow blue. Solutions of powders C-2 and C-5 will glow yellowish green, and solutions of powders C-3 and C-6 will glow purple.

Powder C-1, C-2, or C-3 can also produce a "tornado" effect. To observe this effect, pour 500 ml of distilled water into a 600-ml beaker and place the beaker on a magnetic stirrer. Place a stirring bar in the beaker and adjust the stirring rate to form a vortex of about 0.5 cm in depth. Dim the room lights. Add a scoopula of one of the powders to the vortex and sprinkle a second scoopula of the same powder near the walls of the beaker. The glow will be most apparent in the center of the beaker.

PROCEDURE D

See Hazards section before handling dimethylsulfoxide.

Wear rubber gloves. Cautiously pour 100 ml of DMSO into a 250-ml Erlenmeyer flask. Add 10 ml of 3M potassium hydroxide and 0.05 g of luminol. Stopper the flask and swirl until the luminol is dissolved. To initiate the reaction, shake the stoppered flask vigorously to aerate the mixture [2, 3, 4]. An intense blue glow will be observed.

Alternatively, the mixture can be aerated in a gas washing bottle using a pressurized air supply.

The addition of about 0.01 g of fluorescein will result in an intense yellow glow upon aeration.

HAZARDS

Copper compounds are harmful if taken internally, and dust from copper compounds can irritate mucous membranes.

A 3% solution of hydrogen peroxide is a topical antiseptic and cleaning agent. It decomposes rapidly with evolution of oxygen upon exposure to light, heat, or foreign materials.

Rhodamine B has been shown to be carcinogenic in animal tests. Toxic and carcinogenic properties of luminol and fluorescein are not known.

Potassium hydroxide can cause severe burns of the eyes and skin. Dust from solid potassium hydroxide is very caustic.

Dimethylsulfoxide (DMSO) is a colorless, odorless, very hygroscopic liquid. It is readily absorbed through the skin. Wear rubber gloves when working with DMSO. Since DMSO is a good solvent, it can transport dissolved materials through the skin. Contact with skin or eyes must be avoided.

DISPOSAL

All substances used in this demonstration are water soluble and should be flushed down the drain.

Before disconnecting the coil assembly used in Procedure A, flush it with distilled water.

DISCUSSION

One of the most well-known examples of chemiluminescence is that of luminol (3-aminophthalhydrazide or, in IUPAC nomenclature, 5-amino-2,3-dihydrophthala-zine-1,4-dione), compound **I**:

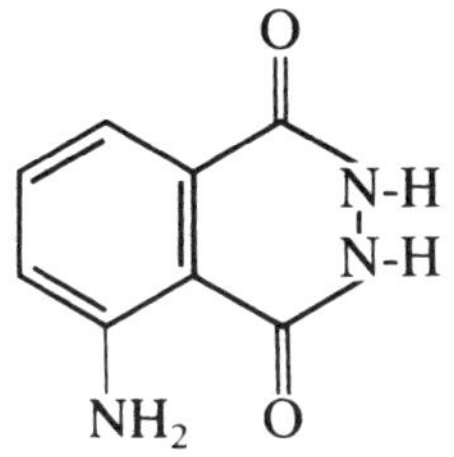

I

The bright blue emission produced by chemiluminescent reactions of luminol is readily visible in a darkened room. In Procedure A, the glowing reaction mixture is passed through tubing, thereby increasing the surface area, so that the emission is visible to all observers, even in very large rooms.

In Procedure B, a blue luminous "tornado" effect is produced. The addition of fluorescein to the reaction mixture produces yellow emission as well as blue. The effects in Procedure B are not as readily visible in a large room as those in Procedure A.

Procedure C describes the preparation of powders which produce a variety of emission colors when sprinkled into a beaker of distilled water. These powders can easily be stored for prolonged periods of time. A "tornado" effect can also be produced with these powders.

Procedure D differs from the previous procedures in that dimethylsulfoxide (DMSO) rather than water is used as solvent. In DMSO, chemiluminescence results from the reaction of luminol with molecular oxygen. Thus, emission can be produced simply by aerating the solution either by shaking it in a stoppered flask or by bubbling air through it in a gas washing bottle.

The first report of chemiluminescence from luminol was made by Albrecht in 1928 [5]. Soon thereafter, in 1934, it was adapted for use as a lecture demonstration [6]. Since then, this chemiluminescent reaction has been studied extensively.

The chemiluminescent reactions of luminol are all oxidations [7], as represented by equation 1:

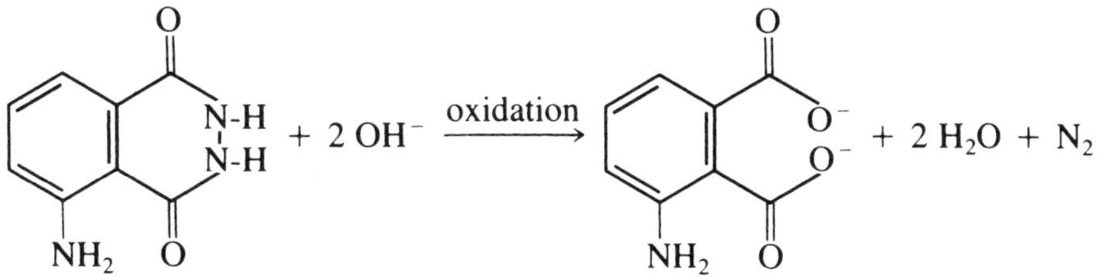

(1)

I   **II**

The reaction can be carried out in a variety of media including protic solvents such as water or an alcohol, and aprotic solvents such as dimethylsulfoxide (DMSO) or di-methylformamide (DMF). The reaction mechanism varies with the solvent. Different oxidants are required and slightly different chemiluminescent spectra are observed in

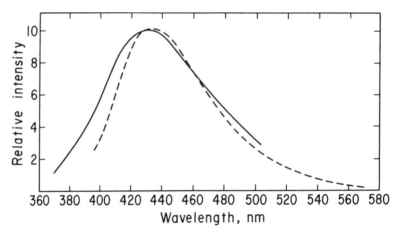

Figure 2. Chemiluminescence spectrum of luminol, *solid line,* and fluorescence spectrum of aminophthalate ion, *dashed line.*

different media. In aprotic media, only molecular oxygen and a strong base are needed to produce chemiluminescence from luminol, and the wavelength of maximum emission occurs at 485 nm [8]. In protic solvents, the reaction requires a strong base, either molecular oxygen or a peroxide, and an auxiliary oxidant such as hypochlorite, ferricyanide, or persulfate ions. The wavelength of maximum emission in protic solvents occurs at 425 nm [7, 9].

In both protic and aprotic solvents, the emitting species has been identified as a form of the product aminophthalate ion, **II**. This identification is made on the basis of the correspondence between the chemiluminescent spectrum of luminol and the fluorescent spectrum of the aminophthalate ion. As shown in Figure 2, the fluorescent spectrum of the aminophthalate ion in water and the spectrum of the chemiluminescence from the aqueous oxidation of luminol are superimposable [8]. However, in aprotic media, such as DMSO, the emitting species is the quinoidal form of the anion, **III** [7]:

III

Thus, the chemiluminescent reaction can be represented in general terms by equation 2, which indicates that the light emission comes from the aminophthalate ion:

$$ \text{oxidation} \atop \text{base} \qquad (2) $$

Considerable data have been accumulated on the mechanism of the oxidation of luminol in both protic and aprotic solvents. The protic and aprotic mechanisms probably differ only in their early steps where the protic system requires a free radical species and the aprotic system utilizes a luminol dianion [10]. Since the conditions required for the reaction are milder in the aprotic solvents, the mechanism in such solvents is better understood. No definitive luminol mechanism is known, but much information about the intermediate species in the oxidation has been gathered [7].

An important intermediate in the reaction is the dianion of luminol, **IV**, which is formed in a stepwise reaction with base, as shown in equation 3 [7]:

$$ \text{(3)} $$

Most recently proposed mechanisms are constructed around this dianion, which may undergo several different reactions [7]. One of these is the formation of an azaquinone, **V**, as represented in equation 4 [11]:

$$ \text{(4)} $$

Azaquinone is believed to be an intermediate in the light-producing pathway in the oxidation of luminol [12, 13]. One such pathway, proposed by White [7], is outlined below:

Another mechanism, suggested by McCapra [12], is based on electron transfer and involves the loss of nitrogen gas much earlier in the reaction than does White's mechanism:

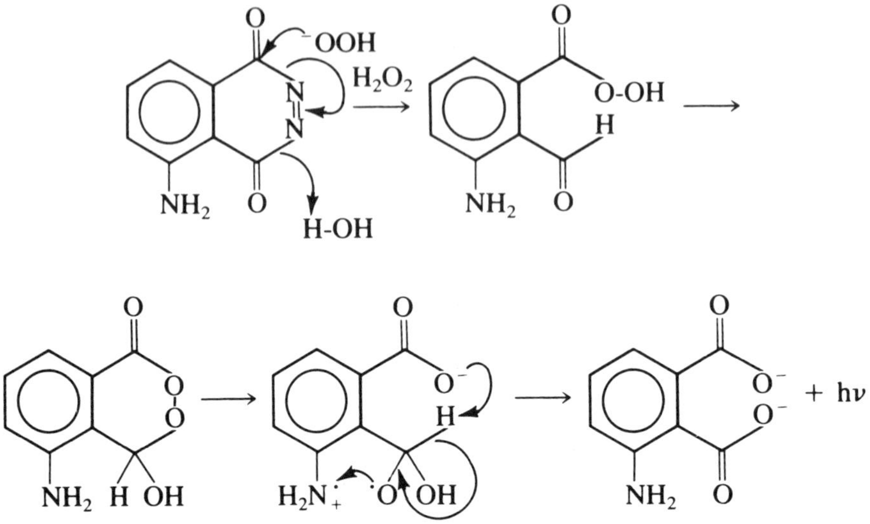

Both of these mechanisms apply to aprotic media. The reaction in aprotic media is less complicated than the reaction in protic solvents since no auxiliary oxidant, or activator, is involved. In aqueous systems an activator is always required [14]. These activators, which are frequently transition-metal ion complexes, may act either as catalysts or as co-oxidants. The transition-metal ions require complexing agents to prevent the precipitation of the metal hydroxide in the strongly alkaline environment required for the chemiluminescent oxidation of luminol [14].

The species required to initiate the oxidation of luminol in aqueous media is the superoxide radical anion, $O_2^{\cdot-}$ [13]. This species is formed in aqueous media by the decomposition of hydrogen peroxide. In highly alkaline aqueous systems, this decomposition is catalyzed by the oxidation activators, such as the hexacyanoferrate(III) ion [14]. The transition-metal complex ions may also be involved in an electron-transfer step in the oxidation of luminol, as illustrated in the first step of the mechanism below for the aqueous oxidation of luminol [15]:

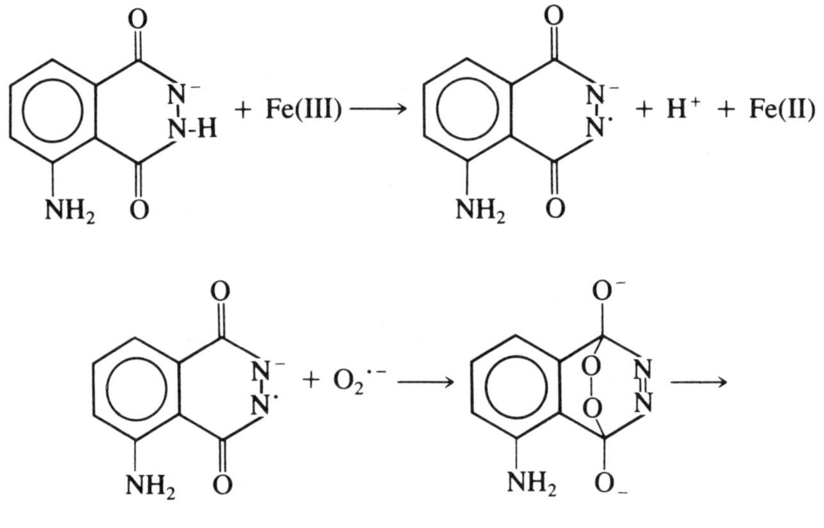

$$N_2 + \left[\begin{array}{c}\text{(isophthalate dianion excited state)}\end{array}\right]^* \longrightarrow \begin{array}{c}\text{(isophthalate dianion)}\end{array} + h\nu$$

Each system that is initiated by transition-metal ions appears to have its own reaction sequence [16]. Unfortunately, little work has been done on these varying reaction mechanisms; instead, interest has focused on analytical applications of luminol chemiluminescence. Presumably, the tetramminecopper(II) complex used in this demonstration (Procedure A) serves to catalyze the formation of the superoxide radical ion,

Table 1.   Summary of Powder Composition and Emission Colors

Powder	Composition	Oxidant	Activator complex	Sensitizer	Emission color
B-1	luminol, hemoglobin, carbonate-bicarbonate	3% H_2O_2	Fe(II)	none	blue
B-2	luminol, hemoglobin, carbonate-bicarbonate	3% H_2O_2	Fe(II)	fluorescein	yellowish green
C-1	luminol, hemoglobin, phosphate	$NaBO_3 \cdot 4H_2O$	Fe(II)	none	blue
C-2	luminol, hemoglobin, phosphate	$NaBO_3 \cdot 4H_2O$	Fe(II)	fluorescein	yellowish green
C-3	luminol, hemoglobin, phosphate	$NaBO_3 \cdot 4H_2O$	Fe(II)	rhodamine B	purple
C-4	luminol, Cu(II), NH_4^+, carbonate-bicarbonate	$NaBO_3 \cdot 4H_2O$	Cu(II)	none	blue
C-5	luminol, Cu(II), NH_4^+, carbonate-bicarbonate	$NaBO_3 \cdot 4H_2O$	Cu(II)	fluorescein	yellowish green
C-6	luminol, Cu(II), NH_4^+, carbonate-bicarbonate	$NaBO_3 \cdot 4H_2O$	Cu(II)	rhodamine B	purple

and perhaps is involved in an electron-transfer step similar to that above. Longer lasting emission results when $Cu(NH_3)_4^{2+}$ is used as the activator instead of $Fe(CN)_6^{3-}$.

In Procedures B and C, which use powders, the activator is either an iron(II) complex found in bovine hemoglobin or a $Cu(II)/NH_3$ complex. Powders C-1, C-2, and C-3 in Procedure C also contain sodium perborate, sodium orthophosphate, and sucrose. The sucrose is presumably a carrier for the mixture, the sodium perborate serves as the oxidant, and the sodium orthophosphate produces the alkaline conditions needed for the reaction. Hemoglobin from other sources including human whole blood can also be the activator. Powders C-4, C-5, and C-6 contain Cu(II) species as the activator, components of a carbonate buffer, and an ammonia source.

In Procedures B and C the reactants are used in powder form to facilitate their storage and to produce colorful "tornado" effects. The powder mixtures in Procedure C contain a solid peroxide source ($NaBO_3 \cdot 4H_2O$). Table 1 summarizes the composition of each mixture and the observed colors.

The chemiluminescent quantum yield of luminol oxidation reactions depends on several factors, including the solvent and the pH of the mixture [17, 18]. Table 2 illustrates the effects of these factors. The reaction is most efficient in producing light in aqueous media with a pH of 11 [17]; at that pH level, the efficiency reaches 2%. Although the reaction is most efficient at a pH of 11, its apparent brightness reaches a maximum at a pH of about 9 using Cu(II) as the activator. In the aqueous oxidation of luminol, hydroxide ions are consumed, as shown in equation 1. To yield maximum brightness, Procedure A employs a carbonate buffer ($0.04M$ CO_3^{2-}, $0.02M$ HCO_3^- in solution A-1) which maintains a pH of about 9.

Table 2. Quantum Yields of Luminol Oxidations in DMSO and Water [17, 18]

Solvent	ϕ_c
DMSO	0.01
Water (pH)	
9.6	0.01
11	0.02
12	0.01
13	0.006
14	0.002
15	0.0004

See references 19–22 for additional information about luminol oxidations by chlorine species. The procedure of Fuchsman and Young [20] is not as effective as the procedures described in this demonstration.

REFERENCES

1. Alyea, H. N.; Dutton, F. B., Eds. "Tested Demonstrations in Chemistry," 6th ed.; Journal of Chemical Education: Easton, Pennsylvania, 1965; pp 81, 186.
2. White, E. H. *J. Chem. Educ.* **1957,** *34,* 275.
3. Wilson, M.; Wood, T. *School Sci. Review* **1972,** *54,* 524.
4. Alyea, H. N.; Dutton, F. B., Eds. "Tested Demonstrations in Chemistry," 6th ed.; Journal of Chemical Education: Easton, Pennsylvania, 1965; p 130.

5. Albrecht, H. O. *Z. Physik. Chem. Unterricht* **1928**, *136*, 321.

6. Huntress, E. H.; Stanley, L. N.; Parker, A. S. *J. Chem. Educ.* **1934**, *11*, 142.

7. Rosewell, D. F.; White, E. H. *Meth. Enzym.* **1978**, *57*, 409.

8. White, E. H.; Bursey, M. M. *J. Am. Chem. Soc.* **1964**, *86*, 941.

9. Lee, J.; Seliger, H. H. *Photochem. Photobiol.* **1970**, *11*, 247.

10. McCapra, F. *Prog. Org. Chem.* **1971**, *8*, 231.

11. Gunderman, K. D. *Angew. Chem. Intl. Ed. Engl.* **1965**, *4*, 566.

12. McCapra, F.; Leeson, P. D. *Chem. Commun.* **1979**, 114.

13. Merenyi, G.; Lind, J. S. *J. Am. Chem. Soc.* **1980**, *102*, 5830.

14. White, E. H. in "Light and Life," McElroy, W. D.; Glass, B., Eds.; Johns Hopkins Press: Baltimore, Maryland, 1961; p 183.

15. Hodgson, E. K.; Fridovich, I. *Photochem. Photobiol.* **1973**, *18*, 451.

16. Burdo, T.; Seitz, W. R. *Anal. Chem.* **1975**, *47*, 1639.

17. Seliger, H. H. in "Light and Life," McElroy, W. D.; Glass, B., Eds.; Johns Hopkins Press: Baltimore, Maryland, 1961; p 200.

18. Lee, J.; Seliger, H. H. *Photochem. Photobiol.* **1972**, *15*, 227.

19. Schneider, H. W. *J. Chem. Educ.* **1970**, *47*, 519.

20. Fuchsman, W. H.; Young, W. G. *J. Chem. Educ.* **1976**, *53*, 548.

21. Marino, D. F.; Ingle, J. D., Jr. *Anal. Chem.* **1981**, *53*, 455.

22. Eriksen, T. E.; Lind, J.; Mereny, G. *Trans. Faraday Soc. 1* **1981**, *77*, 2125.

2.5

Luminol Chemiluminescent Clock Reactions

In a darkened room two colorless solutions are mixed together. After 10–60 seconds a bright blue glow suddenly appears, and considerable gas is evolved [1]. The glow lasts approximately 2 seconds. When the room lights are turned on, the solution is a deep blue. Other mixtures and procedures are described [2, 3] in which the induction period and the glow duration vary.

MATERIALS FOR PROCEDURE A

100 ml 15M aqueous ammonia, NH_4OH

0.20 g luminol (3-aminophthalhydrazide), $C_8H_7O_2N_3$

2 liters of distilled water

1.105 g potassium tricyanocuprate(I), $K_2Cu(CN)_3$ **(See Hazards section before handling cyanide salts.)**

120 ml 3% hydrogen peroxide, H_2O_2

250-ml beaker

2 stirring rods

1-liter volumetric flask with stopper

50-ml volumetric flask with stopper

4 400-ml beakers

5-ml pipette

pipette bulb

4 50-ml beakers

4 magnetic stirrers and stirring bars, or 4 stirring rods

2 100-ml graduated cylinders

MATERIALS FOR PROCEDURE B

3 liters distilled water

4.0 g sodium carbonate (anhydrous), Na_2CO_3

0.2 g luminol (3-aminophthalhydrazide), $C_8H_7O_2N_3$

24.0 g sodium bicarbonate, $NaHCO_3$

0.4 g copper(II) sulfate pentahydrate, $CuSO_4 \cdot 5H_2O$, or 0.25 g copper(II) chloride dihydrate, $CuCl_2 \cdot 2H_2O$

2.5 g ammonium carbonate, $(NH_4)_2CO_3$

0.6 g potassium cyanide, KCN (**See Hazards section before handling cyanide salts.**)

15 ml *fresh* 3% hydrogen peroxide, H_2O_2

copper wire, 20 gauge, 25 cm coiled (optional)

3 1-liter beakers

2 1-liter volumetric flasks

10-ml graduated cylinder

4 250-ml Erlenmeyer flasks

4 100-ml beakers

2 50-ml graduated cylinders

250-ml beaker (optional)

MATERIALS FOR PROCEDURE C

0.08 g DL-cysteine hydrochloride hydrate, $C_3H_7O_2NS \cdot HCl \cdot H_2O$

1500 ml distilled water

5.0 g potassium hydroxide, KOH

0.1 g luminol (3-aminophthalhydrazide), $C_8H_7O_2N_3$

0.25 g copper(II) sulfate pentahydrate, $CuSO_4 \cdot 5H_2O$

10 ml *fresh* 3% hydrogen peroxide, H_2O_2

2 125-ml Erlenmeyer flasks

2 250-ml Erlenmeyer flasks

100-ml volumetric flask

25-ml pipette

1-liter volumetric flask

6 100-ml beakers

5-ml pipette

10-ml Mohr pipette

1-ml Mohr pipette

pipette bulb

3 magnetic stirrers and stirring bars

50-ml graduated cylinder

PROCEDURE A

Solution A-1. Pour 100 ml of 15M aqueous ammonia into a 250-ml beaker. Add 0.20 g of luminol and stir to dissolve. Rinse the solution into a 1-liter volumetric flask and dilute to the mark with distilled water. Stopper the flask and invert several times to mix.

Solution A-2. Place 1.105 g of $K_2Cu(CN)_3$ in a 50-ml volumetric flask. Add 25 ml of distilled water and swirl to dissolve. Dilute to the mark with distilled water. Stopper the flask and invert several times to mix.

Measure 100 ml of solution A-1 into each of four 400-ml beakers. Using the pipette, add 5 ml of solution A-2 and stir. Measure 30 ml of 3% hydrogen peroxide into each of four 50-ml beakers.

These directions are for a 10-second induction time. For other induction times, vary the volume of 3% H_2O_2 solution as indicated:

Approximate induction time (seconds)	Volume of 3% H_2O_2 (ml)
10	30
13	20
25	10
57	5

To perform the demonstration, pour 30 ml of 3% hydrogen peroxide into the first beaker and stir. After 5 seconds pour 30 ml of peroxide into the second beaker and stir. After another 5 seconds add 30 ml of peroxide to the third beaker and stir. After another 5 seconds add 30 ml of peroxide to the fourth beaker and stir. Observe the sequential appearance of the glow in the beakers.

PROCEDURE B

Solution B-1. Pour 500 ml of distilled water into a 1-liter beaker. Add 4.0 g of sodium carbonate and stir to dissolve. Add 0.2 g of luminol and again stir to dissolve. Add 24.0 g of sodium bicarbonate and stir until the solution becomes clear. Add 0.4 g of copper(II) sulfate pentahydrate or 0.25 g of copper(II) chloride dihydrate. Stir to dissolve. In a volumetric flask, dilute to 1 liter with distilled water.

Solution B-2. Pour 500 ml of distilled water into a 1-liter beaker. Add 2.5 g of ammonium carbonate and 0.6 g of potassium cyanide and stir to dissolve. Dilute to 1 liter and mix well. **Caution: This solution contains CN^- ions. Do not acidify this solution (see Hazards section).**

Solution B-3. Measure 885 ml of distilled water into a 1-liter beaker. Add 15 ml of *fresh* 3% hydrogen peroxide and mix well.

To prepare for the demonstration, pour 50 ml of solution B-1 and 10 ml of solution B-2 into each of the four 250-ml Erlenmeyer flasks and stir well. Let the solutions stand without stirring for at least 30 minutes before use. This delay produces a more uniform induction time than is obtained from freshly mixed samples. The freshness and purity of the reagents used will affect the actual induction time. While standing, the color of the solution will change from bluish green to olive green.

To perform the demonstration, measure 50 ml of solution B-3 into each of the four 100-ml beakers. Dim the lights. Add the aliquots of solution B-3 to the solutions in the Erlenmeyer flasks at 5-second intervals, swirling the flask after each addition to insure mixing. An initial dim glow will appear in the mixture. The induction period is about 10 seconds for the amounts specified. For other induction periods, consult the list below.

Approximate induction time (seconds)	Volume of solution B-2 (ml)
5	5
10	10
20	15
55	20

As an optional procedure, prepare solutions B-1, B-2, and B-3 as described, but omit adding Cu(II) to solution B-1. In a 250-ml beaker, mix 50 ml of solution B-1, 10 ml of solution B-2, and 50 ml of solution B-3. Immerse the copper wire in this mixture and dim the lights. Within a minute or less, a blue glow will appear near the wire. As more copper dissolves, the entire solution will glow brightly.

PROCEDURE C

Solution C-1. Place 0.08 g of DL-cysteine hydrochloride in a 125-ml Erlenmeyer flask. Add 50 ml of distilled water and swirl to dissolve. This solution has a limited shelf life (about 1 day) and should be prepared as needed.

Solution C-2. Place 5.0 g of potassium hydroxide pellets in a 250-ml Erlenmeyer flask. Add 100 ml of distilled water and swirl to dissolve. Add 0.1 g of luminol and swirl to dissolve.

Solution C-3. In a 125-ml Erlenmeyer flask, dissolve 0.25 g of copper(II) sulfate pentahydrate in about 50 ml of distilled water. Transfer the solution to the 100-ml volumetric flask and dilute to the mark with distilled water. With a pipette, transfer 25 ml of this solution to the 1-liter volumetric flask and dilute to the mark. The solution in the 1-liter volumetric flask is solution C-3.

Into each of three 100-ml beakers, place a magnetic stirring bar, a 5.0-ml aliquot of solution C-1, and a 5.5-ml aliquot of solution C-2.

Immediately before the demonstration, measure 40 ml of solution C-3 into each of the remaining 100-ml beakers and add to each a 0.5-ml aliquot of *fresh* 3% hydrogen peroxide.

To perform the demonstration, place the three beakers containing solutions C-1 and C-2 on the magnetic stirrers and set all three stirrers for approximately the same stirring rate. Dim the room lights. Add the contents of one of the beakers containing the hydrogen peroxide and solution C-3 mixture to the first beaker on the magnetic stirrer. After 10 seconds, add the contents of the second beaker containing solution C-3 to the second beaker on the magnetic stirrer. After another 10 seconds, add the contents of the third beaker containing solution C-3 to the third beaker on the magnetic stirrer. Observe the sequential emission of light from the contents of the three beakers.

The induction time can be varied by changing the amounts of cysteine hydrochloride, copper(II) sulfate, or hydrogen peroxide. The induction time is directly proportional to the amount of cysteine hydrochloride and inversely proportional to the amount of copper(II) sulfate or hydrogen peroxide. Approximate induction times for various amounts of copper(II) sulfate and hydrogen peroxide follow on the next page.

Approximate induction time (seconds)	Solution C-3 (ml)	H_2O_2 (ml)
100	40	0.1 of 3%
70	40	0.2 of 3%
45	40	0.5 of 3%
35	50	0.5 of 3%
30	60	0.5 of 3%
55	30	0.5 of 3%
70	20	0.5 of 3%
30	40	0.1 of 30%

HAZARDS

Cyanide salts, their solutions, and hydrogen cyanide gas produced by the reaction of cyanides with acids are all extremely poisonous. Hydrogen cyanide is among the most toxic and rapidly acting of all poisons. The solutions and the gas can be absorbed through the skin. Solutions are irritating to the skin, nose, and eyes. Cyanide compounds and acids must not be stored or transported together. An open bottle of potassium cyanide can generate HCN in moist air.

Early symptoms of cyanide poisoning are weakness, difficult breathing, headache, dizziness, nausea, vomiting; these may be followed by unconsciousness, cessation of breathing, and death.

Anyone exposed to hydrogen cyanide should be removed from the contaminated atmosphere immediately. Amyl nitrite should be held under the person's nose for not more than 15 seconds per minute, and oxygen should be administered in the intervals. If the person is not breathing, artificial resuscitation by the Silvester method (not mouth to mouth) should be attempted immediately.

The toxicity or carcinogenicity of luminol and cysteine are not known.

Concentrated aqueous ammonia solutions can cause burns and are irritating to the skin, eyes, and respiratory system.

Potassium hydroxide can cause severe burns of the eyes and skin. Dust from solid potassium hydroxide is very caustic.

Copper compounds are harmful if taken internally. Dust from copper compounds can irritate mucous membranes.

A 3% solution of hydrogen peroxide is a topical antiseptic and cleaning agent, but it decomposes rapidly with evolution of oxygen upon exposure to light, heat, or foreign materials.

DISPOSAL

Solutions containing cyanide should be mixed with an excess of sodium hypochlorite solution (household bleach) and allowed to stand for a few hours. The resulting solution should be flushed down the drain with water.

All the other solutions used in this experiment are water soluble and should be flushed down the drain with water.

DISCUSSION

This demonstration is based on the aqueous chemiluminescent system containing copper species, hydrogen peroxide, luminol, and either cyanide ion or cysteine. Copper complexes of cyanide ion, ammonia, or cysteine play a critical role in affecting the "clock" nature of the reaction. After all the cyanide ions or cysteine are oxidized, the Cu(II) species can function as an activator in the oxidation of luminol by hydrogen peroxide, resulting in chemiluminescence (see Demonstration 2.4).

Procedure A employs the copper-cyanide species, $Cu(CN)_3^{2-}$, hydrogen peroxide, ammonia, and luminol. In fact, copper can be added as copper wire [1]. The mechanism of the reaction is not clearly established, but White proposed the following scheme [1] which accounts for all observations:

$$Cu(CN)_3^{2-} + H_2O_2 + NH_3 \longrightarrow Cu(NH_3)_4^{2+} + CO_3^{2-} + CNO^- + OH^- \text{ (slow)(1)}$$

$$Cu(NH_3)_4^{2+} + CN^- \text{ or } Cu(CN)_3^{2-} \longrightarrow Cu(I) \text{ complex} + (CN)_2 \qquad \text{(fast) (2)}$$

$$Cu(I) \text{ complex} + H_2O_2 + NH_3 \longrightarrow Cu(NH_3)_4^{2+} + OH^- \qquad \text{(fast) (3)}$$

$$Cu(NH_3)_4^{2+} + H_2O_2 + luminol \longrightarrow products + h\nu \qquad (4)$$

$$H_2O_2 + Cu(NH_3)_4^{2+} \longrightarrow O_2 + H_2O \qquad (5)$$

The equations are not necessarily elementary steps, nor are they stoichiometric; they merely represent a scheme consistent with the observations. Equations 1, 2, and 3 account for the "clock" nature of the system, while equation 4 accounts for the chemiluminescence and equation 5 is a side reaction. The identity of the Cu(I) complex is not known.

Procedure B employs $Cu(NH_3)_4^{2+}$, CN^-, H_2O_2, luminol, and a CO_3^{2-}/HCO_3^- buffer mixture. The carbonate buffer mixture controls the pH of the solution to maximize the chemiluminescent intensity (see Discussion section of Demonstration 2.4). The reaction scheme is presumed to be similar to what is outlined above [2], with equations 2 and 3 representing the changes taking place until all the free CN^- is consumed. As long as CN^- is present, no significant amount of Cu(II) can accumulate to activate the hydrogen peroxide–luminol oxidation represented by equation 4.

If copper wire is lowered into a beaker containing hydrogen peroxide, ammonia, and luminol, only the surface of the wire glows as the copper begins to dissolve. When enough copper has dissolved, the glow is produced throughout the solution as in Procedure A of Demonstration 2.4. Cyanide in the form of either CN^- or $Cu(CN)_3^{2-}$ stops both the emission of light and the evolution of O_2 gas from decomposition of hydrogen peroxide. The relative concentrations of cyanide species and hydrogen peroxide determine the induction period prior to light emission and evolution of O_2 gas. The induction period can be lengthened by *decreasing* the hydrogen peroxide concentration or by *increasing* the $Cu(CN)_3^{2-}$ concentration. Mixtures with large amounts of $Cu(CN)_3^{2-}$ have a long induction period prior to a flash of light and subsequent evolution of O_2 gas. Mixtures with small amounts of $Cu(CN)_3^{2-}$ have a short induction period, but the duration of light emission is longer. Mixtures with large amounts of hydrogen peroxide emit light for a longer period of time. When 30% H_2O_2 is used, large volumes of gaseous O_2 are released resulting in a chemiluminescent "fountain effect" [1].

Procedure C employs an alkaline solution of cysteine, $H_2N\text{-}\overset{\displaystyle H}{\underset{\displaystyle CH_2SH}{C}}\text{-COOH}$, Cu^{2+}(aq),

and luminol. The mechanism of the clock reaction involves the oxidation of the thiol cysteine in an alkaline medium in the presence of copper ion [3]. Hydrogen peroxide is produced as a by-product of this oxidation. Copper forms a complex with cysteine, and, as long as cysteine is present, no significant amount of Cu(II) can accumulate to activate the hydrogen peroxide–luminol oxidation. When all the cysteine is consumed, Cu(II) acts as a co-oxidant in the chemiluminescent reaction. The induction period can be shortened by increasing the concentration of copper(II) or decreasing the concentration of cysteine. However, decreasing the concentration of copper(II) or increasing the concentration of cysteine will lengthen the duration of light emission. Rather than depend on the production of hydrogen peroxide from the cysteine oxidation [3], we add a small amount of H_2O_2 to the reaction mixture to shorten the induction period and provide better control of its reproducibility.

We have initiated a study of several factors and conditions that influence the behavior of the chemiluminescent clock reaction systems of Procedure A and Procedure B.† The time elapsed between mixing all reagents and the onset of chemiluminescence is the induction period. The induction period is inversely proportional to the first power of the hydrogen peroxide concentration, directly proportional to approximately the sixth power of CN^- concentration, and inversely proportional to the second power of the Cu(I) concentration. The dependence of the induction period on $Cu(CN)_3^{2-}$ concentration is complicated: low $Cu(CN)_3^{2-}$ concentrations increase the induction period, but high $Cu(CN)_3^{2-}$ concentrations decrease the induction period.

In the absence of ammonia and in alkaline solution, chemiluminescence does not occur. This suggests that the $Cu(NH_3)_4^{2+}$, rather than Cu^{2+}(aq), acts as the activator. Interestingly, the visual appearance of the blue color of $Cu(NH_3)_4^{2+}$ occurs after chemiluminescence is observed.

When $Ag(CN)_2^-$ is used in place of $Cu(CN)_3^{2-}$ or Cu^{2+}(aq)/CN^-, chemiluminescence does not occur. Nor is chemiluminescence observed when a silver wire (or an iron wire) is used in place of a copper wire. Addition of diphenylpicrylhydrazyl (DPPH), which is a radical scavenger, does not alter the induction period, nor does it quench the chemiluminescent reaction. However, addition of ethylenediaminetetraacetic acid (EDTA) diminishes the intensity of chemiluminescence and increases the induction period. Very large concentrations of EDTA quench the blue emission. These observations support the scheme proposed by White [1].

REFERENCES

1. White, E. H. *J. Chem. Educ.* **1957**, *34*, 275.
2. Young, K. E. *J. Chem. Educ.* **1974**, *51*, A124.
3. Young, K. E. *J. Chem. Educ.* **1974**, *51*, A122.

†We thank Mark Fink, who conducted this study.

2.6

Two-Color Chemiluminescent Clock Reaction

In a darkened room, a colorless liquid is poured into a solution in a large beaker. The solution emits a red glow for a few seconds, foams, and then emits a blue glow for a few more seconds [1].

MATERIALS

40 ml distilled water

0.8 g sodium hydroxide pellets, NaOH

0.005 g luminol (3-aminophthalhydrazide), $C_8H_7N_3O_2$

25.0 g potassium carbonate, K_2CO_3

1.0 g pyrogallol (pyrogallic acid) (1,2,3-trihydroxybenzene), $C_6H_6O_3$
   (See Hazards section before handling pyrogallol.)

10 ml 40% formaldehyde, CH_2O

30 ml 30% hydrogen peroxide, H_2O_2

gloves, plastic or rubber

2 100-ml graduated cylinders

250-ml beaker

magnetic stirrer and stirring bar, or stirring rod

1-liter beaker

shallow pan or 2-liter beaker to catch overflow

PROCEDURE

Wearing gloves, measure 40 ml of distilled water into a 250-ml beaker. Add 0.8 g of sodium hydroxide and stir to dissolve. Add 0.005 g of luminol, 25.0 g of potassium carbonate, and 1.0 g of pyrogallol. Stir to dissolve. Add 10 ml of 40% formaldehyde. Pour the solution into a 1-liter beaker, and place the beaker in a 2-liter beaker or shallow pan.

To perform the demonstration, dim the room lights and add 30 ml of 30% hydrogen peroxide to the 1-liter beaker. Stirring is not necessary. The reaction mixture glows dull red for several seconds and then bright blue for a few more seconds. The solution foams extensively and becomes hot.

175

HAZARDS

Plastic or rubber gloves should be worn, because pyrogallol (1,2,3-trihydroxyben-zene or pyrogallic acid) is poisonous and can be absorbed through the skin.

Because 30% hydrogen peroxide is a strong oxidizing agent, contact with skin and eyes must be avoided. In case of contact, immediately flush with water for at least 15 minutes; get immediate medical attention if the eyes are affected.

Avoid contact between 30% hydrogen peroxide and combustible materials. Avoid contamination from any source, because any contaminant, including dust, will cause rapid decomposition and the generation of large quantities of oxygen gas. Store 30% hydrogen peroxide in its original closed container, making sure that the container vent works properly.

Sodium hydroxide can cause severe burns of the eyes and skin. Dust from solid sodium hydroxide is very caustic.

Formaldehyde vapors are extremely irritating to mucous membranes. Skin contact can cause dermatitis. Extended exposure to high concentrations of vapor can have chronic effects, such as laryngitis, bronchitis, conjunctivitis, and skin problems. Preliminary data "have indicated the development of nasal cancers in rats exposed to 15 ppm formaldehyde for 18 months" [2].

The toxicity and carcinogenicity of luminol are not known.

DISPOSAL

All solutions used in this demonstration are water soluble and should be flushed down the drain with water.

DISCUSSION†

This two-color chemiluminescent clock reaction involves two oxidation reactions occurring sequentially. The red glow is attributed to singlet molecular oxygen formed during the oxidation of pyrogallol and formaldehyde by alkaline hydrogen peroxide [3]. The blue glow is caused by the subsequent oxidation of luminol. The sequence of the red glow, followed by vigorous frothing and heat evolution, and finally the blue glow of the luminol oxidation, is noteworthy. The triggering of the luminol reaction by the residue of the red chemiluminescent reaction suggests strongly that free radicals are involved in this reaction sequence.

The chemiluminescent reaction of pyrogallol, formaldehyde, and alkaline hydrogen peroxide was reported by Trautz and Schorigin in 1905 [3]. Much more recently, it has been studied in some detail as a sensitive, quantitative method of determining formaldehyde and as a method of determining small amounts of certain metal ions which catalyze the reaction and enhance the chemiluminescence [4–6]. Much of this work has been done using gallic acid, **II**, in place of pyrogallol, **I**, because the former

†We are grateful to Professor Earle Scott of Ripon College for his contributions toward understanding the behavior of this system.

gives more reliable analytical results, but the understanding of the reaction(s) involving gallic acid appear to transfer directly to the pyrogallol system.

These studies have established that either pyrogallol or formaldehyde gives a very faint chemiluminescence in alkaline hydrogen peroxide, but that the analytically useful chemiluminescence requires the presence of both. Some oxidation products of each of these two reactants are thought to be the species actually involved in the chemiluminescent reaction [4]. Formaldehyde does react with the hydroperoxide ion to give compounds **III** and **IV**:

Further oxidation may lead to formic acid and CO_2 and water.

The oxidation of pyrogallol is more complicated and does not result in a single product. One of the most frequently obtained products is purpurogallin, **V**:

A mechanism for the formation of this product has been proposed [7]:

This reaction sequence may form semiquinones and free radicals in reactions between compounds **I** and **VI** and compounds **VII** and **VIII**, and from the oxidation of compound **V**.

The nature of the products obtained from the Trautz-Schorigin reaction has not been definitively described. The normal products of oxidation of pyrogallol are colored, but the solutions obtained at the end of this demonstration are usually colorless. When we used the intensely colored purpurogallin, **V**, as starting material instead of pyrogallol, the color was largely bleached. No red chemiluminescence was observed, although there were normal gas evolution, heat generation, and blue emission.

The reaction of gallic acid, formaldehyde, and alkaline hydrogen peroxide evolves CO_2, H_2, and O_2 [4]. In this demonstration, the evolution of CO_2 in a solution containing large quantities of carbonate ions is not surprising because the oxidative reaction generates acids. Conceivably, some molecular oxygen is formed by the decomposition reaction of hydrogen peroxide, but that reaction does not generate chemiluminescence, so other routes for the generation of singlet oxygen should be sought to account for it. Slawinska and Slawinski [4] propose that hydrogen gas is generated by the decomposition of the dimethylol peroxide:

$$HOCH_2\text{-}O\text{-}O\text{-}CH_2OH + 2\ NaOH \longrightarrow 2\ HCO_2Na + 2\ H_2O + H_2$$

They also propose methods of generating oxygen from a methylolperoxide radical dismutation, but this reaction cannot be the source of singlet oxygen, because the pyrogallol moiety is essential to that reaction.

Although the source of the singlet oxygen is not clearly established, the red chemiluminescence is undoubtedly caused by singlet oxygen. Taken as a whole, the evidence for this assignment is convincing [4, 5]: known quenchers of singlet oxygen quench this emission, dyes that fluoresce in the presence of singlet oxygen do so in these reactions, the intensity of the chemiluminescence is directly related to the concentration of dissolved oxygen in the solution, and the emission spectra can be rationalized in terms of perturbed singlet oxygen, but not in terms of the fluorescence spectra of the reaction products.

The reaction sequence also defies easy explanation. In the singlet oxygen demonstration involving hypochlorite and hydrogen peroxide in base, the addition of luminol permits both the singlet oxygen emission and the luminol emission to occur simultaneously (see Demonstration 2.1). In this demonstration, however, the oxygen chemiluminescence is finished before the luminol emission begins, at least as far as the eye can detect. These emissions are largely separated by the phase of vigorous gas evolution, during which the luminol emission begins to be detectable. If the pyrogallol is excluded from the mixture used in this demonstration, the luminol chemiluminescence is generated after moderate heating for a couple of minutes. Conceivably, the luminol reaction is triggered by the heat evolved by the oxidation of pyrogallol, and thus the two mechanisms may be totally isolated from each other. More work is obviously required to explain the subsystems in this demonstration and their possible relationships to each other.

REFERENCES

1. Adey, H.; Britton, G. C. *School Sci. Review* **1975**, *57*, 314.
2. "Prudent Practices for Handling Hazardous Chemicals in Laboratories." *Natl. Res. Counc.* **1981**; Committee on Hazardous Substances in the Laboratory; p 131.
3. Trautz, M.; Schorigin, P. *Z. Wiss. Photogr. Photochem.* **1905**, *3*, 121.
4. Slawinska, D.; Slawinski, J. *Anal. Chem.* **1975**, *47*, 2101.
5. Slawinska, D. *Photochem. Photobiol.* **1978**, *28*, 453.
6. Stieg, S.; Nieman, T. *Anal. Chem.* **1977**, *49*, 1322.
7. Barten, D.; Ollis, W. D. "Comprehensive Organic Chemistry"; Pergamon Press: Oxford, 1979; Vol. I, p 789.

2.7

Hydrogen Peroxide Oxidation of Lucigenin

In a darkened room, a colorless liquid and a fluorescent yellow liquid are poured simultaneously into a large glass funnel with a spiral delivery tube emptying into a large glass container. The mixed liquids emit a green chemiluminescent glow as they flow through the tube and into the container [1]. The glow persists for approximately 2 minutes, and its color changes from green to blue. When the room lights are turned on, the mixture in the receiving vessel is golden brown.

Various fluorescent dyes produce different emission colors.

MATERIALS FOR PROCEDURE A

8.0 g sodium hydroxide, NaOH

2 liters distilled water

300 ml ethanol, C_2H_5OH, or 300 ml acetone, C_3H_6O

50 ml 30% hydrogen peroxide, H_2O_2

0.2 g lucigenin (bis-N-methylacridinium nitrate), $C_{28}H_{22}O_6N_4$

0.05 g fluorescein disodium salt,[†] $Na_2C_{20}H_{10}O_5$, or 0.05 g rhodamine B, $C_{28}H_{31}ClN_2O_3$ (optional)

2 1-liter Erlenmeyer flasks with rubber stoppers or 2 1-liter round-bottomed flasks with rubber stoppers and cork support rings

gloves, plastic or rubber

glass funnel (15–20 cm in diameter with long stem to fit tubing)

ring stand (1 m or taller)

ring with clamp holder to support funnel

2 m transparent Tygon tubing (0.5 cm in diameter) or glass spiral assembly (See footnote in Procedure A.)

metal wire to fasten tubing to funnel

6–8 extension clamps and clamp holders

3-liter Erlenmeyer flask or 3-liter round-bottomed flask with cork support ring

MATERIALS FOR PROCEDURE B

0.01 g fluorescein disodium salt, $Na_2C_{20}H_{10}O_5$

or one of the following:

0.01 g rhodamine B, $C_{28}H_{31}ClN_2O_3$

†See pp 202–4 for structures and sources.

0.01 g rhodamine 110, $C_{20}H_{15}ClN_2O_3$

0.01 g rhodamine 6G, $C_{27}H_{29}ClN_2O_3$

0.01 g eosin Y disodium salt, $Na_2C_{20}H_6Br_4O_5$

10 ml ethanol, C_2H_5OH

200 ml solution A-2 (see Procedure A)

200 ml solution A-1 (see Procedure A)

500-ml Erlenmeyer flask with rubber stopper

gloves, plastic or rubber

PROCEDURE A

Solution A-1. In a 1-liter flask dissolve 8.0 g of sodium hydroxide in 650 ml of distilled water. Wearing gloves, add 300 ml of ethanol and 50 ml of 30% hydrogen peroxide. Swirl the flask to mix the solution.

Solution A-2. Wearing gloves, dissolve in the other 1-liter flask 0.2 g of lucigenin in 1000 ml of distilled water.

Place the funnel in the ring and clamp it as high as possible on the ring stand. Slip the Tygon tubing over the funnel stem. If the fit is loose, tighten it by placing a metal wire around it. Place 6–8 clamps at about equal distance from each other along the ring stand. Form a coil with the Tygon tubing around the ring stand and hold the tubing in place with the clamps†. Place the lower end of the tube in the 3-liter receiving flask (see Figure 1).

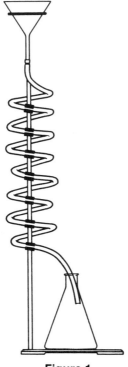

Figure 1.

†At the University of Wisconsin-Madison, we use a glass spiral assembly made of 18-mm glass tubing. The spiral has 4½ turns with a separation of about 7 cm between adjacent turns. The diameter of the coil is 14 cm and its height is 50 cm. One end of the coil is 6 cm long from the turn while the other is 9 cm long. The shorter end of the coil is connected to a 200-mm glass funnel (200-mm top diameter, 18-mm stem diameter). The longer end of the coil is placed directly into the receiving flask.

A variety of shapes can be constructed with the Tygon tubing, as long as the liquid flow from the funnel to the receiving flask is not restricted. For example, numerals and letters can be produced by bending the tubing and spraying appropriate parts with black paint.

To perform the demonstration, dim the room lights but allow sufficient light for the demonstrator to see the flasks and coil assembly. Wearing gloves, slowly pour solutions A-1 and A-2 simultaneously into the funnel. A greenish blue glow will appear. Continue pouring until both flasks are empty. The glowing liquid runs through the spiral and collects in the receiving vessel. The solution will continue to glow for approximately 2 minutes after leaving the spiral.

To enhance the brightness of the emission, add 0.01 g of fluorescein to solution A-2 prior to performing the demonstration. The mixture produces a bright yellowish green glow. If rhodamine B is used instead of fluorescein, a red glow is produced.

PROCEDURE B

Wearing gloves, place 0.01 g of the fluorescein or any of the other fluorescent dyes in the 500-ml Erlenmeyer flask. Dissolve the dye in 10 ml of ethanol. Add 200 ml of solution A-2 and thoroughly mix the contents of the flask.

To perform the demonstration, mix 200 ml of solution A-1 with the contents of the 500-ml Erlenmeyer flask.

Various fluorescent dyes can be used to provide different colors: fluorescein (yellowish green), rhodamine B (red), rhodamine 110 (lime green), rhodamine 6G (orange, changing to pale blue), and eosin Y (orange). The chemiluminescence lasts for 1–2 minutes, depending on the dye used.

HAZARDS

Rhodamine B and rhodamine 6G are cancer-suspect agents.

The toxic and carcinogenic properties of lucigenin, eosin, and rhodamine 110 are not known.

Since both ethanol and acetone are flammable, this demonstration should not be performed near an open flame.

Sodium hydroxide can cause severe burns of the eyes and skin. Dust from solid sodium hydroxide is very caustic.

Since 30% hydrogen peroxide is a strong oxidizing agent, contact with skin and eyes must be avoided. In case of contact, immediately flush with water for at least 15 minutes; get immediate medical attention if the eyes are affected.

Avoid contact between 30% hydrogen peroxide and combustible materials. Avoid contamination from any source, since any contaminant, including dust, will cause rapid decomposition and the generation of large quantities of oxygen gas. Store 30% hydrogen peroxide in its original, closed container, making sure that the container vent works properly.

DISPOSAL

All substances used in this demonstration are water soluble and should be flushed down the drain. The coil assembly used in Procedure A should be rinsed with water.

DISCUSSION

Gleu and Petsch first synthesized lucigenin (bis-N-methylacridinium nitrate), **I**, in 1935 [2]:

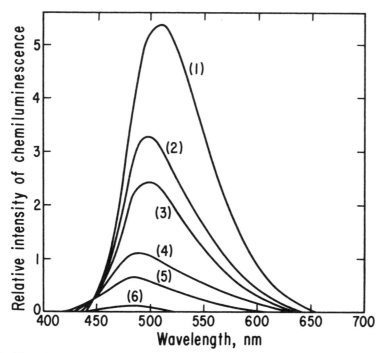

I

They noted the chemiluminescence of lucigenin in alkaline hydrogen peroxide solutions. The chemiluminescence is either blue or green depending upon the concentration of lucigenin (Figure 2) [3]. The emitting species for the blue chemiluminescence has been identified as N-methylacridone [3], and the initial green luminescence has been ascribed to the fluorescence of lucigenin [1].

Figure 2. The chemiluminescence spectra of lucigenin at 23°C [3]. Concentration of lucigenin (wavelength of the maximum):

(1) 2.0×10^{-4}M (510 nm)   (4) 1.0×10^{-5}M (488 nm)
(2) 1.0×10^{-4}M (498 nm)   (5) 5.0×10^{-6}M (485 nm)
(3) 5.0×10^{-5}M (498 nm)   (6) 1.0×10^{-6}M (485 nm)

A mechanism has been proposed to account for the apparent color change observed in the chemiluminescence [1, 4]. In this mechanism, the bis-N-methylacridinium ion, **I**, reacts with peroxide in an alkaline medium to produce the dioxetane, **II**.

$$\xrightarrow[\text{OH}^-]{\text{H}_2\text{O}_2}$$

I   **II**

The dioxetane decomposes spontaneously to form two molecules of N-methylacridone, **III**, in its first singlet excited state, S_1.

$$\longrightarrow 2 \qquad\qquad\qquad (2)$$

II   **III**

The excited singlet **III***(S_1) can return to the ground state via release of a photon, producing blue emission.

$$\mathbf{III}^*(S_1) \longrightarrow \mathbf{III}(S_0) + h\nu \quad \text{(blue emission)} \tag{3}$$

In addition, the excited singlet of **III** can be quenched by energy transfer to a ground state molecule of the starting bis-N-methylacridinium ion, **I**(S_0), producing its first excited singlet, **I***(S_1), as represented in equation 4. This excited state, **I***(S_1), may return to its ground state by emission of a photon of green light.

$$\mathbf{III}^*(S_1) + \mathbf{I}(S_0) \longrightarrow \mathbf{III}(S_0) + \mathbf{I}^*(S_1) \tag{4}$$

$$\mathbf{I}^*(S_1) \longrightarrow \mathbf{I}(S_0) + h\nu' \quad \text{(green emission)} \tag{5}$$

This mechanism explains the gradual change in the color of the emission from green to blue. Initially, since the concentration of **I** is high, it efficiently quenches **III***(S_1) by the process represented in equation 4. The resulting **I***(S_1) decays, producing the green emission. As the reaction proceeds, the concentration of **I** diminishes, resulting in less efficient quenching of **III***(S_1) and therefore in more pronounced blue emission of **III***(S_1). Additional mechanistic details (rate law, effect of pH, quantum yields, etc.) are described in the literature [5–8]. The behavior of both lucigenin and luminol systems is described in reference 8.

The transfer of energy to fluorescent dyes (Procedures A and B) results in emissions of different colors. The mechanism of energy transfer presumably involves either the excited singlet **III***(S_1) or the excited singlet **I***(S_1) and a low-lying singlet state of the acceptor molecule. The spectra of selected dyes are shown on pages 142–44. For some of these dyes, the apparent change in the color of the emission may be due to partial absorption of the light emitted by the lucigenin chemiluminescence. The role of the solvent and its effect on the brightness of emission are not understood.

REFERENCES

1. Aimet, R. G. *J. Chem. Educ.* **1982**, *59*, 163.
2. Gleu, K.; Petsch, W. *Angew. Chem.* **1935**, *48*, 57.
3. Maeda, K.; Hayashi, T. *Bull. Chem. Soc. Japan* **1967**, *40*, 169.
4. McCapra, F.; Richardson, D. G. *Tetrahedron Letters* **1964**, *43*, 3167.
5. Totter, J. R. *Photochem. Photobiol.* **1964**, *3*, 231.
6. Maskiewicz, R.; Sogah, D.; Bruice, T. C. *J. Am. Chem. Soc.* **1979**, *101*, 5347.
7. Maskiewicz, R.; Sogah, D.; Bruice, T. C. *J. Am. Chem. Soc.* **1979**, *101*, 5355.
8. Totter, J. R.; Philbrook, G. E. *Photochem. Photobiol.* **1966**, *5*, 177.

2.8

Air Oxidation
of White Phosphorus

When an oily mixture is rubbed over gloved hands in a darkened room, the gloved hands glow for several seconds.

MATERIALS

0.5 g white (yellow) phosphorus

40 ml mineral oil

test tube, 25 mm × 200 mm

gloves, plastic or rubber, preferably disposable

tongs

large evaporating or crystallizing dish to hold bulk phosphorus for cutting

knife or spatula

50-ml beaker

magnetic stirrer and 13-mm (½-inch) stirring bar

#4 rubber stopper

Additional Materials for Alternate Preparation

5 ml carbon disulfide, CS_2

tank of nitrogen with valve

20 cm of 6-mm glass tubing

1 m of rubber tubing with inside diameter of 6 mm

PROCEDURE

The mixture of phosphorus in mineral oil should be prepared at least 1 day prior to the demonstration.

Place 40 ml of mineral oil in the large test tube. Wearing gloves and using tongs, place the bulk phosphorus in a large dish of water. With a knife or spatula, carefully cut the phosphorus into small pieces. Add these pieces to a tared beaker of water on a balance until about 0.5 g of phosphorus is obtained. Decant all the water from the phosphorus pieces in the beaker and then add the pieces to the oil in the test tube. Place a stirring bar in the test tube and place on a magnetic stirrer to stir for several hours. After a few hours, enough phosphorus will be dissolved in the oil to perform the demonstration, but the phosphorus will not completely dissolve for several days.

Perform this demonstration only in a room with good ventilation.

To perform the demonstration, wear plastic or rubber gloves and pour a small amount (about 1 ml) of the phosphorus–mineral oil mixture into the palm of one glove. In a completely darkened room, quickly rub your gloved hands together to coat palms and backs of both gloves with the mixture. The gloves will glow in the dark for about 20 seconds. The demonstration may be repeated, if desired. The rate of exposing phosphorus to air affects the duration of glow.

Caution: a noticeable warmth will be felt through the gloves and white fumes of phosphorus oxides will be produced. Avoid inhaling the fumes.

Remove the gloves and place them in a hood away from all combustible materials. Wash the hands thoroughly with soap to remove any residues.

Alternate Preparation of Phosphorus-Oil Mixture

Obtain 0.5 g of phosphorus pieces as described above. Decant the water from the phosphorus pieces and, with a paper towel, *quickly* blot the pieces to remove excess water. Add the pieces to 5 ml of carbon disulfide in the large test tube. Stir or swirl to dissolve the pieces. When the phosphorus is dissolved, add 40 ml of mineral oil to the test tube. In a hood, connect the piece of glass tubing with the rubber tubing to the nitrogen supply and slowly bubble nitrogen gas through the carbon disulfide–phosphorus–oil mixture until the carbon disulfide has evaporated. This will produce a somewhat cloudy dispersion of phosphorus in oil, which may clear up in a few hours. Do not use the mixture until it is free of carbon disulfide.

Either method of preparation provides enough phosphorus-oil mixture to perform the demonstration approximately 40 times. We have stored the tightly stoppered test tube containing the mixture for several years. It should be stored in a cool place away from all combustible materials, preferably in a closed metal cabinet.

HAZARDS

White phosphorus is spontaneously flammable in air. Combustion in air produces phosphorus pentoxide. Both white phosphorus and phosphorus pentoxide are very poisonous and can produce severe burns. Chronic effects can result from continued absorption of small amounts. Exposure of the bulk phosphorus to air should be very brief, and all cutting should be done under water.

In this demonstration, reaction of the phosphorus with air produces phosphorus pentoxide and possibly phosphorus trioxide. Both of these oxides are poisonous and irritating to the skin and mucous membranes. The demonstration should be performed in a well-ventilated room. To avoid inhaling the fumes, the demonstrator should keep the gloves away from the face.

Carbon disulfide is extremely flammable and toxic. The explosive range is 1–50% (v/v) in air [1].

Wear gloves when preparing or using the phosphorus-oil mixture or carbon disulfide. The tightly stoppered test tube should be stored in a cool place in a fireproof cabinet away from all combustible materials.

DISPOSAL

If disposable gloves are used, they should be placed in a metal pan in a hood until all the phosphorus has reacted. The gloves should then be rinsed with warm water and discarded. If re-usable gloves are used, they should be washed with soap to remove the oily residue.

Scraps of phosphorus from the cutting procedure should be allowed to dry and burn in a metal pan in a hood or outdoors away from all combustible materials.

If it is necessary to dispose of a large quantity of the phosphorus-oil mixture, the mixture should be poured into a large flat pan and stirred to mix thoroughly with air. The oil should then be burned.

DISCUSSION

The most important reaction of elemental phosphorus is oxidation [2]. There are two distinct oxidation reactions, both accompanied by emission of light. If rapid oxidation occurs, a yellow flame and more than enough heat to kindle combustible materials are produced (see Demonstration 1.31). Slow oxidation occurs when phosphorus vapor in the immediate vicinity of the solid reacts with air and emits a greenish glow. The sublimation pressure of white phosphorus is 0.043 mm of Hg at 25°C [2].

In this demonstration, phosphorus is dispersed in mineral oil to help provide uniform exposure to air. Slow oxidation ensues and produces the greenish glow. In mineral oil the solubility of phosphorus is 1.45% at 25°C, and in carbon disulfide the solubility is about 12 g/ml at 5°C [2]. A study of the oxidation of white phosphorus by Dainton and Bevington [3] revealed a definite boundary between the glow region and the flame region. Variations of these regions with the pressure of air and with temperature are shown in the figure.

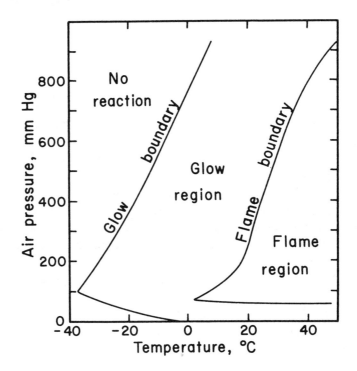

The mechanism proposed by Dainton and Kimberley [4] for the oxidation of phosphorus involves branching chains. The proposed scheme is

$$P_4 + O_2 \longrightarrow P_4O + O \qquad \text{initiation} \qquad (1)$$

$$P_4 + O + M \longrightarrow P_4O + M \qquad \text{propagation} \qquad (2)$$

$$P_4O_n + O_2 \longrightarrow P_4O_{n+1} + O \qquad \text{chain branching} \qquad (3)$$

$$O + O_2 + M \longrightarrow O_3 + M \qquad \text{termination} \qquad (4)$$

$$O + X \longrightarrow \text{stable product} \qquad \text{termination} \qquad (5)$$

$$O + \text{wall} \longrightarrow \text{adsorption} \qquad \text{termination} \qquad (6)$$

In this mechanism, M denotes a third body which may be a molecule of phosphorus, oxygen, or other material, n is an integer ranging from 1 to 9, and X is an inhibitor. Dainton and Kimberley [4] suggest that oxygen atoms are present in the glow reaction. Ozone was detected, but P_4O was not isolated. The greenish glow is emission from an excited state of a PO molecule which is produced by the vapor-phase reaction of phosphorus and oxygen atoms [5]:

$$P(^4S) + O(^3S) + PO(^2\Pi) \longrightarrow PO^* + PO(^2\Pi)$$

The final product of the exothermic reaction is primarily P_4O_{10} ($\Delta H_f^\circ = -3010$ kJ/mole) [6], plus some phosphorus trioxide, P_4O_6, as well as small amounts of other oxides and the various oxyacids of phosphorus produced by traces of moisture. Under the conditions of the demonstration, the temperature will remain low, although the demonstrator will notice a warming effect.

REFERENCES

1. Windholz, M., Ed. "The Merck Index," 9th ed.; Merck and Co.: Rahway, New Jersey, 1976; p 231.
2. Van Wazer, J. R. "Phosphorus and Its Compounds, Vol. I: Chemistry"; Interscience Publishers, John Wiley and Sons: New York, 1958; Ch. 4.
3. Dainton, F. S.; Bevington, J. C. *Trans. Faraday Soc.* **1946**, *42*, 377.
4. Dainton, F. S.; Kimberley, H. M. *Trans. Faraday Soc.* **1950**, *46*, 629.
5. Peck, D. R. in "Mellor's Comprehensive Treatise on Inorganic and Theoretical Chemistry," Vol. VIII, Supplement III, Phosphorus; Interscience Publishers, John Wiley and Sons: New York, 1971; Section V.
6. "Selected Values of Chemical Thermodynamic Properties." *Natl. Bur. Stand. (U.S.)* **1952**; Circ. 500; p 73.

2.9

Air Oxidation of Tetrakis(dimethylamino)ethylene

When a vial containing a yellow liquid is opened momentarily in a darkened room, the contents of the vial emit a bright bluish green glow. The glow gradually fades over a period of approximately 15 minutes.†

MATERIALS

4–7 ml tetrakis(dimethylamino)ethylene (TMAE), $C_{10}H_{28}N_4$ [1]

ca. 25-ml vial of colorless glass with a foil- or plastic-lined screw cap

gloves, plastic or rubber

disposable pipette or dropper with bulb

Optional Materials

small test tube

rubber bulb from dropper (to fit securely over mouth of test tube)

stopcock grease or petrolatum

PROCEDURE

Wear gloves. To prepare for the demonstration, fill the vial about ⅕ full with tetrakis(dimethylamino)ethylene (TMAE), using the pipette or dropper. Ideally, TMAE should be transferred in an inert atmosphere of argon or nitrogen, because it reacts slowly with the oxygen in air. However, the transfer can be accomplished in air if it is done quickly under a fume hood and if all containers are tightly sealed immediately afterward. The cap used to seal the vial should have a foil or plastic liner rather than a paper liner; we have found that TMAE attacks paper liners, causing the vial to leak. A liquid to air volume ratio of 1:4 in the vial allows the chemiluminescent reaction to proceed for about 15 minutes, yet provides insufficient oxygen to consume all the TMAE in one demonstration. If the TMAE is transferred to the vial in air, the contents of the vial will luminesce until all the oxygen gas in the vial has been consumed. Therefore, allow 15–30 minutes for the reaction to be completed before performing the demonstration.

To perform the demonstration, darken the room and open the vial for only 1–2 seconds. Avoid inhaling TMAE vapors. Immediately replace and tighten the cap. Gently shake the vial a few times. The contents of the vial will emit a bright bluish

†We are grateful to Professor T. C. Werner of Union College, Schenectady, New York, for calling this demonstration to our attention.

190

green glow, which slowly fades over a period of about 15 minutes. The same vial can be used repeatedly until it fails to produce a sufficiently bright glow.

A "pen" which will allow one to write luminescent messages can be fashioned from a small test tube and a rubber bulb that fits securely over the mouth of the test tube. With a razor blade, make a slit about 1 cm long in the tip of the rubber bulb. Wearing gloves, place about 2 ml of TMAE into the test tube and insert the mouth of the test tube into the base of the bulb. Placing stopcock grease or petrolatum on the base of the bulb may facilitate inserting the mouth of the test tube into the bulb. Invert the bulb and test tube assembly, and write or draw on paper or poster board. Experiment to find the best way to hold the "pen." Be sure to wear gloves and to use the "pen" in a well-ventilated area.

HAZARDS

The toxic and carcinogenic properties of tetrakis(dimethylamino)ethylene are unknown. Consequently, skin contact with the liquid, as well as inhalation of its vapors, should be avoided. Rubber or plastic gloves should be worn when handling TMAE, and, if it is handled in the air, a fume hood should be used.

DISPOSAL

Once the contents of the vial fail to produce a sufficiently bright glow, the contents of the vial should be flushed down a drain with rapidly running water, and the vial discarded in a waste receptacle.

The paper or board used with the TMAE "pen" should be discarded in a waste receptacle after the message or drawing has stopped glowing.

DISCUSSION

Tetrakis(dimethylamino)ethylene is a yellow liquid with a normal freezing point of $-4°C$, a boiling point of $59°C$ at 0.9 mm of Hg, and a density of 0.86 g/ml.

The first synthesis of tetrakis(dimethylamino)ethylene, TMAE, **I**, was reported in 1950, at which time it was noted that TMAE is strongly chemiluminescent in air [2].

$$
\begin{array}{cc}
(CH_3)_2N & N(CH_3)_2 \\
 & C=C \\
(CH_3)_2N & N(CH_3)_2 \\
 & \mathbf{I}
\end{array}
$$

Although TMAE's strong chemiluminescence in air was noted by Pruett et al. [2], the exact nature of the chemiluminescent reaction was not systematically investigated until the mid-60s.

The major products of the oxidation of TMAE in air are tetramethylurea, **II**, tetramethyloxamide, **III**, and tetramethylhydrazine, **IV** [3]:

$$
\underset{\mathbf{II}}{(CH_3)_2N-\overset{\overset{\displaystyle O}{\|}}{C}-N(CH_3)_2} \qquad \underset{\mathbf{III}}{(CH_3)_2N-\overset{\overset{\displaystyle O\ O}{\|\ \|}}{C-C}-N(CH_3)_2} \qquad \underset{\mathbf{IV}}{(CH_3)_2N-N(CH_3)_2}
$$

These products appear to result from two reaction paths, as indicated in equations 1 and 2 [*4*]:

$$TMAE + O_2 \longrightarrow 2\ (CH_3)_2N\text{-}\overset{\displaystyle O}{\overset{\displaystyle \|}{C}}\text{-}N(CH_3)_2 \qquad 70\text{--}80\% \qquad (1)$$

$$TMAE + O_2 \longrightarrow (CH_3)_2N\text{-}\overset{\displaystyle O}{\overset{\displaystyle \|}{C}}\text{-}\overset{\displaystyle O}{\overset{\displaystyle \|}{C}}\text{-}N(CH_3)_2 + 2\ (CH_3)_2N\cdot \qquad 30\text{--}20\% \qquad (2)$$

Several minor products have also been found, and these have been ascribed to reactions of the amino radicals produced by reaction 2 [*4*].

Virtually all investigations into the mechanism of TMAE chemiluminescence have involved solutions of TMAE rather than the pure compound. Therefore, much of what is known about the mechanism of the chemiluminescent reaction does not apply directly to the reaction as it occurs in this demonstration. However, since the main features of the pure compound emission are probably quite similar to those of the reaction in solution, it is worth describing what is known about the solution reaction.

Since the fluorescence spectrum of TMAE is nearly equivalent to its chemiluminescence spectrum (see figure) and since the oxidation products of TMAE do not fluoresce, the emitting species in the chemiluminescent reaction has been identified as an electronically excited TMAE molecule (TMAE*) [*5*]. The fluorescence quantum yield of TMAE is 0.35, while the chemiluminescence quantum yield is about 0.004 [*4*]. The fluorescence and chemiluminescence quantum yields indicate that about 1% of the oxidized TMAE molecules result in the production of TMAE*. This yield contrasts with the 50% or more of luminol molecules that are oxidized to excited aminophthalate ions [*6*].

Studies of TMAE in hydrocarbon solution indicate that TMAE reversibly forms a weak complex with molecular oxygen [*7*]. No further reaction between oxygen and TMAE occurs in the absence of hydroxylic substances, such as alcohols or water, and no chemiluminescence can be observed [*8*]. In the presence of such substances, however, TMAE reacts with molecular oxygen, producing its oxidation products and exhibiting chemiluminescence. If the TMAE-oxygen complex is formed first and then a hydroxylic compound is added, chemiluminescence is produced. In addition, calori-

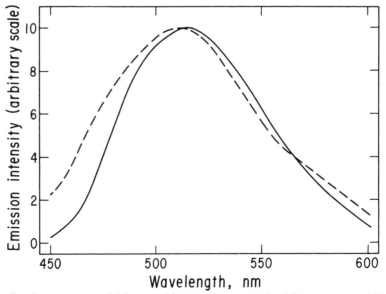

Chemiluminescence, *solid line,* and fluorescence, *dashed line,* spectra of TMAE.

metric studies have shown that the overall reaction to form major products releases 580 kJ/mole [4], considerably more than the 300 kJ needed to produce one einstein of 400 nm emission. Therefore, the mechanism for the production of chemiluminescence must involve the TMAE-oxygen complex and the hydroxylic compound, and the oxidation of one molecule of TMAE is sufficient to produce one excited molecule (TMAE*).

One of the most recently proposed mechanisms to fit these and other facts regarding the chemiluminescent reaction of TMAE is represented by equations 3 through 7, in which E represents TMAE and HA represents a hydroxylic molecule [4]:

$$E + O_2 \longrightarrow E \cdot O_2 \tag{3}$$

$$E \cdot O_2 + HA \longrightarrow (E^{2+} \cdot O_2H^- \cdot A^-) \tag{4}$$

$$(E^{2+} \cdot O_2H^- \cdot A^-) + E \longrightarrow (^+E\text{-}E^+ \cdot O_2H^- \cdot A^-) \tag{5}$$

$$(E^{2+} \cdot O_2H^- \cdot A^-) \longrightarrow products + HA \tag{6}$$

$$(^+E\text{-}E^+ \cdot O_2H^- \cdot A^-) \longrightarrow E^* \text{ or } E + products + HA \tag{7}$$

Equations 6 and 7 explain the two major sets of oxidation products. The coupling step, equation 5, and the chemielectronic step, equation 7, explain the production of TMAE* utilizing the energy provided by the oxidation of TMAE.

Among the known quenchers of the chemiluminescence from TMAE oxidation are tetramethylurea and tetramethyloxamide [5], which are two of the oxidation products of TMAE. For this reason, the chemiluminescence from the oxidation of TMAE disappears before all the TMAE has been oxidized. If the products of the oxidation were removed, more light would be produced by the oxidation reaction.

Since TMAE is commercially available, there is no need to synthesize it, but it can be synthesized by the reaction of 1,1-dimethoxytrimethylamine with dimethylamine [9]. Under distillation conditions, methanol is eliminated:

$$(CH_3)_2NCH(OCH_3)_2 + (CH_3)_2NH \longrightarrow (CH_3)_2NCHOCH_3 + CH_3OH \tag{8}$$

with $N(CH_3)_2$ substituent on the carbon

$$2\ (CH_3)_2NCHOCH_3 \longrightarrow \underset{(CH_3)_2N}{\overset{(CH_3)_2N}{>}} C = C \underset{N(CH_3)_2}{\overset{N(CH_3)_2}{<}} + 2\ CH_3OH \tag{9}$$

REFERENCES

1. Tetrakis(dimethylamino)ethylene can be obtained from Aldrich Chemical Company, stock no. 23,423–0.
2. Pruett, R. L.; Barr, J. T.; Rapp, K. E.; Bahner, C. T.; Gibson, J. D.; Lafferty, R. H., Jr. *J. Am. Chem. Soc.* **1950**, *72*, 3646.
3. Urry, W. H.; Sheeto, J. *Photochem. Photobiol.* **1965**, *4*, 1067.
4. Fletcher, A. N.; Heller, C. A. *J. Phys. Chem.* **1967**, *71*, 1507.
5. Winberg, H. E.; Downing, J. R.; Coffman, D. D. *J. Am. Chem. Soc.* **1965**, *87*, 2054.
6. White, E. H.; Bursey, M. M. *J. Am. Chem. Soc.* **1964**, *86*, 941; see also Discussion section of Demonstration 2.3 in this volume.
7. Heller, C. A.; Fletcher, A. N. *J. Phys. Chem.* **1965**, *69*, 3313.
8. Fletcher, A. N.; Heller, C. A. *J. Catalysis* **1966**, *6*, 263.
9. Eilingsfeld, H.; Seefelder, M.; Weidinger, H. *Chem. Ber.* **1963**, *96*, 2671.

2.10

Chemiluminescence of Tris(2,2′-bipyridyl)ruthenium(II) Ion

In a darkened room, drops of a pale green solution are added to the vortex of a magnetically stirred clear solution. Where the two solutions mix, bright orange flashes appear [1]. If the green solution is injected below the surface of the clear solution, a glowing orange jet develops.

MATERIALS

30 ml 1.0M sulfuric acid, H_2SO_4

0.021 g tris(2,2′-bipyridyl)ruthenium(II) chloride hexahydrate, $Ru(C_{10}H_8N_2)_3Cl_2 \cdot 6H_2O$

0.01 g lead dioxide, PbO_2

50 ml distilled water

0.3 g sodium hydroxide, NaOH

5.0 g sodium borohydride, $NaBH_4$

2 100-ml beakers

glass stirring rod

Whatman #2 filter paper

55-mm glass funnel

ring stand and ring (to hold funnel)

250-ml beaker

250-ml Erlenmeyer flask

magnetic stirrer and stirring bar

dropping pipette and bulb

PROCEDURE

Solution 1. (*Prepare no more than 30 minutes before use.*) Pour 30 ml of 1.0M H_2SO_4 into a 100-ml beaker. Add 0.021 g of tris(2,2′-bipyridyl)ruthenium(II) chloride hexahydrate and stir to dissolve. The solution will be reddish orange in color. Add 0.01 g of lead dioxide and stir. The solution turns yellowish green. Let stand 2 minutes; then filter through Whatman #2 filter paper held in the supported funnel into a clean 100-ml beaker.

Solution 2. (*Prepare no more than 30 minutes before use.*) Pour 50 ml of distilled water into a 250-ml beaker. Add 0.3 g of NaOH and stir to dissolve. Add 5.0 g of sodium borohydride and stir.

To perform the demonstration, pour solution 2 into a 250-ml Erlenmeyer flask. Place the Erlenmeyer flask on a magnetic stirrer and stir at a sufficient rate to cause a vortex. Dim the lights. Using a dropping pipette, add solution 1 to the vortex. Orange flashes will appear. The solution will foam, and heat will be evolved. Violent foaming will occur if solution 1 is added in greater quantities.

If solution 1 is injected with the pipette below the surface of solution 2, an orange jet will be seen.

HAZARDS

The toxicity and carcinogenicity of tris(2,2′-bipyridyl)ruthenium(II) chloride hexahydrate are not known.

Lead and its compounds are toxic. Lead poisoning, either acute or chronic, can result from exposure to dust from the solid.

Sodium hydroxide can cause severe burns of the eyes and skin. Dust from solid sodium hydroxide is very caustic.

Sodium borohydride is irritating to skin, eyes, and respiratory system. Acidifying a solution of sodium borohydride releases hydrogen gas.

Since sulfuric acid is a strong acid and a powerful dehydrating agent, it can cause burns. Spills should be neutralized with an appropriate agent, such as sodium bicarbonate, and then rinsed clean.

Because of the evolution of H_2 gas, the demonstration should be conducted in a well-ventilated area.

DISPOSAL

All the solutions used in this demonstration are water soluble and should be flushed down the drain.

DISCUSSION

The manner in which the two reactant solutions are mixed determines the visual effect obtained in this demonstration. If the two solutions are mixed by drops, flashes of orange chemiluminescence are produced. If the two solutions are mixed in a continuous stream, a continuous orange luminescence is created. The two solutions should not be mixed all at once, since considerable frothing occurs which reduces the visibility of the luminescence and may cause the mixture to overflow its container.

In the preparation of solution 1, the reddish orange tris(2,2′-bipyridyl)ruthenium(II) ion, $Ru(bipy)_3^{2+}$, is oxidized to the yellowish green tris(2,2′-bipyridyl)ruthenium(III) ion, $Ru(bipy)_3^{3+}$, in a heterogenous reaction with lead dioxide, PbO_2, [1]. This reaction is indicated in equation 1:

$$2\ Ru(bipy)_3^{2+} + PbO_2(s) + 4\ H^+ \longrightarrow 2\ Ru(bipy)_3^{3+} + Pb^{2+} + 2\ H_2O \quad (1)$$

The excess PbO_2 is filtered off to prevent any interferences in subsequent reactions. The concentration of the $Ru(bipy)_3^{3+}$ solution is about 10^{-3}M. This solution must be

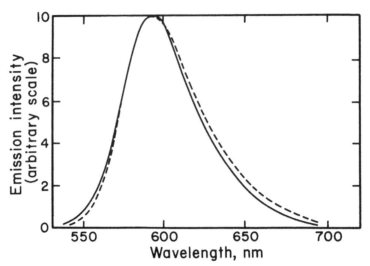

Figure 1. Photoluminescence, *solid line,* and chemiluminescence, *dashed line,* spectra of tris(2,2'-bipyridyl)ruthenium(II) ion.

prepared no more than 30 minutes before use, since the $Ru(bipy)_3^{3+}$, is slowly reduced by water to $Ru(bipy)_3^{2+}$. The solution half-life of $Ru(bipy)_3^{3+}$, which depends on the pH of the solution, is approximately 2 hours in the acidic conditions used in this demonstration [1]. Solution 2, which contains tetrahydridoborate ions, BH_4^-, must also be prepared within 30 minutes of use, since the BH_4^- ions undergo acid hydrolysis to hydrogen gas and borate ions. The sodium hydroxide added to the solution inhibits this hydrolysis [1].

The reaction which produces the chemiluminescence observed in this demonstration is the reduction of $Ru(bipy)_3^{3+}$ to $Ru(bipy)_3^{2+}$ by the powerful reducing agent BH_4^- [1]. Although the details of the mechanism of this reaction are not known, the origin of the luminescence is quite well documented. As indicated in Figure 1, the chemiluminescence spectrum of this reaction is virtually identical to the phosphorescence of $Ru(bipy)_3^{2+}$ [2]. On this basis, the excited-state intermediate formed in the reaction has been identified as an excited state of $Ru(bipy)_3^{2+}$ [1].

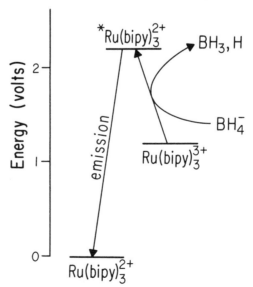

Figure 2. Energy-level diagram for various tris(2,2'-bipyridyl)ruthenium species [1].

The energetic relationships involved in the reaction can be represented as in Figure 2 [*1*]. The energy difference between Ru(bipy)$_3^{2+}$ and Ru(bipy)$_3^{3+}$ has been determined by electrochemical means to be 1.2 volts [*3*]. The energy difference between the excited-state Ru(bipy)$_3^{2+}$ and its ground state is revealed by the emission, which covers a wavelength range of 525 to 700 nm. This corresponds to an energy range of 210 to 160 kJ/mole, which is equivalent to 2.2–1.7 volts. Therefore, the emissive excited state of Ru(bipy)$_3^{2+}$ is as much as 1.0 volt above the ground state of Ru(bipy)$_3^{3+}$. The reaction in which Ru(bipy)$_3^{3+}$ is reduced to Ru(bipy)$_3^{2+}$ must provide this energy, which shows why a powerful reducing agent such as BH$_4^-$ is used to produce chemiluminescence.

A mechanism for the luminescence of the tris(2,2′-bipyridyl)ruthenium(II) ion has been proposed [*4*]. Before considering this mechanism, however, it is necessary to consider several properties of Ru(bipy)$_3^{2+}$ itself.

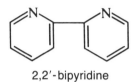

2,2′-bipyridine

The Ru(bipy)$_3^{2+}$ complex ion contains three bidentate 2,2′-bipyridine ligands coordinated to a ruthenium(II) ion. Ruthenium is a second-row transition element, below iron in the periodic table, and the Ru^{2+} ion has a [Kr]4d^6 electronic configuration. In Ru(bipy)$_3^{2+}$, the Ru^{2+} ion is bonded to six nitrogen atoms in an octahedral arrangement, but overall the Ru(bipy)$_3^{2+}$ ion has D$_3$ symmetry. (Cotton [*5*] provides an excellent explanation of symmetry and its applications in chemistry.) In this symmetry, the energies of the d-orbitals of the Ru^{2+} ion are split into three groups by the influence of the nitrogen atoms, as indicated on the left side of Figure 3 [*4*]. In the ground state of Ru(bipy)$_3^{2+}$, the e(d) and a$_1$(d) orbitals are filled by the six d electrons, while the e(d*) orbitals are empty. The orbital energies of the bipyridine ligands are indicated on the right side of Figure 3. The completely filled bonding sigma (σ) and pi (π) orbitals are indicated along with the empty antibonding pi (π*) orbitals.

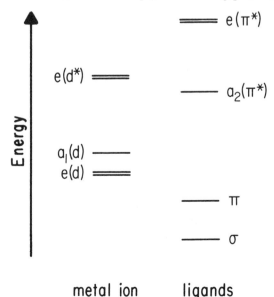

Figure 3. Orbital energy diagram for tris(2,2′-bipyridyl)ruthenium(II).

The relative energies of these orbitals have been determined by a thorough investigation of the absorption spectra of $Ru(bipy)_3^{2+}$ and related complexes [4]. These spectra can contain four types of absorbances, which can be categorized with reference to Figure 3. The first type is the absorbances caused by the ligand-centered transitions, $\pi \rightarrow a_2(\pi^*)$ and $\pi \rightarrow e(\pi^*)$. The energies of these absorbances in the $Ru(bipy)_3^{2+}$ complex are quite similar to those in the protonated ligand [4], and therefore can be easily determined. The second type is the metal-centered absorbances, $a_1(d) \rightarrow e(d^*)$ and $e(d) \rightarrow e(d^*)$. These absorbances can be identified by comparing the spectra of Ru^{2+} complexes containing a variety of similar ligands; the bands at comparable energies in all spectra can be assigned to metal-centered transitions.

The third type of absorbances is caused by the transitions from the metal to the ligand: $a_1(d) \rightarrow a_2(\pi^*)$, $e(d) \rightarrow a_2(\pi^*)$, $a_1(d) \rightarrow e(\pi^*)$, and $e(d) \rightarrow e(\pi^*)$. These transitions are called metal-to-ligand charge transfer (MLCT) transitions and can be identified from absorption spectra taken at low temperature (77 K). At low temperature, the vibrational motion of the ligand is sufficiently slow that some fine structure due to various vibrational levels may appear in the absorption bands that result from MLCT transitions. MLCT transitions occur primarily in complexes having both a metal ion that is easily oxidized, such as Ru^{2+}, and ligands with low-energy unoccupied π^* orbitals, such as bipyridine.

The fourth type of transition is called a ligand-to-metal charge transfer (LMCT). These bands occur at low energy in complexes containing both a metal in a high oxidation state and ligands that are easily oxidized. Since this is not the case in $Ru(bipy)_2^{3+}$, such a transition has not been identified in its spectrum [4].

The emission from $Ru(bipy)_3^{2+}$ has been identified as phosphorescence from the lowest triplet MLCT excited state [4]. The energy of the emission suggests that it arises from an MLCT excited state as identified in the absorption spectrum. The lifetime of the emission has a measured value of 4 microseconds. This lifetime is rather long for fluorescence, which generally has a lifetime in the range of 10^{-9} to 10^{-6} seconds, but it is also somewhat short for phosphorescence, which usually has a lifetime of several milliseconds or more. Long lifetimes are typical of phosphorescence, since it results from transitions between states of differing spin and these transitions are quantum-mechanically forbidden. The rather short lifetime of the phosphorescence of $Ru(bipy)_3^{2+}$ has been ascribed to spin-orbit coupling [4]. This coupling, which is common in compounds of relatively heavy elements such as Ru, tends to reduce the applicability of spin selection rules, thus making the triplet to singlet transition more facile.

Since the photoluminescent and chemiluminescent spectra of Figure 1 are identical, the emission process in the chemiluminescent reduction of $Ru(bipy)_3^{3+}$ must involve the MLCT triplet state of $Ru(bipy)_3^{2+}$. This state may be formed if the reducing agent transfers an electron into a π^* orbital of one of the bipyridine ligands rather than directly to the Ru ion [1]. If this is the case, then the MLCT excited state would be produced directly and could decay to the ground state, thus producing the same luminescence as obtained from photo-excitation.

The exact mechanism of the reaction used in this demonstration is not known. For instance, it is not known whether the reductant that produces the excited $Ru(bipy)_3^{2+}$ is BH_4^- or a hydrogen atom resulting from the one-electron oxidation of BH_4^-. In either case, the generalized explanation of the reaction as shown in Figure 2 is valid and illustrates how spectral, thermochemical, and electrochemical data can be used to understand the energetics of the reaction. The demonstration illustrates how chemi-luminescence can be used to probe the energetics and mechanism of a chemical reaction [*1*].

REFERENCES

1. Gafney, H. D.; Adamson, A. W. *J. Chem. Educ.* **1975**, *52*, 480.
2. Lytle, F. E.; Hercules, D. M. *Photochem. Photobiol.* **1971**, *13*, 123.
3. Buckingham, D. A.; Sargenson, A. M. in "Chelating Agents and Metal Chelates," Dwyer, F. P.; Mellor, D. P., Eds.; Academic Press: New York, 1964; Ch. 6.
4. Lytle, F. E.; Hercules, D. M. *J. Am. Chem. Soc.* **1969**, *91*, 253 and references therein.
5. Cotton, F. A. "Chemical Applications of Group Theory," 2nd ed.; Interscience Publications, John Wiley and Sons: New York, 1971.

2.11

Explosive Reaction
of Nitric Oxide
and Carbon Disulfide

When a mixture of nitric oxide gas and carbon disulfide vapor is ignited at the top of a large glass tube, the reaction travels down the tube, producing a bright blue light and a loud noise. This is described in Demonstration 1.44.

Unsuitable Demonstrations

We are aware of the following systems, and we judge them unsuitable as lecture demonstrations because they are either hazardous or ineffective.

Ozonolysis of Safranine
> Ayers, R. P. *School Sci. Review* **1935**, *17*, 236.

Permanganate Oxidation of Siloxane
> Ayers, R. P. *School Sci. Review* **1935**, *17*, 236.

Oxidation of Lophine
> Cottman, E. W. *J. Chem. Educ.* **1937**, *14*, 236.

Oxidation of Lima Beans, Peanuts, Etc.
> Johnson, L. D. *J. Chem. Educ.* **1940**, *17*, 295.

Reaction of Chloropicrin with Phenylmagnesium Bromide
> Weedy, W. G. *J. Chem. Educ.* **1944**, *21*, 142.

Micro Porphyrin Test for Blood
> Chu, E. J.; Chu, T. C. *J. Chem. Educ.* **1953**, *30*,178.

Chemiluminescence in the Action of Chlorine on Ammonia Solution
> Ormerod, M. B. *School Sci. Review* **1966**, *48*, 792.

Sources and Structures of Sensitizers

Cresyl violet acetate
  FW 321.34
  Aldrich 86,098–0

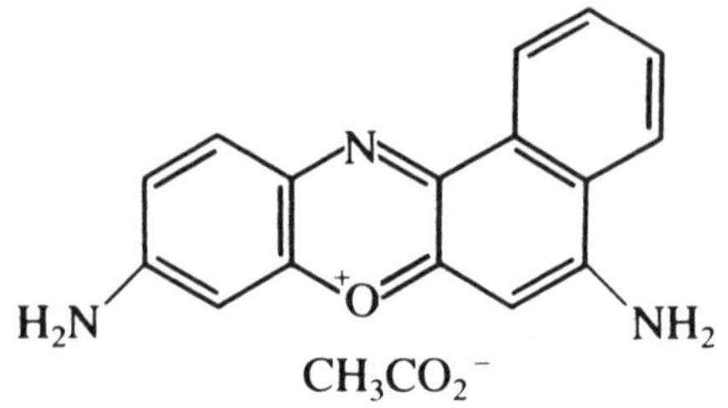

13,13′-Dibenzanthronyl
  FW 482.68
  Aldrich S40,739–9
  (special chemical not
  in 1981–82 catalog)

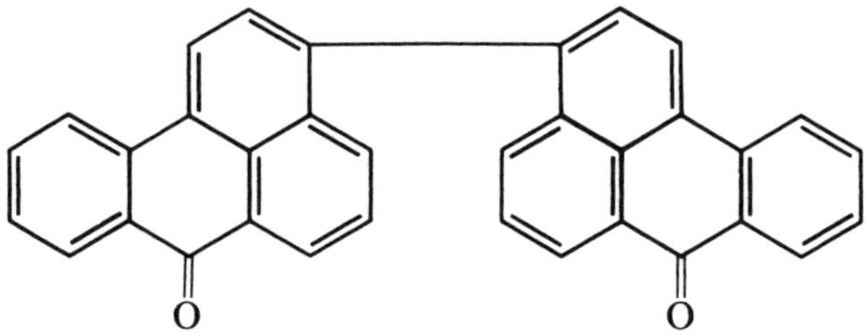

9,10-Diphenylanthracene
  FW 330.43
  Aldrich D20,500–1

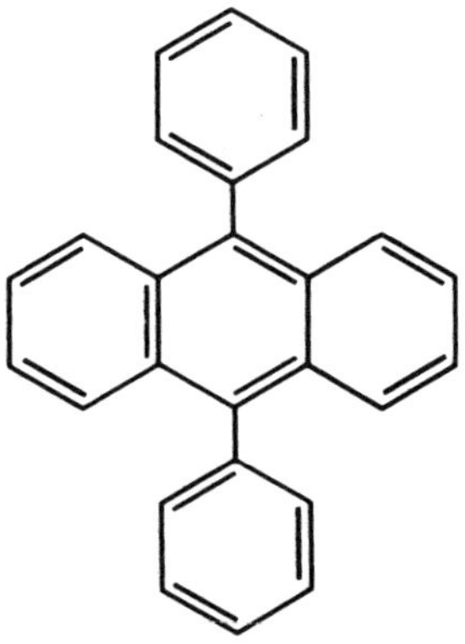

Eosin Y
  FW 691.88
  Aldrich 11,983–0

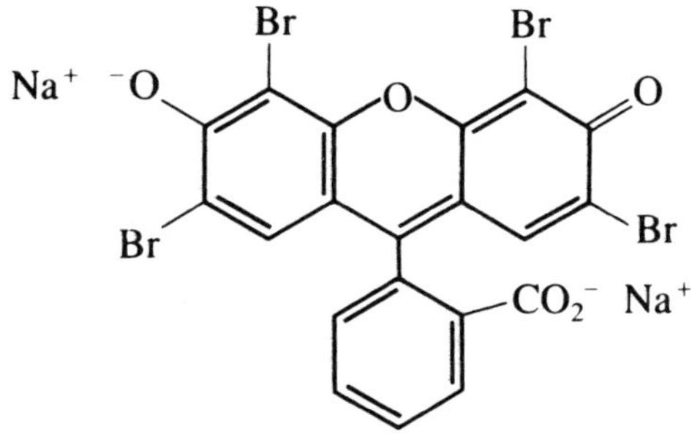

Fluorescein disodium salt
  FW 376.28
  Aldrich 16,630–8

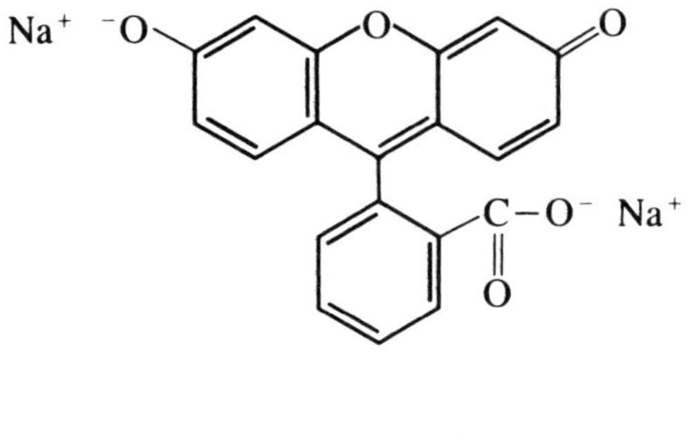

Isoviolanthrone
 FW 456.51
 ICN Pharmaceuticals 13668

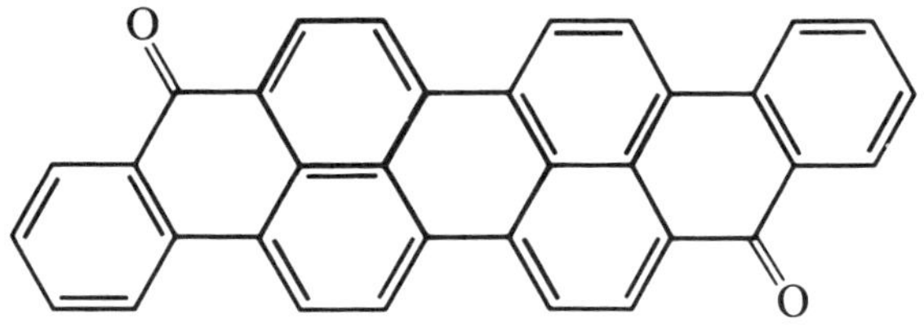

Phenanthrenequinone
 FW 208.22
 Aldrich Pl, 180–8

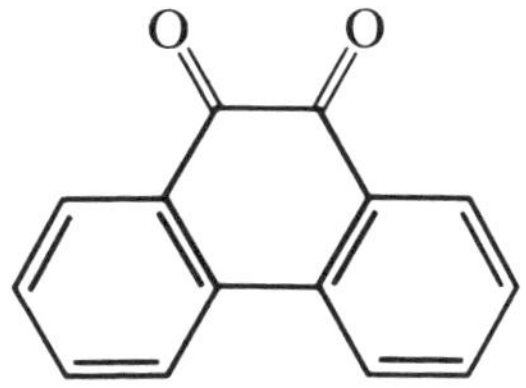

Rhodamine B
 FW 479.00
 Aldrich R–95–3

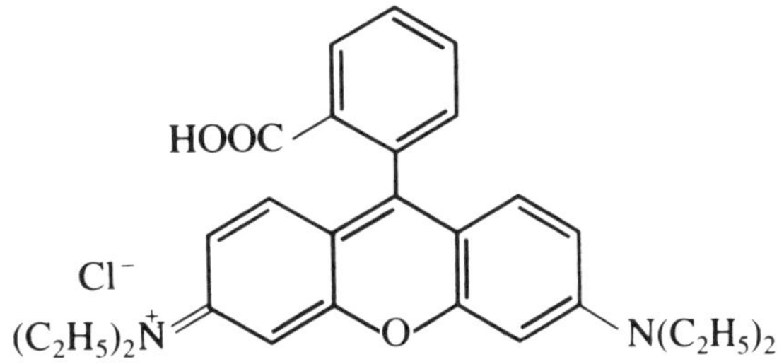

HOOC

Cl^-
$(C_2H_5)_2\overset{+}{N}$ $N(C_2H_5)_2$

Rhodamine 6G
 FW 479.02
 Aldrich 20,132–4

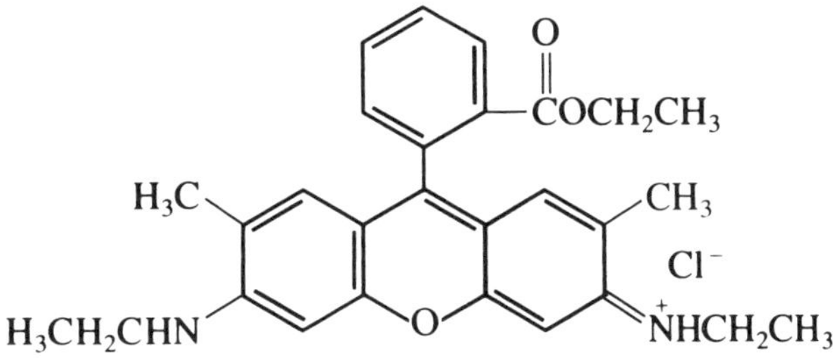

$\overset{O}{\overset{\|}{C}}OCH_2CH_3$
H_3C CH_3
Cl^-
H_3CH_2CHN $\overset{+}{N}HCH_2CH_3$

Rhodamine 110
 FW 366.80
 Kodak 11927

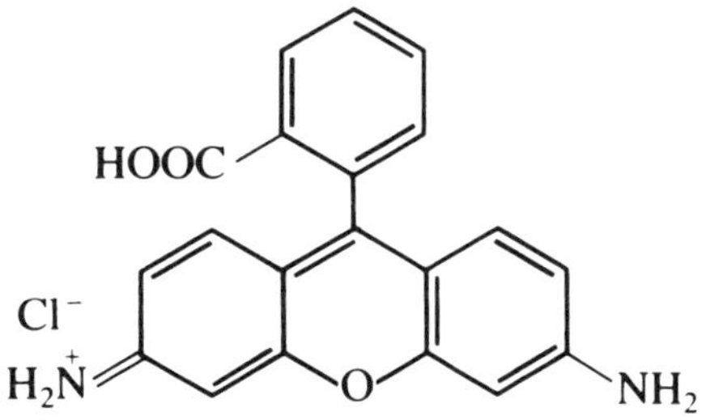

HOOC

Cl^-
$H_2\overset{+}{N}$ NH_2

Rubrene (9,10,11,12-Tetraphenylnapthacene)
 FW 532.69
 Aldrich R220–6

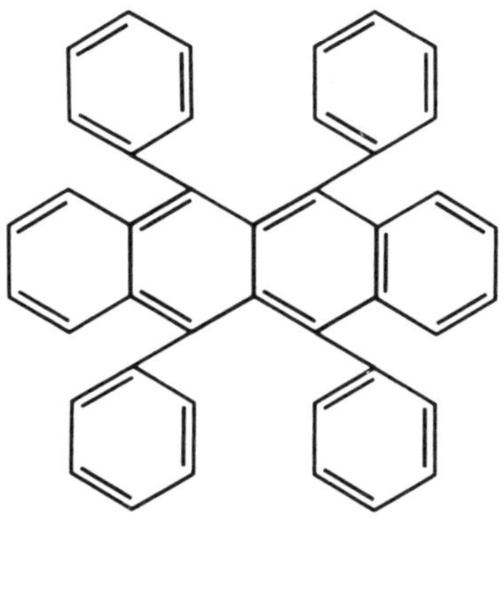

Tetracene (2,3-Benzanthracene)
FW 228.29
Aldrich B240–3

Violanthrone
FW 456.51
ICN Pharmaceuticals 6687

SOURCE ADDRESSES

ICN Pharmaceuticals, Inc.
K & K Labs Division, Life Sciences Group
121 Express Street
Plainview, NY 11803
1981 Catalog

Aldrich Chemical Company
940 W. Saint Paul Avenue
Milwaukee, WI 53233
1981–82 Catalog

Eastman Kodak Company
Kodak Laboratory & Specialty Chemicals
Rochester, NY 14650
1981 Catalog

3

Polymers

Glen E. Dirreen and Bassam Z. Shakhashiri

Polymer chemistry and polymer technology have become prominent in terms of their advances and uses. Over forty percent of professional chemists are engaged in polymer science. Yet, many teachers tend not to discuss polymers in their courses. This chapter contains fourteen demonstrations which can serve as a focus to introduce students to polymers and to discuss their properties and applications. Polymer chemistry must become an integral part of high school and college curricula. Teachers are urged to read the excellent series of articles presented in the "State of the Art Symposium: Polymer Chemistry" [1] and the articles by Mark [2], Marvel [3], and Quirk [4]. Also, college teachers are referred to the article by Seymour [5] detailing the recommended ACS syllabus for introductory courses in polymer science.

Polymers are substances with large molecular weights. Their properties and characteristics vary depending on their composition and method of preparation. Two main types of polymerization methods are used to convert monomers into polymers: chain-reaction or chain-growth polymerization (addition polymerization) and step-reaction or step-growth polymerization (condensation polymerization). In both methods, substances of small molecular weight are transformed to substances of large molecular weight. The major distinctions between the two methods are based on the differences in the kinetics of polymerization.

CHAIN-REACTION POLYMERIZATION (ADDITION)

In this method of synthesis, monomers with carbon-carbon double bonds are converted to polymers in a chain reaction. The mechanism has three steps: Initiation, Propagation, and Termination.

$$ROOR \xrightarrow{\text{heat}} 2\ RO^{\cdot} \quad (1)$$

$$RO^{\cdot} + CH_2{=}CH_2 \longrightarrow RO\text{-}CH_2\text{-}CH_2^{\cdot} \quad (2a)$$

$$RO\text{-}CH_2CH_2^{\cdot} + CH_2{=}CH_2 \longrightarrow RO\text{-}CH_2\text{-}CH_2\text{-}CH_2\text{-}CH_2^{\cdot} \quad (2b)$$

$$\text{etc.}$$

$$RO{-}(CH_2)_n CH_2^{\cdot} + RO{-}(CH_2)_m CH_2^{\cdot} \longrightarrow RO{-}(CH_2)_n CH_2\text{-}CH_2{-}(CH_2)_m OR \quad (3)$$

In the Initiation Step (1) a highly reactive species is generated, usually a free radical. In the Propagation Step (2) the newly generated free radical adds to a carbon-carbon double bond. Addition occurs so as to generate a new free radical which contains the

205

initial monomer (2a). This new free radical adds to another monomer to generate still another new free radical (2b). The propagation step is repeated so that the chain continues to grow. In the Termination Step (3) the growing chain terminates by reacting with another growing chain or by reacting with another species in the polymerization mixture. Typical polymers that are prepared commercially by this method include polyethylene, polystyrene, poly(vinyl chloride), and poly(methyl methacrylate).

Chain-reaction polymerization has the following general characteristics [1]:

(a) Once initiation occurs, the polymer chain forms rapidly (less than 0.1 second).
(b) The concentration of free radicals is very low (about 10^{-8}M). The polymerization mixture consists mostly of the newly formed polymer and unreacted monomer.
(c) The polymerization reaction is usually exothermic.
(d) The polymers formed have high molecular weights (10^4 to 10^7).
(e) Polymers containing secondary chains (branches) attached to the main chain (backbone) can be formed.
(f) Cross-linked polymers can form where all the primary chains are linked to secondary chains.

Table 1 lists typical polymers produced by chain-reaction polymerization.

When two or more different monomers are used in a chain-reaction polymerization, they yield a polymer containing the corresponding repeating units. The process is called copolymerization, and the product is called a copolymer. A wide variety of copolymers can be produced by varying the copolymerization technique and the amounts of each monomer. Many synthetic polymers used today are copolymers.

Table 1. Typical Addition Polymers

Name	Repeating unit	Trade name	Uses
polyethylene	$+CH_2\text{-}CH_2+$	Polythene	film, housewares
polytetrafluoroethylene	$+CF_2\text{-}CF_2+$	Teflon	nonstick surfaces
polystyrene	$+CH_2\text{-}CH+$ (phenyl)	Styrofoam	insulation, coatings
poly(vinyl chloride)	$+CH_2\text{-}CH+$ | Cl	PVC	floor tile, pipe
poly(methyl methacrylate)	CH_3 | $+CH_2\text{-}C+$ | $COOCH_3$	Plexiglas	automotive parts
polyacrylonitrile	$+CH_2\text{-}CH+$ | CN	Orlon	fibers, textiles

STEP-REACTION POLYMERIZATION (CONDENSATION)

In step-reaction polymerization, two monomers with different functional groups combine to form a dimer. The chain growth occurs when a functional group on the dimer reacts further with another functional group on one of the monomers to produce

Step-reaction polymerization [*1*].

a trimer. This process is repeated until all the monomers are converted to low molecular weight units called oligomers, which react further with each other through their free functional groups (see figure). The product does not have as high a molecular weight as that produced in a chain-reaction polymerization.

Step-reaction polymerization has the following general characteristics [*1*]:

(a) The polymer chain forms slowly, sometimes requiring several hours to several days.
(b) Monomers are quickly converted to low molecular weight oligomers.
(c) High temperatures are often required to bring about polymerization.
(d) The polymers formed usually have moderate molecular weights (less than 10^4).
(e) Branching or cross-linking does not occur unless a monomer with three or more functional groups is used.

Table 2 on the following page lists typical polymers produced by step-reaction polymerization.

GENERAL PROPERTIES OF POLYMERS

Polymers exhibit remarkable physical and chemical properties. These properties depend on the molecular weight and chemical structure. Flame-resistance properties, crystallinity, thermal stability, resistance to chemical action, mechanical properties, and other properties determine the usefulness of a polymer. Stereochemical effects play an important role in determining polymer properties.

The properties of a polymer are determined primarily by the extent of interactions among its chains. Some polymers are stiff and strong, some are soft and flexible, and others can withstand considerable stress without breaking. Such mechanical properties are peculiar to the polymer and not characteristic of the monomer(s) from which it was

Table 2. Typical Condensation Polymers

Trade name	Monomers	Repeating unit	Uses
Dacron (polyester)	dimethyl terephthalate and HO-CH$_2$-CH$_2$-OH ethylene glycol		fibers, films, textiles
Lexan (polycarbonate)	bisphenol A and Cl-C-Cl phosgene		tool cases, tele-phone housings
Nylon 66 (polyamide)	HO-C-(CH$_2$)$_4$-C-OH adipic acid and H$_2$N-(CH$_2$)$_6$-NH$_2$ hexamethylenediamine	$\{C(CH_2)_4C\text{-}NH(CH_2)_6NH\}$	fibers, textiles
Perlon U (polyurethane)	HO(CH$_2$)$_4$OH 1,4-butanediol and OCN(CH$_2$)$_6$NCO hexamethylene diisocyanate	$\{O\text{-}(CH_2)_4O\text{-}C\text{-}N(CH_2)_6N\text{-}C\}$	flooring, wood and fabric coat-ings, flotation applications and packaging

made. The magnitude of the interactions between chains in a polymer depends on the nature of intermolecular bonding forces, the molecular weight, and the nature of the chain packing. The secondary bonding forces in polymers are the weak Van der Waals forces and the dipole-dipole interactions. The dipole-dipole interactions result from the presence of different functional groups on the polymer chains and from hydrogen bonding between chains. For nonpolar polymers such as polyethylene, the very weak intermolecular forces are compensated by a relatively high molecular weight and very close packing. Although the molecular weight of a polymer controls the polymer's flow properties, it does not affect other properties such as color, hardness, density, and electrical properties.

The nature of the chain packing is reflected by the extent of disorder in a polymer. Linear polymer chains pack together in disordered (amorphous) and ordered (crystal-line) fashions. Amorphous polymers such as poly(methyl methacrylate) consist of randomly coiled and entangled chains. In crystalline polymers the chains have a highly ordered orientation. Actually, polymers can be divided into two classes: those that are completely amorphous under all conditions, and those that are semicrystalline. The thermal behavior of both types of polymers varies and can be used to characterize each type. The term thermoplastic refers to any material that softens when heated. The following chart compares the thermoplastic characteristics of amorphous and crystalline polymers:

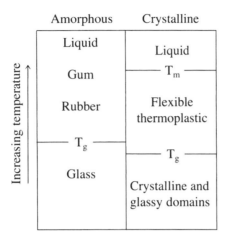

T_g is a transition point known as the glass transition temperature, and T_m is the melting temperature. The glass transition temperature is one of the fundamental properties of polymers. Below T_g the polymer is a hard glassy material, and above it the polymer is a viscous liquid. Above T_g amorphous polymers behave differently from crystalline polymers. As the temperature of an amorphous polymer is raised, the rubbery phase *gradually* becomes soft, then gummy, and finally turns into a liquid. No sharp transitions occur from one phase to another. In contrast, crystalline polymers retain their rubbery properties above T_g until the temperature reaches the melting temperature, T_m, at which point the material liquifies and loses its crystalline characteristics. The amorphous and crystalline behavior described here is characteristic of linear and branched polymers and copolymers, but is not generally exhibited by cross-linked polymers.

A few typical glass transition temperatures are

Polymer	T_g,°C
polystyrene	100
trans-poly-1,3-butadiene	−18
Nylon 66	57
poly(vinyl chloride)	81
polyacrylonitrile	105
cis-polyisoprene	−70
trans-polyisoprene	−50

For more detailed discussions of the preparation and properties of polymers, consult references 6–12.

SELECTED PROPERTIES OF INDUSTRIAL POLYMERS

In addition to the fourteen demonstrations in this chapter, we list here several suggestions and references to demonstrations which can be used to illustrate various properties of polymers. For example, chewing gum contains a rubbery material with a glass transition temperature, T_g, near the normal body temperature of 37°C. When chewing gum is removed from its wrapper and chewed, it becomes soft and rubbery. If an ice cube or ice-cold water is placed in the mouth along with the chewing gum, the gum becomes stiff and remains hard as long as the temperature is significantly lower than body temperature. If hot water or a hot drink is placed in the mouth along with the chewing gum, the gum becomes very soft and sticky.

Another way to show the difference in properties of a polymer above and below the glass transition temperature, T_g, is to freeze a piece of rubber, such as a bouncy ball, in liquid nitrogen (boiling point: −196°C). When the frozen rubber is struck against a hard surface, it shatters. A direct practical application of this property is now commercially used to recycle old automobile tires [13]. Tires are frozen in liquid nitrogen and crushed. The rubber component of the tire is separated from the other minor components and is then mixed with asphalt.

The observation that a piece of rubber contracts when heated and expands when cooled was reported by James Prescott Joule in the 19th century. When a rubber band is stretched and then released, the temperature change is easily detected by using one's cheek, forehead, or lips. Several variations of this procedure and other uses are described in references 14–21. Another experiment [22], which requires about 1 hour, deals with the swelling of a rubber band when covered with toluene.

To examine polymer properties due to crystallinity, two polarizing filters and a light source are required [23, 24]. The most convenient light source to use in a classroom is an overhead projector. Place one polarizer on the overhead projector. Dim the room lights and project the polarized light onto a screen. Hold the other polarizer about 10–15 cm directly above the first one. Rotate either filter and observe on the screen the position at which the filters appear dark. This is the *extinction position,* where the directions of polarization of the two filters are at right angles to each other. Take a transparent sample of polymer (about 5 cm × 5 cm), such as polyethylene, cellophane, or a piece from a plastic lunch bag, and hold it between the crossed polarizers in the extinction position. Rotate the polymer sample and observe the changes in brightness. The phenomenon of *birefringence* is characteristic of crystalline material. In general, the greater the degree of crystallinity in a polymer, the greater the brightness in crossed-polarized light. The angle of brightest appearance, which is the angle formed between the edge of the polymer sample and the extinction position, can be estimated. With the help of another person, stretch the polymer sample while it is in the position of maximum brightness. A striking display of colors, called *interference colors,* will then be observed. By stretching the polymer sample, a nonuniform change in polymer film thickness results. When placed between crossed polarizers, a film of nonuniform thickness exhibits interference colors. The successive sequence of interference colors are called *orders,* and they can be related quantitatively to the polymer thickness. Totally amorphous polymers, such as those prepared according to the procedures in this chapter, do not usually exhibit birefringence or interference colors.

The molecular weight of a polymer can be determined by viscometry. The viscosity of a polymer solution is related to the average molecular weight of the polymer. Molecular weights in the 200,000–400,000 range can be determined by using a simple viscometer [25].

Finally, we call attention to a colorful series of experiments dealing with polydiacetylenes [26], which Patel and Yang [27] claim to be suitable for students from high school to graduate school levels. Patel [28] presented a series of fourteen experiments using test tubes, filter paper, a hair dryer, and a 20-watt short ultraviolet lamp. The experiments revealed striking color changes during polymerization, degradation, phase transitions, dissolution, gelation, ion exchange, and conformation changes. The starting material,

$$CH_3(CH_2)_3NH\overset{O}{\overset{\|}{C}}O(CH_2)_4\text{-}C\equiv C\text{-}C\equiv C\text{-}(CH_2)_4O\overset{O}{\overset{\|}{C}}NH(CH_2)_3CH_3$$

can be synthesized as described by Patel [29].

A WORD OF CAUTION

Most of the chemicals in the following fourteen demonstrations are hazardous. Avoid inhalation of fumes and direct contact with skin. When called for, use a properly functioning fume hood and follow all directions carefully.

REFERENCES

1. Harris, F. W.; et al. *J. Chem. Educ.* **1981**, *58*, 836–955.
2. Mark, H. *J. Chem. Educ.* **1981**, *58*, 527.
3. Marvel, C. S. *J. Chem. Educ.* **1981**, *58*, 535.
4. Quirk, R. P. *J. Chem. Educ.* **1981**, *58*, 540.
5. Seymour, R. B. *J. Chem. Educ.* **1982**, *59*, 652.
6. Allcock, H. R.; Lampe, F. W. "Contemporary Polymer Chemistry"; Prentice-Hall: New York, 1981.
7. Seymour, R. B.; Carraher, C. E., Jr. "Polymer Chemistry: An Introduction"; Marcel Dekker: New York, 1981.
8. Stevens, M. P. "Polymer Chemistry: An Introduction"; Addison-Wesley Publishing Co.: Reading, Massachusetts, 1975.
9. Lenz, R. W. "Organic Chemistry of Synthetic High Polymers"; Interscience Publishers, John Wiley and Sons: New York, 1967.
10. Reich, L.; Schindler, A. "Polymerization by Organometallic Compounds"; Interscience Publishers, John Wiley and Sons: New York, 1966.
11. Sorenson, W. R.; Campbell, T. W. "Preparative Methods of Polymer Chemistry"; Interscience Publishers, John Wiley and Sons: New York, 1968.
12. Billmeyer, Fred W., Jr. "Textbook of Polymer Science"; Interscience Publishers, John Wiley and Sons: New York, 1971.
13. Braton, N. R. "Cryogenic Recycling and Processing"; CRC Press: Cleveland, Ohio, 1980.
14. Calingaert, G. *J. Chem. Educ.* **1952**, *29*, 405.
15. Siegelman, I.; Schmuckler, J. S. *J. Chem. Educ.* **1952**, *38*, 597.
16. Kouhoupt, R. *Popular Sci.* **1972**, 93.
17. Laswick, P. H. *J. Chem. Educ.* **1972**, *49*, 469.
18. DeLorenzo, R. *J. Chem. Educ.* **1973**, *50*, 124.
19. Smith, D. D. *J. Chem. Educ.* **1977**, *54*, 701.
20. Dole, M. *J. Chem. Educ.* **1977**, *54*, 754.
21. Bader, M. *J. Chem. Educ.* **1981**, *58*, 285.
22. Sperling, L. H.; Michael, T. C. *J. Chem. Educ.* **1982**, *59*, 651.
23. 48 cm × 48 cm filter pieces are available from many sources including the Edmund Scientific Co., 101 E. Gloucester Pike, Barrington, New York 08007, 1982 catalog number K72,219.

24. Alyea, H. N. *J. Chem. Educ.* **1967**, *44*, A273.
25. Shakhashiri, B. Z.; Dirreen, G. E. "Manual for Laboratory Investigations in General Chemistry"; Stipes Publishing Co.: Champaign, Illinois, 1982; p 228.
26. Krieger, J. H. *Chem. Eng. News* **1980**, *58*, 31; p 24.
27. Patel, G. N.; Yang, N. L. *J. Chem. Educ.* **1983,** *60*, 181.
28. Patel, G. N. Sixth International Conference on Chemical Education, University of Maryland, College Park, Maryland, August 1981. Author's affiliation: Allied Chemical Company, Allied Corporation, Syracuse Research Laboratory, Solvay, New York 13209.
29. Patel, G. N. *Polymer Preprints* **1978**, *19*, 154 (published by American Chemical Society Division of Polymer Chemistry).

3.1

Nylon 6-10

A film of nylon is formed at the interface between two immiscible liquids. When the film is lifted from the container, it is continually replaced forming a hollow thread of polymer. The continuous thread or "rope" of nylon can be wound on a windlass until one or the other of the two reactants is exhausted.

MATERIALS

50 ml 0.5M hexamethylenediamine (1,6-diaminohexane), $H_2N(CH_2)_6NH_2$, in 0.5M sodium hydroxide, NaOH (To prepare, dissolve 3.0 g of $H_2N(CH_2)_6NH_2$ plus 1.0 g NaOH in 50 ml distilled water. Hexamethylenediamine can be dispensed by placing the reagent bottle in hot water until sufficient solid has melted and can be decanted. The melting point is 39–40°C.)

50 ml 0.2M sebacoyl chloride, $ClCO(CH_2)_8COCl$, in hexane (To prepare, dissolve 1.5 ml to 2.0 ml sebacoyl chloride in 50 ml hexane.)

gloves, plastic or rubber

250-ml beaker or crystallizing dish

forceps

2 stirring rods or a small windlass

food-coloring dye (optional)

phenolphthalein (optional)

PROCEDURE

Wearing gloves, place the hexamethylenediamine solution in a 250-ml beaker or crystallizing dish. *Slowly* pour the sebacoyl chloride solution as a second layer on top of the diamine solution, taking care to minimize agitation at the interface. With forceps, grasp the polymer film that forms at the interface of the two solutions and pull it carefully from the center of the beaker. Wind the polymer thread on a stirring rod or a small windlass. Wash the polymer thoroughly with water or ethanol before handling.

Food coloring dyes or phenolphthalein can be added to the lower (aqueous) phase to enhance the visibility of the liquid interface. The upper phase can also be colored with dyes such as azobenzene [1], but observation of the polymer film at the interface is somewhat obscured. Some of the dye will be taken up with the polymer but can be removed by washing.

HAZARDS

Hexamethylenediamine (1,6-diaminohexane) is irritating to the skin, eyes, and respiratory system. Sodium hydroxide is extremely caustic and can cause severe burns. Contact with the skin and eyes must be prevented.

Sebacoyl chloride is corrosive and irritating to the skin, eyes, and respiratory system. Hexane is extremely flammable. Hexane vapor can irritate the respiratory tract and, in high concentrations, can be narcotic.

DISPOSAL

Any remaining reactants should be mixed thoroughly to produce nylon. The solid nylon should be washed before being discarded in a solid waste container.

Any remaining liquid should be discarded in a solvent waste container or should be neutralized with either sodium bisulfate (if basic) or sodium carbonate (if acidic) and flushed down the drain with water.

DISCUSSION

The word "nylon" is used to represent synthetic polyamides. The various nylons are described by a numbering system that indicates the number of carbon atoms in the monomer chains. Nylons from diamines and dibasic acids are designated by two numbers, the first representing the diamine and the second the dibasic acid [2]. Thus, 6-10 nylon is formed by the reaction of hexamethylenediamine and sebacic acid. In this demonstration the acid chloride, sebacoyl chloride, is used instead of sebacic acid. The equation is

$$H_2N(CH_2)_6NH_2 \ + \ Cl\overset{\overset{\displaystyle O}{\|}}{C}(CH_2)_8\overset{\overset{\displaystyle O}{\|}}{C}Cl \longrightarrow \left[\overset{\overset{\displaystyle H}{|}}{N}(CH_2)_6\overset{\overset{\displaystyle H}{|}}{N} - \overset{\overset{\displaystyle O}{\|}}{C}(CH_2)_8\overset{\overset{\displaystyle O}{\|}}{C} \right] \ + \ 2\ HCl$$

The method of reaction used in this demonstration has been termed interfacial polycondensation. This method is useful because it is a low temperature process, it is rapid even at room temperature, and it does not depend on exact stoichiometry of reactants [3].

Many diamines and diacids or diacid chlorides can be reacted to make other condensation products that are described by the generic name "nylon." One such product is an important commercial polyamide, nylon 6-6, which can be prepared by substituting adipoyl chloride [4] for sebacoyl chloride in the procedure described here. The equation is

$$H_2N(CH_2)_6NH_2 \ + \ Cl\overset{\overset{\displaystyle O}{\|}}{C}(CH_2)_4\overset{\overset{\displaystyle O}{\|}}{C}Cl \longrightarrow \left[\overset{\overset{\displaystyle H}{|}}{N}(CH_2)_6\overset{\overset{\displaystyle H}{|}}{N} - \overset{\overset{\displaystyle O}{\|}}{C}(CH_2)_4\overset{\overset{\displaystyle O}{\|}}{C} \right] \ + \ 2\ HCl$$

REFERENCES

1. Alyea, H. N.; Dutton, F. B., Eds. "Tested Demonstrations in Chemistry," 6th ed.; Journal of Chemical Education: Easton, Pennsylvania, 1965; p 136.
2. Ravve, A. "Organic Chemistry of Macromolecules"; Marcel Dekker: New York, 1967; Ch. 15.
3. Sorenson, W. R.; Campbell, T. W. "Preparative Methods of Polymer Chemistry"; Interscience Publishers, John Wiley and Sons: New York, 1968; pp 90–93.
4. Alyea, H. N.; and Dutton, F. B., Eds. "Tested Demonstrations in Chemistry," 6th ed.; Journal of Chemical Education: Easton, Pennsylvania, 1965; p 164.

3.2

Polyurethane Foam

When two viscous liquids are mixed, a rigid foam is produced, whose volume is 20–30 times that of the original mixture [1].

MATERIALS FOR PROCEDURE A

40 ml Part A and 40 ml Part B of a two-component polyurethane foam system [2]

gloves, plastic or rubber

2 50-ml beakers or paper cups

paper towels to cover area ca. 0.5 m × 0.5 m

1-liter beaker, preferably tall-form

2 stirring rods, ca. 25 cm

food-coloring dye (optional)

MATERIALS FOR PROCEDURE B

40 ml Part A and 40 ml Part B of a two-component polyurethane foam system [2]

200-ml disposable cup

paper towels to cover area ca. 0.5 m × 0.5 m

gloves, plastic or rubber

2 50-ml beakers or paper cups

stirring rod

food-coloring dye (optional)

PROCEDURE A

Perform this demonstration in a well-ventilated area.

Wearing gloves, place approximately 40 ml of each component in separate small beakers. The exact amount is not critical as long as equal amounts are used. Place the paper towels under the large beaker. Pour the contents of both beakers into the large beaker. With a stirring rod, mix the contents thoroughly. When the foam begins to expand, stop stirring. The volume will increase 20–30 times that of the original mixture. If carefully supported by two glass rods, the rigid foam will form a column about twice the height of the beaker.

Since the freshly prepared foam usually contains unreacted isocyanate, it should not be handled until it has cured several hours in a well-ventilated area.

Foams of different colors can be produced by adding a few drops of a food-coloring dye to the lighter-colored component prior to mixing with the other component.

PROCEDURE B

Place the 200-ml disposable cup in the center of a mat of paper towels 0.5 m × 0.5 m. Wearing gloves, mix equal amounts of Parts A and B in the disposable cup. When the foam begins to expand, stop stirring. As the mixture foams, it rises out of the cup, overflows, and forms a bell-shaped solid that adheres to the paper towels.

Because the freshly prepared foam usually contains unreacted isocyanate, it should not be handled until it has cured several hours in a well-ventilated area.

As in Procedure A, a few drops of a food-coloring dye can be added to the lighter-colored component.

HAZARDS

Isocyanates are irritants to the skin, eyes, and respiratory system. Toluene 2,4-diisocyanate, which has been used in many polyurethane systems, is reported to cause skin irritation, allergic reactions, and bronchial asthma [3, 4]. The foam product should be allowed to cure several hours in a well-ventilated area before handling.

DISPOSAL

Since the individual components are soluble in acetone, they can be dissolved and the solutions flushed down the drain with water. The rigid foam, when cured, can be discarded in a waste container.

The fully cured polyurethane is soluble in solvents such as dimethylformamide. Before it is fully cured, the foam can be removed from the beaker by using a spatula and small amounts of acetone.

DISCUSSION

The polyurethane foam system used in this demonstration consists of two viscous liquids. Part A, light amber in color, contains a polyether polyol, a blowing agent, a silicone surfactant, and a catalyst. Part B, which is dark, contains a polyfunctional isocyanate.

A polyurethane foam is formed by producing a polyurethane (polycarbamate) polymer in the presence of a fluorocarbon blowing agent. The polymer is formed by the reaction of a polyester polyol, HO-R-OH (whose probable molecular weight range is 400–4000), with a polyfunctional isocyanate, OCN-R'-NCO [for example, methylene bis(4-phenylisocyanate)]. Amine compounds or metal salts are used as catalysts. The reaction can be represented as:

$$\text{HO}-\text{R}-\text{OH} + \text{O}=\text{C}=\text{N}-\text{R}'-\text{N}=\text{C}=\text{O} \longrightarrow \left[\text{O}-\text{R}-\text{O}-\overset{\overset{\text{O}}{\|}}{\text{C}}-\overset{\overset{\text{H}}{|}}{\text{N}}-\text{R}'-\overset{\overset{\text{H}}{|}}{\text{N}}-\overset{\overset{\text{O}}{\|}}{\text{C}} \right]$$

The polyfunctional character of the reactants results in a high degree of cross-linking in the product, forming a rigid foam.

REFERENCES

1. Dirreen, G. E.; Shakhashiri, B. Z. *J. Chem. Educ.* **1977**, *54*, 431.
2. We have used a product, Super Foam, marketed by Edmund Scientific Co., 101 E. Gloucester Pike, Barrington, New Jersey 08007. Similar products may be available at paint, hardware, or lumber stores.
3. Hocking, M. B.; Canham, G. W. R. *J. Chem. Educ.* **1974**, *51*, A580.
4. Windholz, M., Ed. "The Merck Index," 9th ed.; Merck and Co.: Rahway, New Jersey, 1976; p 1225.

3.3

Phenol-Formaldehyde Polymer

Concentrated hydrochloric acid is added to a mixture of aqueous formaldehyde, glacial acetic acid, and phenol. After approximately one minute the solution turns pink and solidifies into a cake-like material.

MATERIALS

25 ml 12M formaldehyde, HCHO (Formalin, 37%)

55 ml glacial (17M) acetic acid, CH_3COOH

20 g phenol, C_6H_5OH

55 ml concentrated (12M) hydrochloric acid, HCl

gloves, plastic or rubber

250-ml beaker

100-ml graduated cylinder

stirring rod

PROCEDURE

Perform this demonstration in a hood and wear gloves.

In a 250-ml beaker, mix 25 ml of aqueous formaldehyde, 55 ml of glacial acetic acid, and 20 g of phenol [1]. Quickly add 55 ml of concentrated hydrochloric acid while stirring. After about a minute, the solution will turn pink and solidify. The polymerization is exothermic. If not stirred, the resulting polymer is friable, but continuous stirring during polymerization results in a solid mass more resistant to crumbling.

HAZARDS

All compounds used in this demonstration should be handled in a hood. Plastic or rubber gloves should be worn.

Formaldehyde vapors are extremely irritating to mucous membranes. Skin contact can cause dermatitis. Extended exposure to high concentrations of vapor can cause chronic effects such as laryngitis, bronchitis, conjunctivitis, or skin problems. Preliminary data "have indicated the development of nasal cancers in rats exposed to 15 ppm formaldehyde for 18 months" [2]. Mixing of formaldehyde with hydrogen chloride "could result in the generation of bis(chloromethyl)ether, a potent human carcinogen" [3].

Phenol is toxic and causes burns. It can be absorbed rapidly through the skin. Prolonged inhalation of phenol vapor can have chronic effects.

Both glacial acetic acid and concentrated hydrochloric acid can produce skin burns, eye irritation, and irritation of the respiratory tract.

DISPOSAL

The product contains acid and probably unreacted formaldehyde or phenol; before handling, it should be washed thoroughly with dilute aqueous sodium hydroxide solution and then water. The washed polymer should be discarded in a waste container.

DISCUSSION

In general, phenol-formaldehyde resins are condensation polymers consisting of aromatic rings linked by methylene groups. The product linkages occur primarily in the ortho or para position to the phenolic hydroxyls (see figure). In addition, ether linkages are found under conditions of excess formaldehyde and neutral pH.

Representative methylene and ether linkages in a phenol-formaldehyde polymer [4].

In this demonstration, prepared with a molar excess of formaldehyde and under acidic conditions, the product resin is formed by reactions that lead mostly to methylene bridges. The reaction sequence can be illustrated as follows [4]:

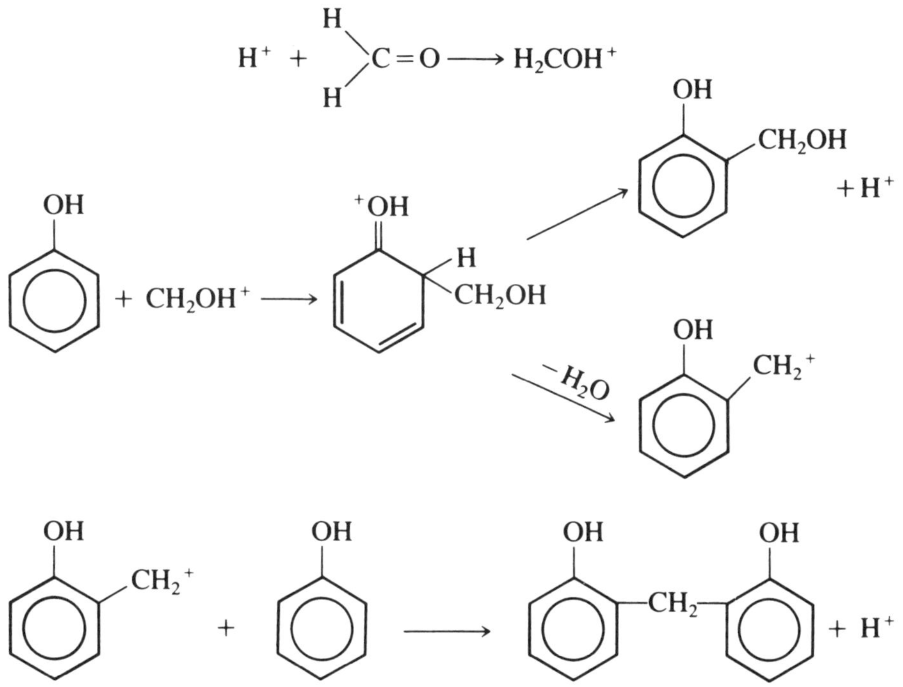

All possible bridges do not form, because all phenolic rings do not lie in the same plane. These gaps in the resin weaken the polymer. The presence of occluded reaction materials and by-products, such as water, causes further weakening.

REFERENCES

1. Wilson, A. S.; Peterson, V. R. *J. Chem. Educ.* **1978**, *55*, 652.
2. "Prudent Practices for Handling Hazardous Chemicals in Laboratories." *Natl. Res. Counc.* **1981**; Committee on Hazardous Substances in the Laboratory; p 131.
3. "Prudent Practices for Handling Hazardous Chemicals in Laboratories." *Natl. Res. Counc.* **1981**; Committee on Hazardous Substances in the Laboratory; p 30.
4. Ravve, A. "Organic Chemistry of Macromolecules"; Marcel Dekker: New York, 1967; Ch. 17.

3.4

Resorcinol-Formaldehyde Polymer

When several drops of potassium hydroxide or hydrochloric acid solution are added to a hot yellow solution containing resorcinol and formaldehyde, an exothermic reaction produces a brittle red solid.

MATERIALS FOR PROCEDURE A

6 g resorcinol (1,3-dihydroxybenzene), $C_6H_4(OH)_2$

10 ml distilled water

8 ml 12M formaldehyde, HCHO (Formalin, 37%)

1 ml 3M potassium hydroxide, KOH (To prepare 10 ml of solution, dissolve 1.7 g of KOH in water and dilute to 10 ml.)

gloves, plastic or rubber

150-ml beaker

10-ml graduated cylinder

hot plate

dropper

MATERIALS FOR PROCEDURE B

6 g resorcinol (1,3-dihydroxybenzene), $C_6H_4(OH)_2$

10 ml distilled water

8 ml 12M formaldehyde, HCHO (Formalin, 37%)

2 ml concentrated (12M) hydrochloric acid, HCl

gloves, plastic or rubber

150-ml beaker

10-ml graduated cylinder

dropper

PROCEDURE A

Perform this demonstration in a hood and wear gloves.

Place 6 g of resorcinol, 10 ml of distilled water, and 8 ml of formaldehyde solution in a beaker [1]. Heat the mixture to boiling and remove from the heat source. Immediately add 5–10 drops of 3M potassium hydroxide. Caution: spattering may occur. Within a few seconds an exothermic reaction produces a red, glassy polymer.

Gas evolution may cause the polymer surface to bubble and become porous. The polymer will darken as it cures and become very brittle. Wash the polymer thoroughly with water before handling.

PROCEDURE B

Perform this demonstration in a hood and wear gloves.

Place 6 g of resorcinol, 10 ml of distilled water, and 8 ml of formaldehyde solution in a beaker. Slowly add, by drops, 2 ml of concentrated hydrochloric acid solution. The resulting polymer is pink.

HAZARDS

All compounds used in this demonstration should be handled in a hood. Plastic or rubber gloves should be worn.

Resorcinol is an irritant to the skin and eyes and can be absorbed through the skin.

Formaldehyde vapors are extremely irritating to mucous membranes. Skin contact can cause dermatitis. Extended exposure to high concentrations of vapor can cause chronic effects such as laryngitis, bronchitis, conjunctivitis, or skin problems. Preliminary data "have indicated the development of nasal cancers in rats exposed to 15 ppm formaldehyde for 18 months" [2]. Mixing of formaldehyde with hydrogen chloride "could result in the generation of bis(chloromethyl)ether, a potent human carcinogen" [3].

Dust from solid potassium hydroxide is very caustic. Hydrochloric acid vapors are extremely irritating to the skin, eyes, and respiratory system. Potassium hydroxide and hydrochloric acid can both cause severe burns of the skin and eyes.

DISPOSAL

The polymer should be washed with water and discarded in a waste container.

DISCUSSION

The reaction of resorcinol (1,3-dihydroxybenzene) with formaldehyde is very similar to the reaction of phenol with formaldehyde (see Demonstration 3.3). Substitution in the meta position by an ortho-para directing group enhances the rate of the reaction. Phloroglucinol (1,3,5-trihydroxybenzene) would react still more rapidly.

Catalysis by a base, as in this demonstration, has been explained by a mechanism in which a phenoxide ion undergoes electrophilic attack by formaldehyde at the ortho or para position [4]:

The reaction can also be catalyzed by hydrochloric acid, in which case the polymerization occurs at room temperature and the product is similar in consistency to the phenol-formaldehyde polymer.

REFERENCES

1. Alyea, H. N.; Dutton, F. B., Eds. "Tested Demonstrations in Chemistry," 6th ed.; Journal of Chemical Education: Easton, Pennsylvania, 1965; p 110.
2. "Prudent Practices for Handling Hazardous Chemicals in Laboratories." *Natl. Res. Counc.* **1981**; Committee on Hazardous Substances in the Laboratory; p 131.
3. "Prudent Practices for Handling Hazardous Chemicals in Laboratories." *Natl. Res. Counc.* **1981**; Committee on Hazardous Substances in the Laboratory; p 30.
4. Ravve, A. "Organic Chemistry of Macromolecules"; Marcel Dekker: New York, 1967; Ch. 17.

3.5

Aniline Hydrochloride–Formaldehyde Polymer

Solutions of aniline hydrochloride and formaldehyde are mixed to produce a red gelatin-like polymer.

MATERIALS

15 ml aniline (aminobenzene), $C_6H_5NH_2$

20 ml 6M hydrochloric acid, HCl (To prepare 20 ml of solution, dilute 10 ml of concentrated [12M] HCl to 20 ml with distilled water.)

30 ml 12M formaldehyde, HCHO (Formalin, 37%)

gloves, plastic or rubber

100-ml beaker

50-ml graduated cylinder

250-ml beaker

stirring rod

PROCEDURE

Perform this demonstration in a hood and wear gloves.

Prepare aniline hydrochloride solution in a 100-ml beaker by adding 15 ml of aniline to 20 ml of 6M HCl solution. Either allow the solution to cool to room temperature before use, or prepare the solution in an ice bath and then allow it to warm up to room temperature.

Place 30 ml of formaldehyde solution in a 250-ml beaker. Quickly add the aniline hydrochloride solution with stirring. Within seconds, the mixture turns red and solidifies into a rubbery mass. The polymerization reaction is exothermic, producing a temperature increase to 40–50°C. As a result the surface of the polymer sometimes expands and becomes porous.

HAZARDS

All compounds used in this demonstration should be handled in a hood. Plastic or rubber gloves should be worn.

Aniline is toxic by inhalation, ingestion, or absorption through the skin. Long exposure to the vapor can have chronic effects.

Hydrochloric acid can produce skin burns, eye irritation, and irritation of the respiratory tract.

Formaldehyde vapors are extremely irritating to mucous membranes. Skin contact can cause dermatitis. Extended exposure to high concentrations of vapor can cause chronic effects such as laryngitis, bronchitis, conjunctivitis, or skin problems. Preliminary data "have indicated the development of nasal cancers in rats exposed to 15 ppm formaldehyde for 18 months" [1]. Mixing of formaldehyde with hydrogen chloride "could result in the generation of bis(chloromethyl)ether, a potent human carcinogen" [2].

DISPOSAL

The rubbery polymer can be removed from the beaker with a spatula. The polymer is likely to be wet with unreacted starting materials and water, and should not be touched until it is rinsed with water and allowed to dry. The dry brittle solid can be discarded in a waste container.

DISCUSSION

Aniline-formaldehyde resins are condensation polymers consisting of aromatic rings linked by methylene groups [3]. The product linkages occur primarily in the ortho or para position to the amine group. The reaction of aniline hydrochloride with formaldehyde [4] is analogous to the reaction of phenol with formaldehyde in an acidic medium (see Demonstration 3.3):

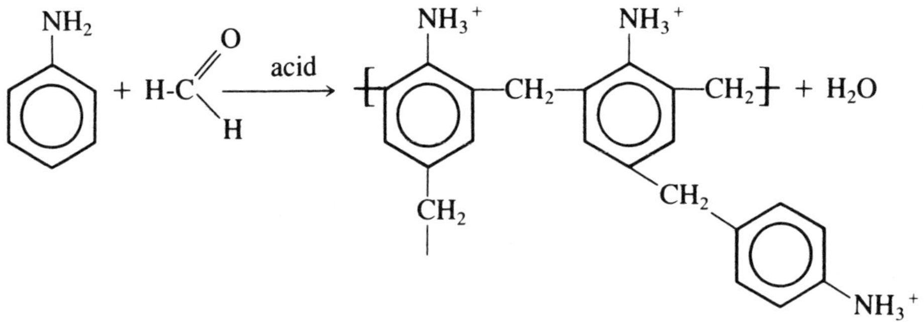

REFERENCES

1. "Prudent Practices for Handling Hazardous Chemicals in Laboratories." *Natl. Res. Counc.* **1981**; Committee on Hazardous Substances in the Laboratory; p 131.
2. "Prudent Practices for Handling Hazardous Chemicals in Laboratories." *Natl. Res. Counc.* **1981**; Committee on Hazardous Substances in the Laboratory; p 30.
3. Ravve, A. "Organic Chemistry of Macromolecules"; Marcel Dekker: New York, 1967; Ch. 17.
4. Walker, J. F. "Formaldehyde," 3rd ed.; Reinhold Book Corp.: New York, 1964; p 370.

3.6

Urea-Formaldehyde Polymer

After several drops of concentrated sulfuric acid are added to a clear solution containing formaldehyde and urea, an exothermic reaction produces a white polymer.

MATERIALS

20 ml 12M formaldehyde, HCHO (Formalin, 37%)

10 g urea, H_2NCONH_2

1 ml concentrated (18M) sulfuric acid, H_2SO_4

gloves, plastic or rubber

100-ml beaker

dropper

stirring rod

PROCEDURE

Perform this demonstration in a hood and wear gloves.

Dissolve 10 g of urea in 20 ml formaldehyde solution. This solution will be nearly saturated. Add 10 drops (approximately ½ ml) of concentrated sulfuric acid. Caution: spattering may occur. In a few seconds a vigorous exothermic reaction produces a white polymeric solid.

HAZARDS

The demonstration should be performed in a hood. Plastic or rubber gloves should be worn.

Formaldehyde vapors are extremely irritating to mucous membranes. Skin contact can cause dermatitis. Extended exposure to high concentrations of vapor can cause chronic effects such as laryngitis, bronchitis, conjunctivitis, or skin problems. Preliminary data "have indicated the development of nasal cancers in rats exposed to 15 ppm formaldehyde for 18 months" [1].

Since concentrated sulfuric acid is a strong acid and a powerful dehydrating agent, it must be handled with great care. Spills should be neutralized with an appropriate agent, such as sodium bicarbonate ($NaHCO_3$), and then wiped up.

DISPOSAL

With some difficulty, the polymer can be scraped from the beaker. The polymer should then be washed with water and discarded in a waste container.

DISCUSSION

Aldehydes undergo condensation reactions with urea and with amines such as melamine and guanidine. The products are referred to as aminoplasts [2]. The reaction mechanism can be considered as an electrophilic attack by formaldehyde on a nucleophilic urea. The structure of the product contains both linear and branched segments:

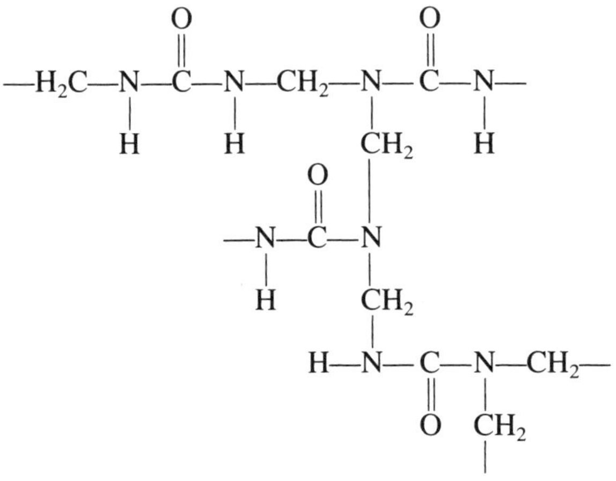

REFERENCES

1. "Prudent Practices for Handling Hazardous Chemicals in Laboratories." *Natl. Res. Counc.* **1981**; Committee on Hazardous Substances in the Laboratory; p 131.
2. Ravve, A. "Organic Chemistry of Macromolecules"; Marcel Dekker: New York, 1967; Ch. 18.

3.7

Phenolphthalein–Terephthaloyl Chloride Polymer

A dark red solution containing phenolphthalein and sodium hydroxide is vigorously stirred with a magnetic stirrer. When a clear solution of terephthaloyl chloride in 1,2-dichloroethane or acetone is added, the red color gradually fades and a white polymeric suspension forms [1].

MATERIALS

200 ml distilled water

3 g phenolphthalein, $C_{20}H_{14}O_4$

0.1 g sodium hydroxide, NaOH

2.1 g terephthaloyl chloride, $C_6H_4(COCl)_2$

30 ml 1,2-dichloroethane, $ClCH_2CH_2Cl$, or acetone, CH_3COCH_3

gloves, plastic or rubber

400-ml beaker

magnetic stirrer and stirring bar

50-ml beaker

PROCEDURE

Perform this demonstration in a hood and wear gloves.

Put 200 ml of water in the 400-ml beaker and add 3 g of phenolphthalein and 1 or 2 pellets (0.1 g) of sodium hydroxide. A deep violet-red suspension results. Place the beaker on a magnetic stirrer and stir vigorously. Add to this mixture a solution of 2.1 g of terephthaloyl chloride in 30 ml of 1,2-dichloroethane. As the reaction proceeds, the color fades and a white polymer precipitates. To achieve a more rapid color change, dissolve the terephthaloyl chloride in 30 ml of acetone instead of 1,2-dichloroethane. After about 5 minutes of stirring, turn off the stirrer and allow the precipitate to settle.

HAZARDS

The demonstration should be performed in a hood. Plastic or rubber gloves should be worn.

Terephthaloyl chloride is corrosive and an irritant to skin and eyes.

Vapors of 1,2-dichloroethane are highly flammable and are irritating to the eyes and respiratory tract. Dichloroethane is a suspected carcinogen.

DISPOSAL

After the solid has settled, decant the solution and rinse the solid with water. Collect the finely divided solid on filter paper and discard in a waste container.

DISCUSSION

When the polymerization is performed with 1,2-dichloroethane as the organic solvent, the demonstration is an example of an interfacial condensation reaction. The equation is

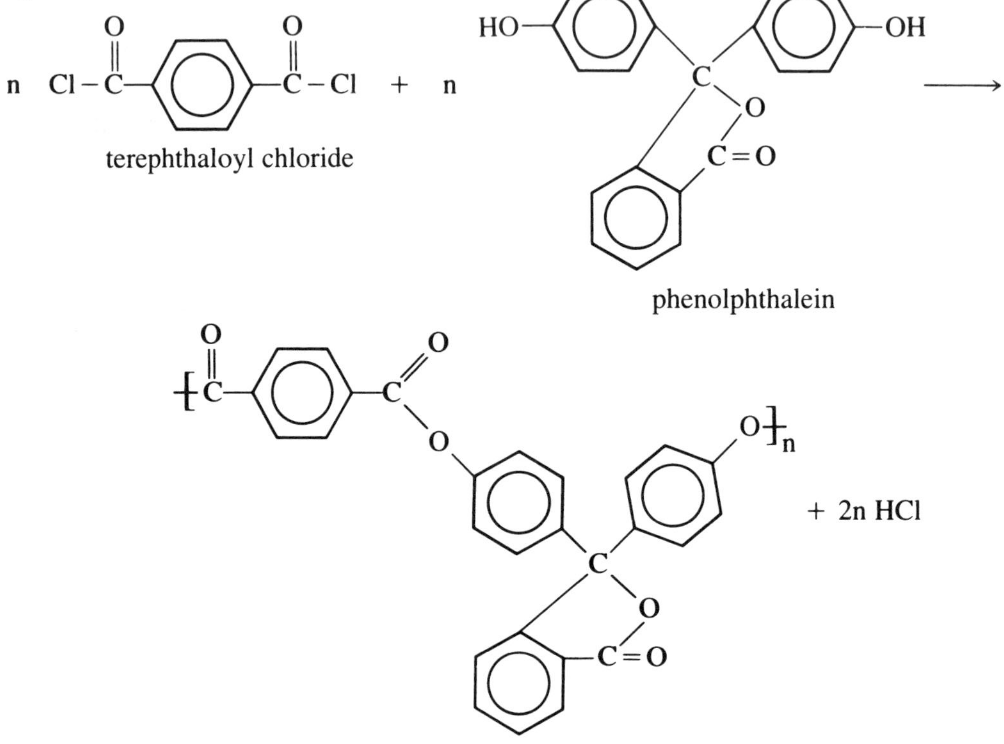

terephthaloyl chloride

phenolphthalein

+ 2n HCl

The reaction is interesting for the color change. The color change of the phenophthalein indicator is

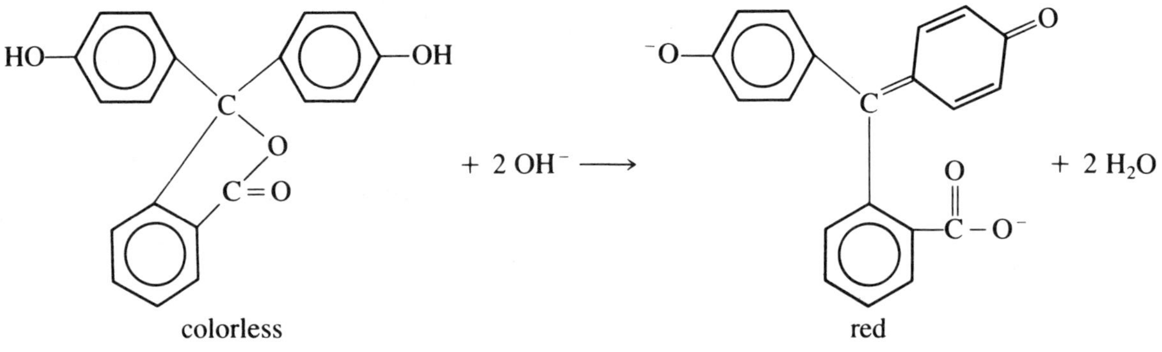

colorless

+ 2 OH⁻ ⟶

red

+ 2 H₂O

REFERENCE

1. McCaffery, E. L. "Laboratory Preparation for Macromolecular Chemistry"; McGraw-Hill: New York, 1970; p 109.

3.8

Polybutadiene
(Jumping Rubber)

A violet suspension of an "alfin" catalyst is added to a bottle containing a solution of 1,3-butadiene in pentane. The bottle is corked and shaken for several seconds. The mixture sets to a gel, and within 2 minutes the contents erupt from the bottle [1].

MATERIALS

Jumping Rubber kit [2]. The kit includes:
    wax-sealed, screw-capped vial containing the alfin catalyst
    small sealed bottle containing 1,3-butadiene dissolved in dry pentane
    cork for bottle

gloves, plastic or rubber

bottle opener

stirring rod

tongs

PROCEDURE

Warning! The alfin catalyst is a fire hazard.
Perform this demonstration in a well-ventilated room and wear gloves.

Open the catalyst vial and stir its contents with the glass rod. Remove the crown seal from the bottle. Quickly add *all* the catalyst to the bottle. Immediately cork the bottle and shake. Do not point the corked bottle at anyone. The temperature increases slightly (to about 50°C), and the pressure increases (perhaps 2–3 atmospheres) until, within 2 minutes, the cork is forced from the mouth of the bottle and a polymer "snake" shoots out. The bottle is left almost dry, and the liquid is trapped in the polymer. Use a pair of tongs to handle the polymer. During the next hour or so, the trapped pentane will evaporate and the polymer will shrink.

HAZARDS

This dramatic demonstration should be performed only by individuals who understand the fire hazard and reactivity of organosodium reagents. We have performed this demonstration over three hundred times without a single accident. Under hot or extremely humid conditions, the catalyst could ignite in air or ignite the butadiene-pentane solution. A carbon dioxide fire extinguisher must be available.

Pentane is a highly flammable and volatile liquid (boiling point: 36°C), which can explode when exposed to heat, sparks, or flame. Avoid inhalation of pentane vapors since they are slightly toxic. In high concentrations pentane is a narcotic.

The compound 1,3-butadiene (boiling point: $-5°C$) can be irritating to skin and mucous membranes and is a narcotic in high concentrations.

DISPOSAL

Since pentane is trapped in the product, the polymer should be kept away from flames. After several hours the pentane will evaporate, and the shrunken polymer can be discarded.

The empty screw-capped vial should be rinsed carefully with water and discarded.

DISCUSSION

Alfin catalysts are a class of heterogeneous catalysts which cause rapid polymerization of butadiene, isoprene, and other monomers resulting in polymers with very high molecular weights (1–2 million or higher). The alfin catalyst used in this demonstration is a solid surface catalyst developed by A. A. Morton [3–5] and co-workers. It is produced by reacting amyl chloride ($C_5H_{11}Cl$) with sodium, which is then reacted with isopropyl alcohol [$(CH_3)_2CHOH$]. The resulting mixture contains amylsodium ($C_5H_{11}Na$), sodium isopropoxide [$(CH_3)_2CHONa$], and sodium chloride. Propylene ($CH_2=CHCH_3$) is added to produce allylsodium ($CH_2=CHCH_2Na$) from amylsodium. The following sequence shows the necessary stoichiometry [6]:

$$1.5\ C_5H_{11}Cl + 3\ Na \longrightarrow 1.5\ C_5H_{11}Na + 1.5\ NaCl$$

$$1.5\ C_5H_{11}Na + (CH_3)_2CHOH \longrightarrow 0.5\ C_5H_{11}Na + (CH_3)_2CHONa + C_5H_{12}$$

$$0.5\ C_5H_{11}Na + 0.5\ CH_2=CHCH_3 \longrightarrow 0.5\ CH_2=CHCH_2Na + 0.5\ C_5H_{12}$$

The sodium isopropoxide-allylsodium combination gives the highest polymer yield. The role of the sodium chloride is not clear, although it could be acting as a support for the catalyst.

The catalyst is the mixture of the three sodium compounds: allyl sodium, sodium isopropoxide, and sodium chloride. All are essential constituents. The name "alfin" was derived from the words "alcohol" and "olefin" because both are involved in the preparation of the reagent.

The alfin catalyst is believed [7, 8] to adsorb and orient the monomer prior to the chain-growth process. Surface effects presumably influence the stereochemistry [6], and the polymer consists mainly of trans-1,4 repeating units. Since one allyl group is incorporated into each polymer chain, the process resembles Ziegler-Natta catalysis [6]:

trans-1,4-polybutadiene

The predominance of 1,4-polymerization has been suggested as evidence for a free radical propagation mechanism. It has been proposed [6] that complexes such as

$$CH_3-CH-CH_3 \qquad\qquad CH_3-CH-CH_3$$

are formed and that the adsorbed monomer displaces the allyl anion from the complex to form an ion pair, which then reacts to form a radical pair:

$$[CH_2{\cdots}CH{\cdots}CH{\cdots}CH_2Na]^+ \; {}^-[CH_2{\cdots}CH{\cdots}CH_2] \longrightarrow$$
$$\cdot CH_2-CH=CH-CH_2{}^-Na^+ \; + \; \cdot CH_2-CH=CH_2$$

The radical anion initiates polymerization, which continues until combination with an allyl radical occurs. This combination does not occur very readily because the allyl radical is bound to the catalyst surface [6], and hence very high molecular weights are obtained.

An anionic mechanism for polymerization with the alfin catalyst has been proposed [9]. According to this mechanism, monomer molecules are inserted into the chain:

The references cited include further discussion of these and other hypotheses.

REFERENCES

1. Shakhashiri, B. Z.; Dirreen, G. E.; Williams, L. G. *J. Chem. Educ.* **1980**, *57*, 738.
2. The Jumping Rubber Kit can be obtained from Organometallics, Inc., Route 111, East Hampstead, New Hampshire 03826.
3. Morton, A. A.; Magat, E.; Letsinger, R. L. *J. Am. Chem. Soc.* **1947**, *69*, 950.
4. Morton, A. A.; Welcher, R. P.; Collins, F.; Penner, S. E.; Combs, R. D. *J. Am. Chem. Soc.* **1949**, *71*, 487.
5. Sorenson, W. R.; Campbell, T. W. "Preparative Methods of Polymer Chemistry," 2nd ed.; Interscience Publishers, John Wiley and Sons: New York, 1968; pp 305–7.
6. Stevens, M. P. "Polymer Chemistry: An Introduction"; Addison-Wesley Publishing Co.: Reading, Massachusetts, 1975; pp 179–80.
7. Morton, A. A.; Lanpher, E. J. *J. Polymer Sci.* **1960**, *44*, 233. (Includes bibliography.)
8. Ravve, A. "Organic Chemistry of Macromolecules"; Marcel Dekker: New York, 1967; p 125.
9. Reich, L.; Schindler, A. "Polymerization by Organometallic Compounds"; Interscience Publishers, John Wiley and Sons: New York, 1966; Ch. 5.

3.9

Poly(methyl acrylate)

Methyl acrylate in an aqueous emulsion polymerizes in the presence of a free radical catalyst. The polymer is coagulated in a concentrated solution of sodium chloride to yield a white product. Using an acetone solution, a film can be prepared [1].

MATERIALS

300 ml 0.6M aqueous (5%) methyl acrylate, $CH_2 = CHCOOCH_3$ (To prepare, mix 16 ml of methyl acrylate with 284 ml of water.)

5 ml 0.1M potassium bromate, $KBrO_3$ (To prepare 25 ml of solution, dissolve 0.4 g of $KBrO_3$ in water and dilute to 25 ml.)

5 ml 0.45M sodium hydrogen sulfite, $NaHSO_3$ (freshly prepared) (To prepare 10 ml of solution, dissolve 0.47 g of $NaHSO_3$ in water and dilute to 10 ml.)

300 ml 5M sodium chloride, NaCl (To prepare, dissolve 88 g of NaCl in water and dilute to 300 ml.)

10 ml acetone, CH_3COCH_3

gloves, plastic or rubber

1-liter Erlenmeyer flask

10-ml graduated cylinder

1-liter beaker

50-ml beaker

2 stirring rods

polyethylene sheet, ca. 15 cm $\times$ 15 cm

PROCEDURE

Perform this demonstration in a hood and wear gloves.

Place 300 ml of 5% methyl acrylate emulsion in a 1-liter Erlenmeyer flask. Add 5 ml of 0.1M potassium bromate and 5 ml of 0.45M sodium hydrogen sulfite. Swirl the flask to mix the contents thoroughly and allow the reaction to proceed for about 15 minutes, shaking the flask occasionally during this time. As the particles grow, the mixture quickly takes on a milky appearance. Pour the emulsion into about 300 ml of 5M sodium chloride solution to coagulate the polymer. Remove the polymer mass. Wash it thoroughly and knead it under water to remove salt and any unreacted monomer. Repeat once or twice with additional fresh water.

To prepare a film, tear or cut the white poly(methyl acrylate) into small pieces and dissolve it in acetone (about 1 g/10 ml). Stir the mixture with a glass rod. The

235

resulting liquid should be thick and viscous. Pour this liquid onto a piece of polyethylene film and allow it to dry for several hours. The film can then be peeled from the polyethylene surface.

HAZARDS

The demonstration should be performed in a hood. Plastic or rubber gloves should be worn.

Methyl acrylate (2-propenoic acid methyl ester) has an acrid odor and is a lachrymator. The monomer is highly irritating to eyes, skin, and mucous membranes. If large concentrations of the vapor are inhaled, they can cause lethargy and convulsions. Because methyl acrylate and acetone are extremely flammable, this demonstration should be performed away from flames or other sources of heat.

DISPOSAL

The washed polymer should be discarded in a waste container.

DISCUSSION

Poly(methyl acrylate) is an addition polymer. The reaction can be represented as:

$$n\ CH_2=CH \xrightarrow[\text{H}_2\text{O}]{\text{free radical catalyst}} \left[CH_2-CH_2\right]_n$$

with pendant groups:
$$\begin{array}{c} | \\ C=O \\ | \\ O \\ | \\ CH_3 \end{array} \qquad \begin{array}{c} | \\ C=O \\ | \\ O \\ | \\ CH_3 \end{array}$$

Because the high rate and heat of polymerization of acrylates make control of bulk polymerization impractical, the most important method of preparation is by emulsion polymerization [2]. Water serves as a moderator and a solvent for the catalyst system. The free radical catalyst is produced by the reaction of bromate ions and hydrogen sulfite ions and is known to produce hydroxyl radicals ($\cdot$OH) and hydrogen sulfite radicals (HSO$_3\cdot$). Other water soluble catalysts, such as ammonium peroxydisulfate or potassium peroxydisulfate (often called persulfate), can be used, but then the mixture must be heated to generate free radicals.

REFERENCES

1. Shakhashiri, B. Z.; Dirreen, G. E. "Manual for Laboratory Investigations in General Chemistry"; Stipes Publishing Co.: Champaign, Illinois, 1982; p 214.
2. Saunders, K. J. "Organic Polymer Chemistry"; Chapman & Hall Ltd.: London, 1973; Ch. 6.

3.10

Poly(methyl methacrylate)

A colorless liquid, methyl methacrylate, is polymerized into a transparent solid by the action of heat and an initiator, benzoyl peroxide. Two alternative procedures yield slightly different products.

MATERIALS FOR PROCEDURE A

10 ml methyl methacrylate, $H_2C = C(CH_3)COOCH_3$

0.05 g benzoyl peroxide, $(C_6H_5CO)_2O_2$

2 g alumina, Al_2O_3

gloves, plastic or rubber

2 600-ml beakers

hot plate

thermometer, $-10°C$ to $+110°C$

test tube, ca. 16 mm $\times$ 150 mm

50-ml beaker

60-mm funnel

11-cm filter paper

MATERIALS FOR PROCEDURE B

dental-grade self-curing acrylic (available from dental suppliers)

gloves, plastic or rubber

10-ml graduated cylinder

disposable cup or small beaker

MATERIALS FOR PROCEDURE C

100 g poly(methyl methacrylate), low molecular-weight beads

1 g benzoyl peroxide, $(C_6H_5CO)_2O_2$

100 ml methyl methacrylate, $H_2C = C(CH_3)COOCH_3$

2 ml N,N-dimethyl-p-toluidine, $CH_3C_6H_4N(CH_3)_2$

gloves, plastic or rubber

2 150-ml beakers

10-ml graduated cylinder

100-ml graduated cylinder

2 stirring rods

disposable cup or small beaker

PROCEDURE A

Perform this demonstration in a hood and wear gloves.

Heat about 400 ml of water to 80–90°C. In a test tube, place about 10 ml of methyl methacrylate, add 0.05 g of benzoyl peroxide, and mix the two reactants thoroughly to dissolve the peroxide. Place the tube in the hot water for about 10 minutes, then transfer the tube to a second beaker of water at 50–60°C. Allow the tube to remain in the bath until the polymer has hardened, approximately an hour. The hardened polymer can be removed by breaking the test tube.

Larger amounts of methyl methacrylate can be polymerized with correspondingly larger quantities of benzoyl peroxide, but local heating caused by the high heat of polymerization often causes bubbles to form in the casting. (The boiling point of the monomer is 100°C.)

If polymerization does not occur, the inhibitor must be removed from the monomer. To do so, slurry about 15 ml of methyl methacrylate with alumina in a small beaker. Filter the monomer into a test tube and repeat the procedure described above.

PROCEDURE B

A dental-grade self-curing acrylic is readily available from dental supply houses. Dental laboratories routinely use the two-component system for repairing and relining dentures and for making various orthodontic appliances. The materials are also sold as a heat-curing acrylic system. They are either pink (flesh-colored) or clear (translucent).

The liquid component contains methyl methacrylate monomer, a small amount of an inhibitor, and a small percentage of a material that acts as an accelerator during the polymerization reaction. The solid component contains poly(methyl methacrylate), an initiator such as benzoyl peroxide, and coloring agents, if desired.

By varying the ratio of monomer to polymer, the setting time can be increased or decreased. For one brand, Duz-all [1], a mixture of 3 ml of the liquid component with 5 g of the solid produced a viscous mass, which set in about 10 minutes at room temperature.

PROCEDURE C

In this procedure we use a two-phase system similar to the commercial self-curing acrylic [2]. The solid phase is a mixture of benzoyl peroxide and poly(methyl methacrylate) beads in a ratio of 1:100. The liquid phase is a mixture of N,N-dimethyl-p-toluidine and methyl methacrylate monomer in a ratio of 2:100.

To prepare a stock mixture of the solid phase, thoroughly mix 1 g of benzoyl peroxide with 100 g of poly(methyl methacrylate) beads. To prepare a stock solution of the liquid phase, add 2 ml of N,N-dimethyl-p-toluidine to 100 ml of methyl methacrylate monomer. It is not necessary to remove the inhibitor that is present in the monomer.

In a disposable cup or small beaker, prepare a slurry of 5 g of the solid component and 3 ml of the liquid component. Stir well. The polymer will set in about 15 minutes.

HAZARDS

The demonstration should be performed in a hood. Plastic or rubber gloves should be worn.

Methyl methacrylate is flammable and its vapors can be harmful. Skin contact can produce irritation.

Benzoyl peroxide (dibenzoyl peroxide) is highly flammable and shock sensitive. It should not be stored in screw-capped bottles, because the friction produced by opening the bottle could cause an explosion.

DISPOSAL

Although the polymer will ignite in a direct flame, it is otherwise inert and should be discarded in a waste container.

DISCUSSION

Methyl methacrylate can be polymerized by bulk, solution, suspension or emulsion techniques. Bulk polymerization is the predominant commercial method [3]. The reaction is

The initiator, benzoyl peroxide, is cleaved by heating into two benzoyl radicals:

Poly (methyl methacrylate) is sold under such trade names as Lucite and Plexiglas and in numerous acrylic paints and sprays. It is also extensively used in the production of dentures.

REFERENCES

1. Duz-all is available from Coralite Dental Products, Chicago, Illinois.
2. Sandler, S. R.; Karo, W. "Polymer Syntheses"; Academic Press: New York, 1974; Vol. I, p 288.
3. Saunders, K. J. "Organic Polymer Chemistry"; Chapman & Hall Ltd.: London, 1973; Ch. 6.

3.11

Polystyrene

Clear, liquid styrene is mixed with an initiator, benzoyl peroxide, and the mixture is then heated in a boiling water bath. The styrene slowly polymerizes to form a transparent solid.

MATERIALS

20 ml styrene, $C_6H_5CH = CH_2$

3 g alumina, Al_2O_3

0.5 g benzoyl peroxide, $(C_6H_5CO)_2O_2$

gloves, plastic or rubber

2 800-ml beakers

stirring rod

11-cm filter paper

60-mm funnel

test tube, 16 mm $\times$ 150 mm, with stopper

hot plate

PROCEDURE

Perform this demonstration in a hood and wear gloves.

To remove the inhibitor from the styrene, place 20 ml of styrene in a small beaker and add 2–3 g of alumina. Stir the mixture well to form a slurry. Filter the mixture through filter paper supported in a small funnel.

Place the styrene in a test tube. Add 0.5 g of benzoyl peroxide; stopper and shake the tube well to dissolve the initiator. Remove the stopper and heat the tube in a boiling water bath for 10–20 minutes. The styrene will become viscous with heating.

If you prefer a mold other than the test tube, check the viscosity perodically during the heating period. When the styrene is still liquid enough to pour, transfer it to another mold to harden. Hardening may take a day or more. Alternatively, if the mold and styrene are placed in an oven at approximately 100°C, the styrene should harden in an hour or two.

You can embed objects in the polystyrene matrix or make castings by keeping the temperature of the mixture at approximately 50°C for several days [1].

HAZARDS

The demonstration should be performed in a hood. Plastic or rubber gloves should be worn.

Styrene is flammable, and it is irritating to the eyes and mucous membranes. In high concentrations, it can be narcotic.

Benzoyl peroxide (dibenzoyl peroxide) is highly flammable and shock sensitive. It should not be stored in screw-capped bottles, because the friction produced by opening the bottle could cause an explosion.

DISPOSAL

Although the polymer will ignite in a direct flame, it is otherwise inert and should be discarded in a waste container.

DISCUSSION

Free radical polymerization of styrene can be carried out by bulk, solution, suspension, or emulsion techniques. The reaction equation can be represented as:

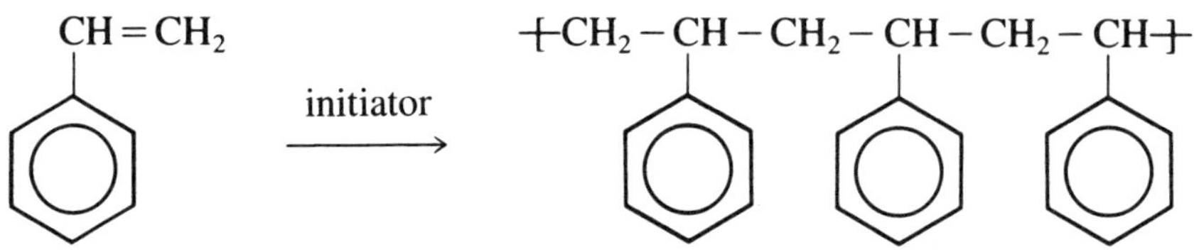

The arrangement of styrene in the polymer is considered to be "head-to-tail."

The initiator, benzoyl peroxide, is cleaved by heat into two benzoyl radicals:

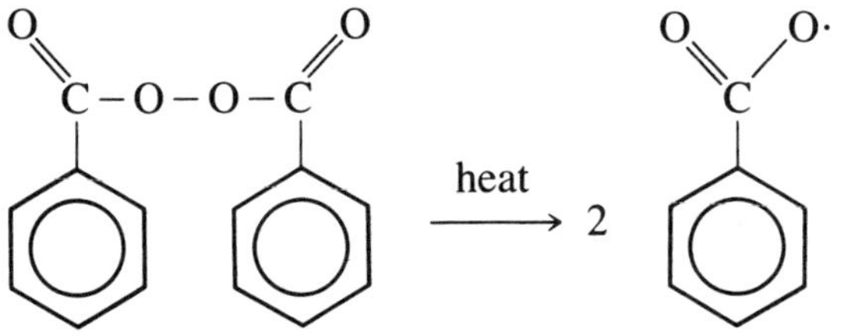

Polystyrene is one of the major commercial plastics widely used in appliances, toys, food containers, packaging, etc. It is a good electrical insulator. Many copolymers of styrene with other monomers have been investigated, and some, such as styrene-acrylonitrile copolymers, are commercially important [2].

REFERENCES

1. Sorenson, W. R.; Campbell, T. W. "Preparative Methods of Polymer Chemistry," 2nd ed.; Interscience Publishers, John Wiley and Sons: New York, 1968; pp 218–19.
2. Saunders, K. J. "Organic Polymer Chemistry"; Chapman & Hall Ltd.: London, 1973; Ch. 3.

3.12

Sulfur Polymer
or Plastic Sulfur

When elemental sulfur is melted and then poured into water, the resulting rubbery mass remains flexible for a varying period of time but eventually returns to the brittle, crystalline form [1].

MATERIALS

50 g sulfur, S_8

test tube, 25 mm $\times$ 200 mm

test-tube clamp

Meker burner

800-ml beaker

tongs

PROCEDURE

Fill the test tube to within 3 cm of the top with sulfur. Under a hood, heat the tube to melt the sulfur and continue to heat until the molten liquid is very dark reddish brown in color. The sulfur often ignites in the flame and burns at the mouth of the tube. Pour the molten sulfur in a thin stream into 600 ml of water in the beaker. If the sulfur has ignited, it will form a burning stream from the test tube to the water bath. A golden brown "string" of sulfur is formed. With tongs, remove the mass of rubbery sulfur from the water and compare it with crystalline sulfur. After it is removed from the water, the polymeric sulfur will revert to the brittle rhombic form, but the time for the change varies from a few minutes to several hours.

HAZARDS

The demonstration should be performed in a hood, because sulfur dioxide (SO_2), a toxic and irritating gas, is produced by the ignition of molten sulfur. For several minutes after it is formed, the plastic sulfur may contain molten sulfur and therefore should be handled with care.

DISPOSAL

The sulfur may be recycled or discarded in a waste container.

DISCUSSION

At room temperature, sulfur is stable in the orthorhombic form and consists of staggered, eight-membered cyclic rings of the monomer, S_8. At temperatures above 160°C, sulfur forms linear polymers with high molecular weights on the order of 1,500,000 or more [2].

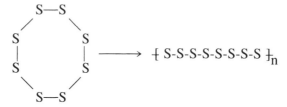

Rhombic sulfur melts at 113°C, and at about 180°C the previously fluid material turns brown and becomes extremely viscous. Viscosity data suggest that the average chain length of the sulfur polymer is about 1 million atoms at about 170°C.

Because rhombic sulfur is the only stable form at room temperature, the polymer chains eventually break and S_8 rings form. The overall heat of reaction for this change should be approximately zero because one S—S bond is broken for every new one formed [3].

REFERENCES

1. Alyea, H. N.; Dutton, F. B., Eds. "Tested Demonstrations in Chemistry," 6th ed.; Journal of Chemical Education: Easton, Pennsylvania, 1965; p 41.
2. Sorenson, W. R.; Campbell, T. W. "Preparative Methods of Polymer Chemistry," 2nd ed.; Interscience Publishers, John Wiley and Sons: New York, 1968; pp 394–95.
3. Schmidt, M. In "Inorganic Polymers"; Stone, F. G. A.; Graham, W. A. G., Eds.; Academic Press: New York, 1962; Ch. 3.

3.13

Thiokol Rubber

When a mixture of ethylene dichloride and a hot sodium polysulfide solution is stirred, a lump of yellow polymer slowly forms [1, 2].

MATERIALS

6 g powdered sulfur, S_8

100 ml 1M sodium hydroxide, NaOH (To prepare, dissolve 4 g of NaOH in water and dilute to 100 ml.)

3 drops liquid detergent

20 ml 1,2-dichloroethane, $ClCH_2CH_2Cl$ (ethylene dichloride)

gloves, plastic or rubber

2 250-ml beakers

25-ml graduated cylinder

hot plate

stirring rod

PROCEDURE

Perform this demonstration in a hood and wear gloves.

Add 6 g of powdered sulfur to 100 ml of 1M NaOH in a 250-ml beaker and boil for 5–10 minutes. If necessary, add a few drops of liquid detergent to reduce surface tension and help mix the chemicals. After several minutes of boiling, the solution should appear dark red and homogeneous. Remove the beaker from the heat source. If small pieces of sulfur remain undissolved, decant the hot solution into a second beaker.

Cautiously add 20 ml of 1,2-dichloroethane to the dark red solution and stir. The boiling point of 1,2-dichloroethane is 83–84°C. The temperature of the solution should be above approximately 80°C for the reaction to proceed. During several minutes of stirring, a rubbery lump of yellowish white polymer will precipitate from the solution. The liquid can be decanted and discarded, preferably in a hood because some of the reaction products are very malodorous. The lump of rubber should be kneaded under water, washed thoroughly with water, and allowed to dry for several minutes before handling. The rubber usually retains a foul odor, probably because of low molecular weight dithiols or cyclic disulfides.

HAZARDS

The demonstration should be performed in a hood. Plastic or rubber gloves should be worn.

Sodium hydroxide is extremely caustic and can cause severe burns. Contact with the skin and eyes must be prevented.

Vapors of 1,2-dichloroethane are highly flammable and are irritating to the eyes and respiratory tract. The compound is a suspected carcinogen.

DISPOSAL

In a hood, the reaction liquid should be flushed down the drain with water. The polymer should be discarded in a waste container.

DISCUSSION

Thiokols are made by condensing a polychloro compound with sodium polysulfide. Thiokol A, the one described here, is made from 1,2-dichloroethane:

$$ClCH_2CH_2Cl + Na_2S_x \longrightarrow \{ CH_2CH_2S_x \} + 2\,NaCl$$

Sodium polysulfides (Na_2S_x where x can range from 2 to 6) are prepared by heating sulfur (S_8) with sodium hydroxide solution. In aqueous solutions, mixtures of anions from S^{2-} to S_6^{2-} may be present, and the sulfur chains are linear [3].

Polysulfide elastomers have few industrial applications. In general, these polymers have outstanding resistance to oils and solvents, a resistance that increases as the sulfur content of the polymer increases. They are attacked by oxidizing acids or strong bases.

REFERENCES

1. Alyea, H. N.; Dutton, F. B., Eds. "Tested Demonstrations in Chemistry," 6th ed.; Journal of Chemical Education: Easton, Pennsylvania, 1965; p 110.
2. Shakhashiri, B. Z.; Dirreen, G. E. "Manual for Laboratory Investigations in General Chemistry"; Stipes Publishing Co.: Champaign, Illinois, 1982, p 219.
3. Saunders, K. J. "Organic Polymer Chemistry"; Chapman & Hall Ltd.: London, 1973; Ch. 17.

3.14

Cuprammonium Rayon

When a blue solution containing copper(II) ions, ammonia, and cellulose (paper) is injected into an acid bath, blue threads of rayon are formed [1].

MATERIALS

25 g cupric sulfate pentahydrate, $CuSO_4 \cdot 5H_2O$

100 ml distilled water

165 ml concentrated (15M) aqueous ammonia, NH_3

300 ml 0.5M sulfuric acid, H_2SO_4 (To prepare, carefully add 8 ml of concentrated [18M] H_2SO_4 to water and dilute to 300 ml.)

250-ml beaker

magnetic stirrer and stirring bar

stirring rods

250-ml graduated cylinder

11-cm Buchner funnel

5 sheets 11-cm filter paper

1-liter filter flask

2 400-ml beakers

10-ml pipette and pipette bulb or 10-ml syringe with #18 or larger needle

PROCEDURE

In a hood, place a 250-ml beaker with stirring bar on a magnetic stirrer. While stirring, dissolve 25 g of cupric sulfate pentahydrate in 100 ml of distilled water in the beaker. Slowly add 13 ml of concentrated aqueous ammonia solution (15M) to the cupric sulfate solution to form bluish green cupric hydroxide, $Cu(OH)_2$. Avoid adding excess ammonia solution.

With a Buchner funnel, one piece of filter paper, and filter flask, filter the mixture to collect the precipitate of cupric hydroxide. Wash the precipitate with several small portions of cold water to remove ammonium sulfate. The filtration will be very slow since the particle size of the precipitate is small.

Place the damp cupric hydroxide in a 400-ml beaker and add 150 ml of 15M ammonia. With the magnetic stirrer, stir to dissolve. Shred four pieces of 11-cm filter paper (or equivalent) and add to the beaker containing the ammoniacal cupric hydroxide. Stir the mixture until the paper is dissolved (30–60 minutes). The mixture should be approximately the same viscosity as light syrup. If it is too viscous, add more 15M ammonia. This solution can be stored in a well-stoppered bottle.

To perform the demonstration, pour about 300 ml of 0.5M H_2SO_4 into a beaker. Fill the pipette or syringe with the blue copper–ammonia–paper solution. Place the tip of the pipette or syringe below the level of 0.5M H_2SO_4 in the beaker and expel the blue solution. On contact with the acid, the solution precipitates a dark blue thread of regenerated cellulose. Wash the thread thoroughly with water before handling.

HAZARDS

Concentrated aqueous ammonia solution can cause burns and is irritating to the skin, eyes, and respiratory system.

Copper compounds are harmful if taken internally. Dust from copper compounds can irritate mucous membranes.

Because sulfuric acid is a strong acid and a powerful dehydrating agent, it can cause burns. Spills should be neutralized with an appropriate agent, such as sodium bicarbonate, and then rinsed clean.

DISPOSAL

The washed thread can be discarded in a waste container. The apparatus used in the preparation of cupric hydroxide can be cleaned by rinsing with 1M hydrochloric acid. Solutions should be flushed down the drain with water.

DISCUSSION

Rayon was the first commercially important man-made fiber. It consists of regenerated cellulose. Cellulose is a natural polymeric material, and the preparation of rayon involves dissolving the cellulose, extruding the solution, and precipitating the rayon as a fiber.

In this demonstration the naturally occurring polymer, cellulose, is dissolved in a solution containing $Cu(NH_3)_4^{2+}$ (Schweizer's reagent). The dissolved polymer swells in concentrated base solution and then is precipitated by neutralizing the base solution. Naturally occurring cellulose has an average molecular weight varying from about 300,000 to over 1,000,000 depending on its source. Cellulose is highly crystalline and is insoluble in water. The polymer product prepared in this demonstration, called cuprammonium rayon, is an example of regenerated cellulose. It differs from the original material because extensive degradation takes place during dissolution. Regenerated cellulose is usually less crystalline than the naturally occurring material.

The chemical structure of the product can be represented by the formula:

```
        H   OH              OH
         \ /                 /
          C————————C
         /           \
        /            H | \
 ┌ O–C–H              C
 ┼                     \
 └    \               /   ] n
       C————————O
      /  \
    H     \
        H-C-OH
          |
          H
```

REFERENCE

1. Alyea, H. N.; Dutton, F. B., Eds. "Tested Demonstrations in Chemistry," 6th ed.; Journal of Chemical Education: Easton, Pennsylvania, 1965; p 111.

4

Color and Equilibria
of Metal Ion Precipitates
and Complexes

Earle S. Scott, Bassam Z. Shakhashiri,
Glen E. Dirreen, and Frederick H. Juergens

This chapter deals primarily with the behavior of transition and post-transition metal ions in aqueous solution and with two phenomena which are classically recognized as indicative of a chemical reaction, namely, the formation or dissolution of a solid and a change in color of solution or solid. Traditionally, these phenomena have been explained in terms of equilibrium constants, such as solubility products and formation constants for complexes. This chapter will follow that tradition, but a few words about the approach used are necessary to avoid confusion. We assume that readers are familiar with solubility principles, complex ion terminology, and simple spectroscopy.

SOLUBILITY AND STABILITY CONSTANTS

It is well known that "equilibrium constants" calculated from the concentrations of species in solution are not constant. In fact, even when corrections are made to obtain activities using standard corrections for ionic strength, the equilibrium constants are still not constant. In some cases, sufficient study of one system with appropriate extrapolation to infinite dilution allows for the determination of a thermodynamic equilibrium constant value, which is of great significance to the theoretical chemist but may not work as well as some other number in describing the system in a beaker during a demonstration. In other cases, the nature of the species under consideration requires a high concentration of salts in the solution, and the determination of a thermodynamic equilibrium constant is impossible. We believe that the "constants" we use should provide a reasonable description of real systems in terms of concentrations and that they should be as internally consistent as possible. For these reasons, we have used values taken from Critical Stability Constants [1] whenever possible, because the authors of that compilation have selected the best values from the literature to give an internally consistent set of constants for solutions of finite ionic strength. The criteria used in selecting the best values are presented in the introduction to each volume of Critical Stability Constants.

In cases where stability constants are not available in the above reference, we have taken values from the most complete compilation of such data, namely Special

Publications 17 and 25 of The Chemical Society [2, 3]. These references contain values for stability constants and formation constants for a vast array of systems, along with an indication of the method of measurement, the nature of the system involved, and literature references. The variation in the value of any one constant provides a problem of selection. In selecting values from these references, we have sometimes been quite arbitrary, while still trying to select numbers that seemed to us to describe real systems in terms of concentrations and to maintain internal consistency.

Special Publications 17 and 25 of The Chemical Society provide not only an entry to the literature of complexes but also the system of notation we use in presenting the equilibrium constants. A summary of that notation system follows. (See the introduction to either of these two references for a more detailed and complete description.)

Equilibrium Constant Notation

Simple Solubility

Traditionally, the solubility of ionic substances has been expressed by using the ion product or solubility product, as in the case of silver iodide:

$$AgI(s) \rightleftharpoons Ag^+(aq) + I^-(aq)$$

$$K_{sp} = [Ag^+][I^-]$$

The square brackets in the second expression imply the molar concentration of the hydrated species. We represent the quantity K_{sp} as K_{s0}. The subscript s indicates that the solid is involved, and the subscript zero indicates that the silver ion in solution has no ligand, I^-, attached to it. Thus, K_{sp} and K_{s0} represent the same quantity.

In a relatively large number of cases, neutral species having the same composition as the solid exist in solution in a finite concentration, and the solid will precipitate only if the limiting concentration of that species is exceeded. For example, copper(I) chloride exists as CuCl(aq) in equilibrium with solid CuCl. The equilibrium constant for the reaction, designated K_{s1}, has the following form:

$$K_{eq} = [CuCl] = K_{s1}$$

Again, the subscript s indicates that the solid is involved, and the subscript 1 indicates that one ligand, Cl^-, is attached to the metal ion in solution.

The relationship between K_{s0}, K_{s1}, and the formation constant for CuCl(aq), K_1, is

$$Cu^+(aq) + Cl^-(aq) \overset{K_1}{\rightleftharpoons} CuCl(aq)$$

$$K_{s0} \diagdown \qquad \diagup K_{s1}$$

$$CuCl(s)$$

Knowledge of the numerical values for any two of the constants allows calculation of the value of the third.

In the corresponding description of the mercury(II) iodide, HgI_2, system, the equation representing the solubility of HgI_2 is

$$HgI_2(s) + H_2O(l) \rightleftharpoons HgI_2(aq)$$

The equilibrium constant, designated K_{s2}, has the following form:

$$K_{eq} = [HgI_2] = K_{s2}$$

Complex Formation

The starting materials for complex formation are assumed to be the hydrated metal ion and the hydrated ligand in the form in which it appears in the complex. Thus, the formation of a complex actually involves the displacement of water from the first coordination sphere of both the metal ion and the ligand, although the equations are written only to show the association of the metal ion and the ligand.

Complexes containing only one metal ion are *mononuclear complexes*. The addition of ligands occurs by steps, and each step is described by a formation constant, K_n, where n is the number of ligands attached to the metal ion in the product:

$$ML_{n-1} + L \longrightarrow ML_n$$

$$K_n = \frac{[ML_n]}{[ML_{n-1}] \cdot [L]}$$

Examples of such complexes are

$$Hg^{2+}(aq) + I^-(aq) \longrightarrow HgI^+(aq)$$

$$K_1 = \frac{[HgI^+]}{[Hg^{2+}][I^-]}$$

$$HgI^+(aq) + I^-(aq) \longrightarrow HgI_2(aq)$$

$$K_2 = \frac{[HgI_2]}{[HgI^+][I^-]}$$

$$HgI_2(aq) + I^-(aq) \longrightarrow HgI_3^-(aq)$$

$$K_3 = \frac{[HgI_3^-]}{[HgI_2][I^-]}$$

and so on up to the maximum possible value of n.

Since each of the steps in the formation of a complex ion is an equilibrium step, an overall equilibrium must also exist. In the case illustrated above, the formation of HgI_3^- is the sum of the three successive additions of iodide ion as shown. This sum can be represented equally well by the equation:

$$Hg^{2+}(aq) + 3\,I^-(aq) \longrightarrow HgI_3^-(aq)$$

The equilibrium constant for this reaction has the form:

$$K_{eq} = \frac{[HgI_3^-]}{[Hg^{2+}][I^-]^3}$$

and is given the symbol β_3. Algebraically, β_3 equals $K_1K_2K_3$.

Similarly, $\beta_1 = K_1$, $\beta_2 = K_1K_2$, and $\beta_n = K_1K_2 \ldots K_n$. These β values are frequently referred to as cumulative stability constants or, simply, stability constants. Only β values are tabulated in Critical Stability Constants [1], whereas stepwise formation constants (K's) are tabulated preferentially in Special Publications 17 and 25 [2, 3].

Complexes that contain more than one metal ion are referred to as *polynuclear complexes*. They can be represented by the generalized formula M_mL_n. For such a complex, the stability constant is represented by the expression:

$$\beta_{nm} = \frac{[M_mL_n]}{[M]^m[L]^n}$$

In the subscripts of β, the number of ligands (n) is specified first, then the number of metal ions (m). This order is consistent with the description of mononuclear complexes if we assume that those would be represented as β_{n1} and that the 1 is understood. Examples of polynuclear complexes are

$$2\,Ag^+(aq) + I^-(aq) \longrightarrow Ag_2I^+(aq)$$

$$\beta_{12} = \frac{[Ag_2I^+]}{[Ag^+]^2[I^-]}$$

$$3\,Cu^+ + 6\,Cl^- \longrightarrow Cu_3Cl_6^{3-}$$

$$\beta_{63} = \frac{[Cu_3Cl_6^{3-}]}{[Cu^+]^3[Cl^-]^6}$$

Formation of Complexes from Solids

When a solid dissolves in an excess of either the metal ion or the ligand, the equilibrium constant is designated by a symbol K carrying the subscript s followed by subscripts n and m, which show the composition of the product in solution. Examples of such notation are

$$AgI(s) + Ag^+(aq) \longrightarrow Ag_2I^+(aq)$$

$$K_{s12} = \frac{[Ag_2I^+]}{[Ag^+]}$$

$$AgI(s) + 2\,I^-(aq) \longrightarrow AgI_3^-(aq)$$

$$K_{s31} = \frac{[AgI_3^-]}{[I^-]^2}$$

Figure 1 is a diagrammatic representation of the Ag^+/I^- system in which each arrow representing a reaction is identified by the symbol for the equilibrium constant for that reaction. Starting from the right of Figure 1, we have the two aqueous ions that can combine in appropriate ratios to form any one of the identified species. The

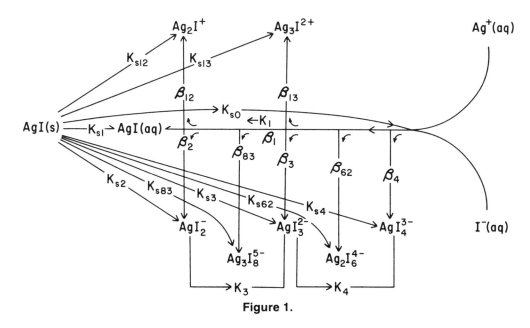

Figure 1.

equilibrium constant for formation directly from the metal ion and ligand may be represented by the appropriate β_{nm}, although β_1 is frequently represented as K_1. The equilibrium constant for the dissolution of the solid AgI, designated K_{s0}, is included in the group representing the dissolution process to form complexes.

The copper(I) chloride–chloride ion system is somewhat analogous to the silver iodide–iodide ion system, although all its polynuclear complexes are negatively charged. A recent paper by Fritz [4] describes this system in a manner that illustrates nicely both the use of the notation for the various equilibrium constants and the difficulties encountered in converting experimental data into thermodynamic equilibrium constants.

If the solid involved contains more than one metal ion, the notation becomes more complicated. In such cases, see reference 1 or 2.

Representation of Weak Acid Equilibria

References 1 and 2 deal with the protonation of the anion of a weak acid as the formation of a complex. The notation is consistent with that used to describe complex formation involving metal ions. Consequently, the phosphoric acid system is described in terms of the following formation constants:

$$H^+(aq) + PO_4^{3-}(aq) \rightleftharpoons HPO_4^{2-}(aq)$$

$$K_{11} \text{ or } K_1 = \frac{[HPO_4^{2-}]}{[H^+][PO_4^{3-}]}$$

Note that K_1 is the reciprocal of the third dissociation constant of phosphoric acid in the conventional treatment. Also, K_{12} is the reciprocal of the conventional K_2:

$$H^+(aq) + HPO_4^{2-}(aq) \rightleftharpoons H_2PO_4^-(aq)$$

$$K_{12} = \frac{[H_2PO_4^-]}{[H^+][HPO_4^{2-}]}$$

In addition, K_{13} is the reciprocal of the conventional first dissociation constant of H_3PO_4:

$$H^+(aq) + H_2PO_4^-(aq) \rightleftharpoons H_3PO_4(aq)$$

$$K_{13} = \frac{[H_3PO_4]}{[H^+][H_2PO_4^-]} = \frac{1}{K_1}$$

See references 1 and 2 for extensive tabulations of stability constants.

Although other forms of equilibrium constants occur in the treatment of complex formation and solubility, those described above cover the main cases of interest. Any special case can be covered by representing the equilibrium constant as K *and* by providing the equation for the reaction under consideration.

Species Distribution Diagrams

For systems in which a metal ion forms a series of complexes, one needs to identify the complexes that predominate in any given solution. Of great help in such cases is the "species distribution diagram" for the metal ion–ligand combination under consideration. Such diagrams show the percentage of the total metal ion concentration that is present as each species, i.e., the hydrated metal ion and each successive complex. The species distribution diagram for the Ni^{2+}/NH_3 system is presented in Figure 2. As we shall see, the percentage values for each species are determined solely by the

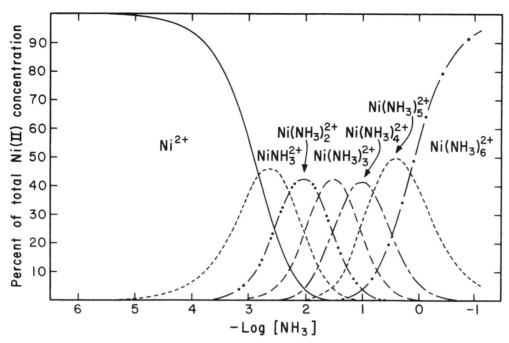

Figure 2. Nickel(II)/ammonia species distribution.

values of the stability constants and the free ligand concentration; therefore, the diagram applies to all solutions as long as no polynuclear complexes form. Because of the utility of these diagrams, the basic relationships involved in their construction are presented below.

Consider the Ni^{2+}/NH_3 system in which complex formation proceeds by successive addition of the ligand until the maximum coordination number of the ion is satisfied by the ligand. Applying the principle of the conservation of matter, we can write the following summation:

$$\text{Total metal ion concentration} = [M] + [ML] + [ML_2] + [ML_3] + [ML_4] + [ML_5] + [ML_6] \quad (1)$$

Using the defining formula for the β values, we can replace each of the terms representing complexes in this summation by its equivalent in terms of β, $[M]$, and $[L]$:

$$\beta_n = \frac{[ML_n]}{[M][L]^n} \qquad \text{so } [ML_n] = \beta_n[M][L]^n$$

Note that because each such term will have $[M]$ in the numerator, $[M]$ can be factored out:

$$\text{Total metal ion concentration} = [M]\left[1 + \beta_1[L] + \beta_2[L]^2 + \beta_3[L]^3 + \beta_4[L]^4 + \beta_5[L]^5 + \beta_6[L]^6\right]$$

This step divorces the resulting diagram from the absolute concentration. Each term within the large brackets still represents the relative concentration of one species, and the sum within these brackets still represents the total metal ion concentration on a relative basis. The fraction of metal ion present as the nth species is known as α_n, with α_o representing the free metal ion. α_n is then equal to $1/[\;]$ for the metal ion or $\beta_n[L]^n/[\;]$ for all other species, where $[\;]$ stands for the sum of terms in the brackets. The percentages are equal to $\alpha_n \cdot 100\%$.

If all the β_n values are known, the values of the individual terms and the value of the sum can be determined for any assigned value of the concentration of free ligand, [L]. Values of α_n and percentages can then be calculated. If a complete diagram is desired, it is convenient to start with a free ligand concentration that makes α_1 just less than 1% of α_0, i.e., $\beta_1[L] < 0.01$. The values of [L] can then be increased in any desired increments until virtually all the metal ions are in the most highly coordinated complex or until the free ligand concentration becomes unreasonably large. We stop calculating when the free ligand concentration is approximately equal to 10M. The drudgery of these calculations is eased greatly with a computer, in which case it is convenient to start with an initial logarithm of the free ligand concentration and then increment that logarithm.

We can just as well write a summation for the total ligand as for the total metal ion:

$$\frac{\text{Total ligand}}{\text{concentration}} = [L] + [ML] + 2[ML_2] + 3[ML_3] + 4[ML_4] + 5[ML_5] + 6[ML_6]$$

(2)

This summation is not used as generally as equation 1, but it is important in at least two ways. It allows us to note that the free ligand concentration employed in the previous calculations is equal to the first term in this summation. It also allows us to visualize the significance of the term "total bound ligand," which is the entire sum minus the first term:

$$\frac{\text{Total bound ligand}}{\text{concentration}} = \frac{\text{Total ligand}}{\text{concentration}} - \frac{\text{Free ligand}}{\text{concentration}}$$

Total bound ligand concentration is used in dealing quantitatively with complexes in solution. It can be obtained because the preparation of a solution gives a value for the total ligand concentration, while the free ligand concentration can often be determined by a spectrophotometric or potentiometric measurement. Using the total bound

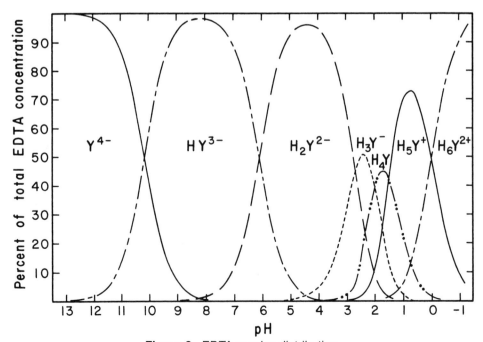

Figure 3. EDTA species distribution.

ligand concentration, we can calculate the average number of ligands bound per metal ion. This quantity is represented by the symbol $\bar{n}$:

$$\bar{n} = \frac{\text{Total bound ligand concentration}}{\text{Total metal ion concentration}}$$

Note that the denominator still contains the term for the free metal ion concentration, so the average obtained is not the average number of ligands in the complexes, although it approaches that rapidly as the concentration of free metal ions diminishes with increasing ligand concentration.

If the appropriate constants are available, precisely the same type of species distribution diagram can be prepared for a polyfunctional acid as a function of pH. In this instance, the anion with the maximum negative charge takes the place of the metal ion, and the hydrogen ion takes the place of the ligand. Such diagrams are important in acid-base chemistry, but they are also important in complex formation where the metal ion and the proton may be competing for the same negatively charged ligand. If only the totally deprotonated ligand can form a complex, it is helpful to know what fraction of the total ligand this represents at various pH values. Figure 3 presents the diagram for ethylenediaminetetraacetic acid (EDTA).

The β values for the addition of protons to the tetranegative ethylenediamine-tetraacetate ion (Y^{4-}) are tabulated below. Alongside these are the calculated, stepwise formation constants (K_n values):

$$\beta_1 = 1.48 \times 10^{10} \qquad K_1 = 1.48 \times 10^{10}$$

$$\beta_2 = 1.91 \times 10^{16} \qquad K_2 = \frac{\beta_2}{\beta_1} = 1.29 \times 10^6$$

$$\beta_3 = 9.12 \times 10^{18} \qquad K_3 = 4.77 \times 10^2$$
$$\beta_4 = 9.12 \times 10^{20} \qquad K_4 = 1.0 \times 10^2$$
$$\beta_5 = 2.89 \times 10^{22} \qquad K_5 = 3.17 \times 10^1$$
$$\beta_6 = 2.89 \times 10^{22} \qquad K_6 = 1.0$$

See Figure 3 for the species distribution diagram calculated from these β values.

The large value of K_1 means that the protonation of Y^{4-} will begin at a very low concentration of hydrogen ions, or at a high pH. Protonation is significant at pH = 12, and equal concentrations of Y^{4-} and HY^{3-} occur at pH = 10.15. This pH range will be the middle buffer region in a titration curve. Perhaps as important is the fact that many EDTA titrations are carried out in a buffer made of ammonia and ammonium ions, which provides a pH of approximately 10. In such cases, roughly half of the uncomplexed EDTA is present as Y^{4-}, the other half being HY^{3-}. Thus, the formation of complexes with Y^{4-} is likely.

Note in Figure 3 that HY^{3-} reaches 98% of the total EDTA at about pH = 8.2. The maximum will occur at the pH at which the curves for Y^{4-} and H_2Y^{2-} intersect, which are the only two other species of significance in that pH range. An intermediate species such as HY^{3-} can achieve such a high percentage of the total only if its formation constant is high and if the formation constant for the next species is on the order of 10^4 times the first formation constant. Although HY^{3-} is the dominant species in solution from pH 10 to pH 6, the second most important species switches at pH = 8.2 from Y^{4-} to H_2Y^{2-}.

Because the formation constant of H_2Y^{2-} (K_2) is 10^4 smaller than K_1 and 10^4 larger than K_3, this species also becomes a dominant species, reaching 96% of the total

EDTA at pH close to 4.5. This result is significant from a practical point of view because the standard source of EDTA is $Na_2H_2Y \cdot 2H_2O$, which should yield a maximum of H_2Y^{2-} in solution when dissolved. The pH of that solution will be about 4.5.

After H_2Y^{2-}, the formation constants differ little. Consequently, no species ever gains dominance, and choosing a species as the one to include in a net ionic equation becomes problematic. Fortunately, in most cases the neutral molecule is relatively insoluble, thus eliminating the problem.

Now let us consider what happens structurally during these changes. Presumably, the first proton attaches to the Y^{4-} at the most basic functional group, one of the nitrogen atoms. The second proton attaches to the second nitrogen, and significant interaction must occur between the first protonated nitrogen and the second for the formation constants to be separated by as much as 10^4. Presumably, there is hydrogen bridge bonding between the two N atoms.

The third proton begins the protonation of the carboxylate anions. That step is thus separated from the preceding one by the usual difference between protonating an amine or protonating a carboxylate ion. From then on, the carboxylate ions are protonated successively and, because these ions do not interact directly to any large degree, the protonation occurs with only slightly increasing difficulty. No species becomes dominant unless the system achieves a concentration of acid above 10M, which is not usually a functional region of the diagram.

To illustrate further how species distribution diagrams are used, we shall look at the Ag^+/I^- system (see also Demonstration 4.4). Figures 4 and 5 are the species distribution diagrams for AgI in excess I^- and in excess Ag^+, respectively. The figures show that in either case the free ion concentration must be approximately $1 \times 10^{-5}M$ before the first complex approaches 1% of the concentration of AgI(aq), which is constant at $5.9 \times 10^{-9}M$. Thus, on either side of the equivalence point, there are approximately 4 decades of ion concentration in which complexing is insignificant. At

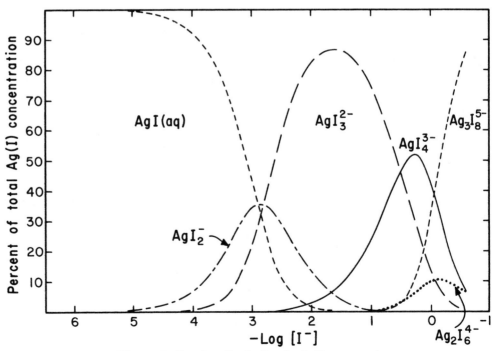

Figure 4. Species distribution for AgI in excess I^-.

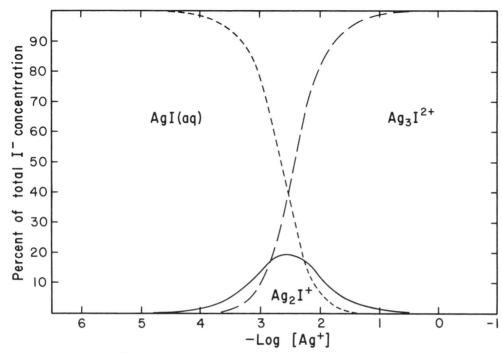

Figure 5. Species distribution for AgI in excess Ag$^+$.

$Ag^+(aq)$ concentrations above $10^{-5}M$, the complexes $Ag_2I^+(aq)$ and $Ag_3I^{2+}(aq)$ form, but the $Ag_3I^{2+}(aq)$ soon begins to dominate (Figure 5). When $[Ag^+]$ is about $3 \times 10^{-3}M$, the concentration of $Ag_3I^{2+}(aq)$ equals the concentration of AgI(aq). By $[Ag^+] = 3 \times 10^{-2}M$, the concentration of $Ag_3I^{2+}(aq)$ has become 100 times the concentration of AgI(aq), and its concentration rises sharply thereafter until it becomes high enough to dissolve all the AgI(s). This appears to happen when $[Ag^+]$ is less than 1M.

When excess iodide is added to the suspension of solid AgI, complex formation becomes significant at $[I^-] = 10^{-5}M$ (Figure 4). When $[I^-]$ is slightly greater than $10^{-3}M$, the concentrations of AgI(aq), $AgI_2^-(aq)$, and AgI_3^{2-} are approximately the same, i.e., about $5.9 \times 10^{-9}M$. The concentration of AgI_3^{2-} continues to grow until it maximizes at about $2.5 \times 10^{-2}M$ I^-. At this point [AgI] is still about 1% of the total, so the AgI(s) will not dissolve. The AgI(s) dissolves as $[I^-]$ approaches 1M. In this concentration range, AgI_4^{3-} is the most important species, but AgI_3^{2-}, $Ag_2I_6^{4-}$, and $Ag_3I_8^{5-}$ are all present in significant concentrations. Although the system of complexes is more complicated on the iodide-rich side of the equivalence point than on the silver-rich side, the net effect of adding excess ions is approximately the same on both sides, and the initially precipitated AgI dissolves as the concentration of free ions approaches 1M.

When a solution of AgI in excess Ag^+ is diluted with distilled water, the free silver ion concentration is reduced and, in effect, the system moves to the left in the diagram in Figure 5. The percentage of the total iodide present as AgI(aq) increases. When the actual concentration of that species exceeds $5.9 \times 10^{-9}M$, silver iodide precipitates. A similar argument explains the precipitation of AgI from its solution in an excess of iodide ions.

Enhanced Solubility Through Complex Formation

Inherent in the discussion of equilibrium constants is the idea that complex formation can compete with precipitation so that the addition of a complexing agent may modify the conditions under which a metal ion will precipitate with a given counterion. (See Demonstrations 4.1, 4.4, 4.5, 4.6, 4.8, 4.9, and 4.11.)

COLOR

One of the most aesthetically pleasing aspects of many demonstrations is the generation of colors or color changes. Fortunately, the simplest possible explanation for color changes turns out to be essentially correct: a change in the color of a solution implies a change in the chemical species in that solution. Although the color we see is merely a complement of the color absorbed by the solution and although the human eye is not a quantitative instrument, training in the recognition of characteristic colors of substances and in the estimation of concentrations is very important. Providing such training is a secondary purpose of every demonstration involving colored substances.

The level of interpretation of a color or a color change depends on one's pedagogical purposes and the sophistication of one's audience. When we add permanganate ions to an acidic solution and watch the color disappear, we may not be concerned with why permanganate ions are colored but only with the disappearance of the color as an indicator of the presence of a reducing agent in the solution. At other times, we may be interested in the reasons for the color, the intensity of the color, or the presence in the absorption spectrum of a series of maxima superimposed on a broad band.

Electromagnetic Radiation

Energy is transmitted through space in the form of electromagnetic radiation, which in a vacuum has a constant rate of propagation, c, and which can be characterized by either its frequency, v, or its wavelength, λ:

$$\lambda v = c$$

where λ is in meters, v in seconds^{-1}, and c in m·s^{-1}. However, traditions in different fields have generated a great variety of units, such as:

Wavelength (λ)		
1 Angstrom	Å	10^{-10} m or 10^{-1} nm
1 nanometer	nm	10^{-9} m
1 millimicron	mμ	10^{-9} m or 1 nm
1 micron	μ	10^{-6} m or 1 micrometer
Frequency (v)		
wave number	cm^{-1}	$1/\lambda$ in cm
frequency	s^{-1}	c/λ

When energy is exchanged between matter and light, the exchange is made in discrete packets of energy known as *photons*. From the standpoint of the light, the photon energy is described by the equation:

$$\text{Energy of the photon } = hv$$

where h is Planck's constant with the value of 6.6×10^{-34} joule seconds. From the standpoint of the chemical system involved in the exchange, the photon energy must correspond to the energy difference between two states of the system. Changes may occur in translational, rotational, and vibrational energy, or in the electric potential energy of the assembly of nuclei and electrons. Only translational energy can vary continuously; all the other energies are quantized, i.e., can have only one of a limited set of values at any one time. As a rule of thumb, microwave radiation ($\lambda \simeq 10^{-2}$ m) brings about changes in rotational states, and infrared radiation changes primarily the vibrational states of matter, but these latter changes are accompanied by changes in rotational states as well. Visible and ultraviolet light primarily produce changes in the electric potential energy of the assembly of charges, but these changes may also be accompanied by incidental changes in vibrational and rotational states. Thus, the emission or absorption of visible light is caused principally by electronic transitions in which emission is accompanied by an electron moving from a higher to a lower energy state, and absorption is accompanied by movement of an electron from a lower to a higher energy state.

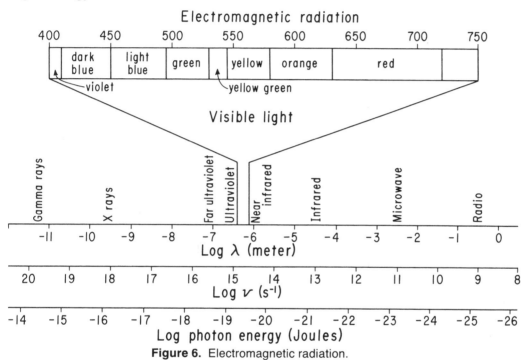

Figure 6. Electromagnetic radiation.

Although the range of wavelengths of electromagnetic radiation is tremendous, the range to which the human eye is receptive is quite narrow, roughly from 400 nm to 750 nm. This range, known as the visible spectrum, spans the range of waves contained

Photon Energies in the Visible Spectrum

Wavelength (nm)	Photon energy (joules)	Energy (kJ/mole)
400	5.0×10^{-19}	300
450	4.4×10^{-19}	260
500	4.0×10^{-19}	240
550	3.6×10^{-19}	220
600	3.3×10^{-19}	200
650	3.0×10^{-19}	180
700	2.8×10^{-19}	170
750	2.6×10^{-19}	160

in greatest intensity in "white light" from the sun. Figure 6 shows the range of electromagnetic radiation in terms of both wavelength and frequency. The names applied to such radiation in the different wavelength ranges are given, and the narrow band known as visible light is expanded out of the continuum to show the relationship between wavelength and color. The table summarizes the energies corresponding to a range of wavelengths of visible light. Note that these photon energies are, indeed, sufficient to bring about drastic changes in electronic arrangements, since they range from 159 kJ/mole (enough energy to rupture the F-F bond in F_2) to 297 kJ/mole (close to the average C-Br bond energy).

Absorption Spectroscopy

Most of the instruments used to quantify light emission or absorption convert a light intensity into an electric current so that the current is proportional to the intensity, or photon density. By the appropriate use of gratings or prisms and slits, such measurements can be made as a function of wavelength, thereby obtaining an emission or absorption spectrum. In most demonstrations, however, we see the light after the chemical species have absorbed whatever they can from the ambient light. Whether we are looking at transmitted or reflected light, we would like to be able to infer something about the nature of an absorption spectrum from the appearance of the solution, and vice versa. What we see is really the complement of the absorption, since the sum of the transmitted or reflected light plus the absorbed light must equal the light impinging on the sample. Because different light sources produce different distributions of frequencies, a colored substance may look quite different in two different lights.

Perhaps the simplest approach to relating the color observed to the color of light absorbed is the so-called "color rosette" (Figure 7). The colors are arranged in a circular disc so that an absorption of one color will leave visible the color opposite the absorbed color on the rosette. If an object absorbs red light, it will appear green, and vice versa. If an object absorbs dark blue light, it will appear yellow, and vice versa.

At least two factors complicate this phenomenon. One is that absorption may occur, to varying degrees, at a number of wavelengths. The other is that our eyes are not equally sensitive to light of all wavelengths. Human eyes are about 5–10 times as sensitive to the green and yellow parts of the spectrum as to the blue and red, and by

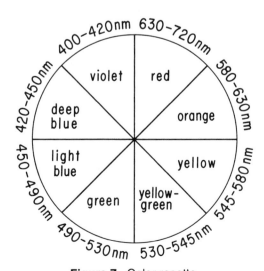

Figure 7. Color rosette.

comparison they hardly detect the violet at all. Therefore, the correlation between what one sees and the absorption spectrum of the substance one observes is not perfect; however, useful guidelines do exist. The absorption spectra of a number of solutions are shown in Figures 8a–l. We suggest that you prepare transparencies or copies of the absorption spectra to accompany the solutions.

The two yellow solutions, iron(III) chloride (Figure 8a) and potassium chromate (Figure 8b), result from the absorption of violet light by the tail of a major absorption peak in the ultraviolet region, which leaves the yellow to be seen. The comparison of dilute and more concentrated solutions of iron(III) chloride shows why the apparent color shifts toward orange at higher concentrations and why, as the concentration increases, the tail of the absorption peak impinges farther and farther into the visible region. The orange of dichromate ion solution, as opposed to the yellow of chromate ion solution, is created by the appearance of a new absorption band which removes blue light.

The green solutions, nickel ion and chlorocuprate complexes, show absorption of violet light and red light. The relative magnitudes of these absorptions determine the shade of green observed. Conversely, the solutions having a purple appearance absorb in the middle of the visible range and allow violet and blue light on one end of the spectrum and red light on the other end to be transmitted (Figures 8d, 8k, and 8l). Again the shade observed is determined by the exact placement and relative magnitude of these absorptions. The clear blue of copper(II) sulfate solution is the result of absorption throughout the red and into the infrared region of the spectrum.

So far, we have described the emission and absorption of light as if they occur only when the emitting or absorbing species moves from one energy state to another. Under this assumption, the absorption of light to elevate an electron from one state to another should always take the same quantity of energy, and the absorption spectra should consist of a series of lines analogous to the Fraunhofer lines in a spectrum generated from sunlight. The spectra in Figures 8a–l, however, show each absorption occurring over a broad range of wavelengths, or photon energies.

If we want to see the line spectra characteristic of a system in which all the energy separations of one kind are identical, we use the spectra of atoms which are isolated from one another in the gas phase. Because each atom is a centrosymmetric system identical to every other atom, the energy levels are identical, the energies of photons absorbed or emitted are the same, and the emission and absorption spectra consist of a series of lines. Emission spectrography and atomic absorption spectrometry depend on this sharpness for their specificity.

When that same atom is considered as part of a molecule, the presence of other atoms alters the energy levels of an electron. Because the energy levels are shifted, the emission or absorption of energy will occur at different wavelengths. Another major change is that all the atoms are no longer identical. Molecules undergo rotation and vibration, and these factors affect the energy states of the molecule. A change in electronic energy state may be accompanied by a change in vibrational or rotational state, or both. Consequently, the photon energies vary over a limited range, and the line in the spectrum broadens, even when the molecules are quite well isolated from one another in the gaseous state.

If we now consider that same molecule, or a molecule-ion, dissolved in a solvent, we realize that in all likelihood all the molecules, or ions, are not the same. Each is interacting with solvent molecules and other species in solution, and all these species are in rapid motion giving rise to the constantly changing geometry of the molecule and of the electric fields the electrons are experiencing. These changes alter the energy

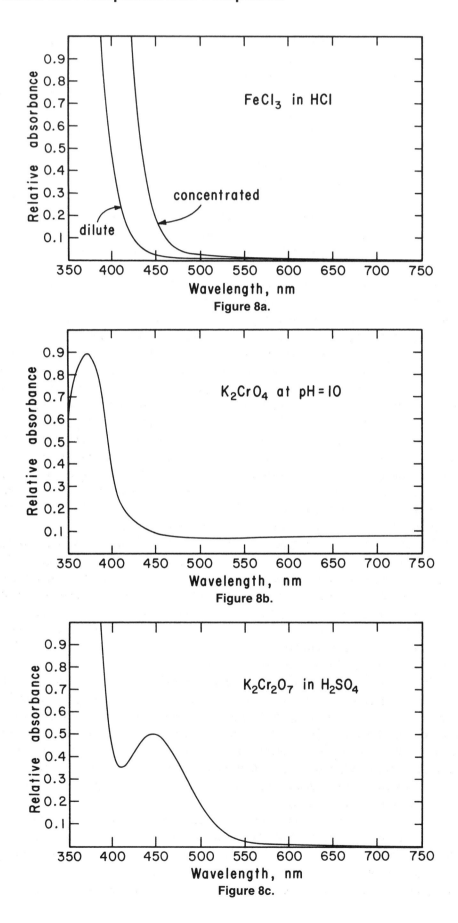

Figure 8a.

Figure 8b.

Figure 8c.

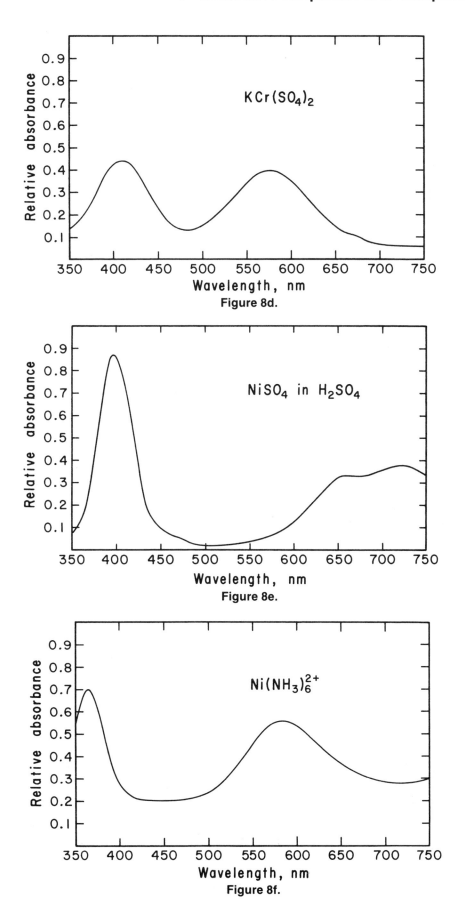

Figure 8d.

Figure 8e.

Figure 8f.

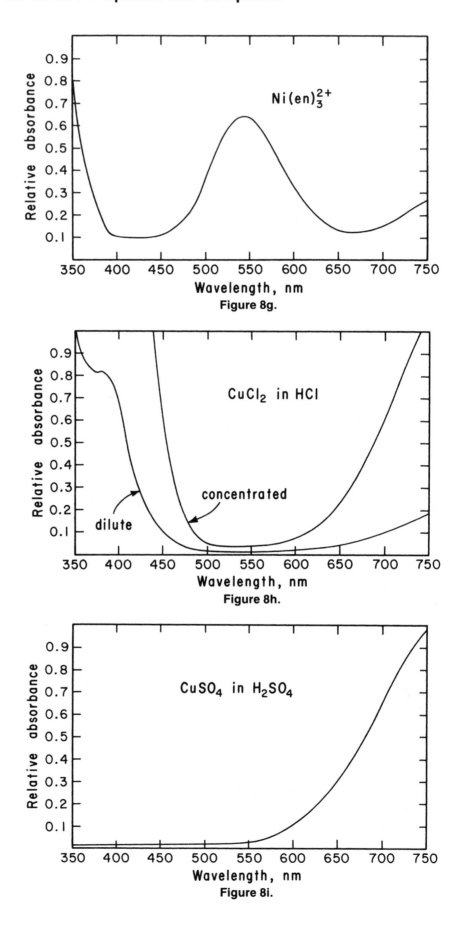

Figure 8g.

Figure 8h.

Figure 8i.

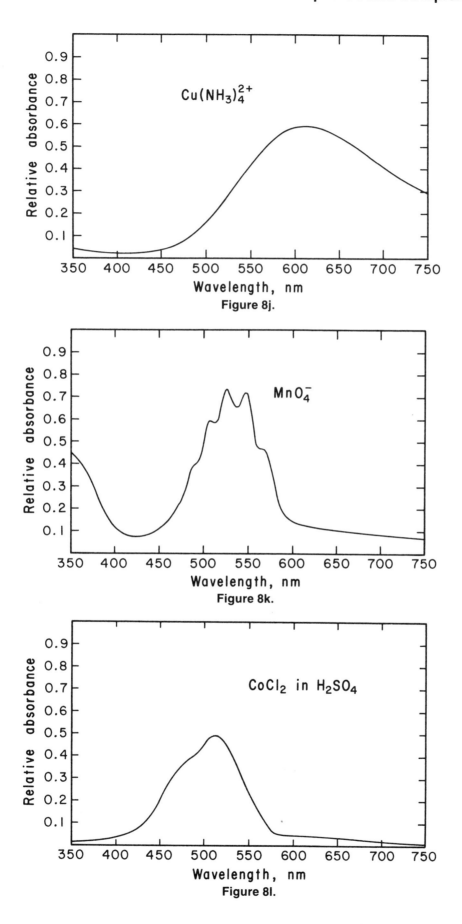

Figure 8j.

Figure 8k.

Figure 8l.

levels available in the molecule or ion, and they broaden even more the range of wavelengths absorbed. Thus, what would be a line absorption in an isolated gaseous atom becomes a broad-band absorption in a molecule in solution (see Figures 8a–l). Nonetheless, each absorption is characterized by a wavelength at which the absorption is maximized; this wavelength is designated as λ_{max}.

An absorption band is also characterized by a proportionality constant, relating the concentration of the absorbing species to the absorbance of the solution. The absorbance, A, is given by the equation:

$$A = \log \frac{I_o}{I} = \epsilon \, c \, \ell$$

where A is the absorbance,
   I_o is the intensity of the incident radiation,
   I is the intensity of the transmitted radiation,
   c is the molar concentration of the absorbing species,
   ℓ is the path length of the radiation through the solution, and
   ϵ is a proportionality constant relating these measurable quantities.

ϵ is a physical constant associated with an absorbing species and has a specific value at each wavelength, with the value at λ_{max} being the one normally tabulated. ϵ is called the molar absorptivity, the molar absorbance coefficient, or the molar extinction coefficient. The values of ϵ range from 0 to over 10^5. Its order of magnitude is useful in determining the type of electronic transition responsible for absorption. A large value of ϵ at λ_{max} indicates that, given the availability of photons, a transition occurs frequently, i.e., that the transition has a high probability of occurring. A low value indicates that the transition has a low probability of occurring.

Selection Rules for Electronic Transitions

The magnitude of the molar absorptivity, ϵ, is useful not only because it allows us to make use of light absorption in quantitative analysis but also because it helps us decide what kind of an electronic transition is responsible for the absorption. Some transitions are said to be *allowed,* and ϵ is large for such transitions. Other transitions are said to be *forbidden,* and ϵ is relatively low for them. But "allowed" and "forbidden" are not absolute terms and should be thought of as "high probability" and "low probability." Described below are some of the factors that control the probability of a transition [5].

1. A transition has a low probability if the electron being excited must undergo a spin inversion. Such transitions are sometimes said to be spin forbidden. Transitions from a singlet to a triplet state fall into this category. In general, changing the number of unpaired electrons in a system is improbable.

2. A transition has a low probability unless the transition involves one symmetric and one antisymmetric orbital. Symmetric and antisymmetric orbitals are designated by a subscript g or u, respectively (from the German *gerade* or *ungerade*). Transitions of the type g $\rightarrow$ u or u $\rightarrow$ g are said to be *parity allowed* or LaPorte allowed. Others are said to be *parity forbidden* or LaPorte forbidden. This rule is similar to the selection rule, $\Delta l = \pm 1$, which allows s $\rightarrow$ p and p $\rightarrow$ d transitions but not s $\rightarrow$ d transitions.

3. A transition has a low probability unless the orbitals involved have a fair degree of overlap. For an isolated atom, the degree of overlap would be determined by Δn, the difference in the principal quantum numbers of the orbitals involved. However, for

bonded species, the degree of overlap depends heavily on the distance of separation of the two atoms, assuming the atomic orbitals involved have the right symmetry characteristics to allow overlap.

4. A transition has a low probability if a highly symmetrical ground state or excited state is involved.

These simplified rules governing electronic transitions help rationalize what we observe experimentally. Tremendous numbers of substances are white or colorless, which means that the energies of their electronic transitions are not in the visible region of the spectrum. The ground state of most molecules is a singlet state, and the transition to the lowest-energy excited state, a triplet state, is spin forbidden. The spin-allowed transition to a higher-energy singlet state requires energy above that of the visible region. Molecules having an unpaired electron, such as NO_2 and ClO_2, tend to be colored. In these molecules, both the ground state and first excited state are triplet states. The transition between these states, which involves energy in the visible region of the spectrum, is spin allowed.

The metal ions that are colored in solution are those that have unfilled d or f orbitals and, therefore, may have unpaired electrons. The constraints of the spin selection rule do not require that these ions be colorless. It might appear that the selection rule $\Delta l = \pm 1$ would require a d electron to be promoted to some orbital outside the d set when energy is absorbed. In fact, this rule does not prohibit the absorption of energy by promoting an electron from one d orbital to another of the same set. The degeneracy of the d orbitals is broken when the centrosymmetric character of the isolated gaseous atom (ion) is lost. Whenever the atom has about it a set of negatively charged species, the d orbitals pointing toward those negative charges will be of higher energy than are those d orbitals not so directly involved with the surrounding negative charges. (These negative charges may be the electron pairs on neutral molecules as well as negative charges on ions.) In a situation of this kind, the electrons tend to reside in the orbitals of lower energy but can be promoted to the higher-energy, unfilled d orbital. The selection rule operating against such a transition simply means that it will be relatively improbable and thus will have a low molar absorptivity. Many of the colors of hydrated transition metal ions are a result of these low probability d-d transitions.

However, experience indicates that not all compounds of the transition metal ions are weak absorbers. Complexes that allow the detection or determination of metal ions at the parts-per-million level are obviously making use of much more probable electronic transitions. These transitions, known as *charge transfer transitions,* involve the transfer of an electron from an orbital that is predominantly a ligand orbital to one that is predominantly a d orbital in the metal ion, or vice versa. Thus, they do not necessarily encounter either the parity or spin restrictions. The molar absorptivities for charge transfer absorptions tend to be approximately 100 times the molar absorptivities of d-d transitions.

Many of the demonstrations in this chapter can be scaled down for presentation by using an overhead projector. However, all the directions given here are for large-scale presentations on the lecture table. Directions for using overhead projection for displaying color and precipitation phenomena will be given in a future volume in this series.

Demonstrations 4.2, 4.5, 4.7, 4.8, and 4.11 deal with four first-row transition metal ions; Demonstrations 4.4 and 4.6 deal with a second-row transition metal ion;

Demonstration 4.1 deals with a third-row transition metal ion; Demonstrations 4.3 and 4.9 deal with post-transition metal ions; and Demonstration 4.10 deals with the chemical behavior of the H_2CO_3/Ca^{2+} system. In all procedures we indicate the necessary amounts and volumes of chemicals to be used. The list of materials for each procedure provides directions for preparing 1 liter of stock solution. In most cases that much solution will not be needed for a single presentation. Users should prepare the volumes they deem necessary for both rehearsing and presenting the demonstration. Unless otherwise indicated, the solutions will not decompose if properly stored and can be used over a period of years. However, unwanted solutions and chemicals should be disposed of properly according to local practice and regulations.

REFERENCES

1. Smith, R. M.; Martell, A. E., Eds. "Critical Stability Constants," Vol. IV, Inorganic Complexes; Plenum Press: New York, 1976.
2. Sillén, L. G.; Martell, A. E., Eds. "Stability Constants of Metal Ion Complexes," Special Publication No. 17; The Chemical Society: London, 1964.
3. Sillén, L. G.; Martell, A. E., Eds. "Stability Constants of Metal Ion Complexes," Supplement No. 1, Special Publication No. 25; The Chemical Society: London, 1971.
4. Fritz, J. J. *J. Phys. Chem.* **1980**, *84*, 2241.
5. Cotton, F. A. "Chemical Applications of Group Theory," 2nd ed.; Interscience Publishers, John Wiley and Sons: New York, 1971.

4.1

Iodo Complexes of Mercury(II): "The Orange Tornado"†

A 4-liter beaker containing 3500 ml of a clear, colorless solution is placed on a magnetic stirrer, which is adjusted to create a smooth, nonturbulent vortex. A clear, colorless solution is added in small increments to the vortex with time to allow for observations after each addition. The initial additions produce a silvery, shifting column like a tornado, which eventually disappears. Subsequent additions increase the quantity of precipitate, and the texture and color of the precipitate change. As more additions are made, the precipitate gradually dissolves. The last precipitate to dissolve is red. A second clear, colorless solution is added to the vortex and generates a yellowish orange "tornado," which may persist for several minutes. The entire process can be repeated. The "tornado" effect provides an aesthetically satisfying demonstration.

MATERIALS

3500 ml distilled water

100 ml 0.10M mercury(II) nitrate, $Hg(NO_3)_2$ (To prepare 1 liter of 0.10M stock solution, dissolve 21.7 g of mercury(II) oxide, HgO, in 15 ml of concentrated, 16M, nitric acid, HNO_3, and dilute with distilled water to 1 liter. Filter if necessary. The mercuric nitrate solution is prepared from HgO, because mercuric nitrate is extremely hygroscopic. If available, $Hg(NO_3)_2$ can be used, but it should be dissolved in dilute nitric acid to avoid hydrolysis.)

100 ml 1.0M potassium iodide, KI (To prepare 1 liter of 1.0M stock solution, dissolve 166 g KI in distilled water and dilute to 1 liter.)

100 ml 0.1M sodium sulfide, Na_2S (optional) (To prepare 1 liter of 0.1M stock solution, dissolve 24.0 g $Na_2S \cdot 9H_2O$ in distilled water and dilute to 1 liter.)

4-liter beaker

magnetic stirrer with 2-inch stirring bar

2 long droppers

2 25-ml graduated cylinders or 2 burets

black backdrop

light source directed to beaker at 90° to direction of viewing (optional)

2 ring stands

2 3-fingered clamps for holding burets

†This demonstration was originally proposed by Professor Richard W. Ramette of Carleton College during the 1980 University of Wisconsin Chemical Demonstrations Workshop.

PROCEDURE A

Place the 4-liter beaker containing 3500 ml of distilled water and the stirring bar on the magnetic stirrer in front of a black backdrop. Adjust the stirring rate to form a smooth, nonturbulent vortex extending approximately 2 cm below the water's surface. If desired, adjust lighting from the side to light the column of liquid above the stirring bar without shining in the eyes of the observers. A small high-intensity desk lamp is satisfactory. Add 35 ml of 0.10M mercuric nitrate solution and allow a couple of minutes for complete mixing.

To perform the demonstration, add approximately 1 ml of 1.0M potassium iodide solution by injecting it vertically into the vortex using either a medicine dropper or a buret. Allow time for observation. When the solution is once more clear and colorless, add a second increment of KI solution. One-ml increments provide a very satisfactory result through the addition of 4–5 ml, by which time the precipitate of mercury(II) iodide, HgI_2, no longer redissolves and each addition of KI solution increases the amount of precipitate. To enhance mixing, inject subsequent additions of KI solution down the side of the beaker instead of into the vortex.

The HgI_2 precipitate dissolves after the addition of about 20 ml of KI solution, but the texture and color of the precipitate undergo almost continuous change throughout the process, so many small additions should be made instead of one large one. Shortly after the precipitate becomes generally dispersed, it takes on the color of orange ice cream and has a silky texture reminiscent of the mercurous chloride precipitate in the Zimmerman-Reinhardt determination of iron. As KI solution is added, the color becomes increasingly red and is frequently caught in the surface tension in the vortex.

When all the mercuric iodide has dissolved, add 2–3 more ml of KI solution to establish some excess iodide ions in solution. Then add to the vortex some mercuric nitrate solution. One ml of $Hg(NO_3)_2$ solution creates a satisfactory tornado of HgI_2. However, 3–5 ml produce a more impressive tornado, which lasts long enough for some of the color changes previously observed to take place. Frequently, the last solid to disappear is red and is visible as a flickering ghost of the tornado for a minute or more.

Slowing the rotation of the stirrer before adding mercuric nitrate solution allows greater freedom of motion to the added ions and frequently creates some beautiful effects. Experimentation is rewarding.

Eventually, the addition of mercuric nitrate solution produces a permanent precipitate, which obscures the tornado. Addition of KI solution down the side of the beaker redissolves the mercuric iodide and sometimes generates another tornado.

Alternate additions of mercuric nitrate and potassium iodide permit the phenomenon to be observed repeatedly.

The demonstration needs no special termination to make it satisfying. However, a black tornado can be generated by pouring very dilute sulfide ion solution into the vortex. The addition of stannous chloride solution forms mercurous iodide but does not produce a good "tornado," probably because the reduction is not fast enough. Either of these procedures illustrates useful chemistry and helps prepare the mercury solution for disposal.

PROCEDURE B

This procedure is designed to permit semiquantitative treatment of the precipitation and dissolution of mercuric iodide. Follow the directions in the first paragraph of Procedure A but place the magnetic stirrer on the base of a large ring stand on which two 50-ml burets are mounted independently so that either can be centered over the vortex or shifted to the side of the beaker. Fill one buret with 0.10M mercuric nitrate and mount it over the side of the beaker. Establish the meniscus at 0.00. Fill the other buret with 1.0M KI solution and mount it over the vortex. Side lighting is especially advantageous to allow for observation of the first solid to appear and the last solid to disappear as the appropriate reagents are added. A small, high-intensity desk lamp is satisfactory.

If you wish to retain the aesthetic advantage of the tornado, add KI solution to the vortex. Useful results can be obtained if 1.00-ml increments are made. Note carefully the volume added that just produces a permanent precipitate. Alternatively, add KI solution in smaller increments to a region closer to the side of the beaker and note the volume of KI solution required just to create a permanent precipitate.

After establishing this first "endpoint," add KI solution rapidly up to about 20 ml, by which time the precipitate should be largely redissolved. Again, if the aesthetic quality of the textural and color changes of the precipitate is important, make this addition in several steps with modest, nonturbulent stirring. As the solution clears, increase the stirring rate until the vortex contacts the stirring bar to guarantee that all solid has an opportunity to redissolve. Then reduce the stirring rate to facilitate observation. The objective is to determine the volume of KI solution that is just sufficient to dissolve the mercuric iodide. Useful results can be obtained by making the additions of KI solution in 1.00-ml increments. If the increments are too small, it becomes difficult to see the endpoint. Record the volume of KI solution when the endpoint is reached and then add 3–5 ml more KI solution.

Add 0.10M $Hg(NO_3)_2$ solution, once more recording the volume that just produces a permanent precipitate. Then add several ml beyond this endpoint.

Once more, back titrate with KI until the precipitate disappears. Each time this cycle of precipitation and solution is repeated, you obtain two endpoints for determining what occurs in the beaker when the precipitate dissolves. Obtain at least five sets of endpoints. See Discussion section for a method of data analysis. The demonstration can be enhanced by creating tornadoes, although this takes additional time.

HAZARDS

Mercury and all of its compounds are poisonous, and toxic effects can result from inhalation, ingestion, or skin contact. Chronic effects can result from exposure to small concentrations over an extended period of time. The dust from salts of mercury is quite poisonous and can irritate the skin and eyes.

Nitric acid is a strong acid and a powerful oxidizing agent. Contact with combustible material can cause violent and explosive reactions. The liquid can cause severe burns to the skin and eyes. The vapor is irritating to the eyes and to the respiratory system.

DISPOSAL

Add ammonium sulfide or sodium sulfide solution to the reaction mixture to precipitate mercury(II) sulfide. Adjust the pH until basic to litmus by addition of aqueous ammonia. The precipitate of mercury(II) sulfide should be collected, washed, and dried. Disposal should be in a landfill designed for toxic chemicals. The remaining solution can be flushed down the drain with water.

Mercury and mercury compounds should not be allowed to contaminate ground water supplies. Current regulations in Madison, Wisconsin, permit the disposal of no more than 2 micrograms of mercury per liter of effluent into the sewer system. Local conditions and regulations may dictate other methods of disposal.

DISCUSSION

When mercuric ions coordinate with iodide ions, the first two iodide ions are very tightly bound, and the third and fourth are relatively loosely bound. The neutral species, HgI_2, exists in finite concentration in solution but is relatively insoluble. The following set of equilibrium constants is based on those reported in "Stability Constants" [1, 2], but we have made minor changes in values to generate an internally consistent set of constants. None of the constants is in serious contradiction with reported values except at the research level. The list provides the equations representing the various reactions between Hg^{2+} and I^-, the equilibrium constant expressions for those reactions, the selected values for the logarithms of the constants, and the values of the constants.

Equation	K_{eq} expression	Log K	K	
$Hg^{2+}(aq) + I^-(aq) \longrightarrow HgI^+(aq)$	$K_1 = \dfrac{[HgI^+]}{[Hg^{2+}][I^-]}$	13.3	2×10^{13}	(1)
$HgI^+(aq) + I^-(aq) \longrightarrow HgI_2(aq)$	$K_2 = \dfrac{[HgI_2(aq)]}{[HgI^+][I^-]}$ †	11.0	1×10^{11}	(2)
$Hg^{2+}(aq) + 2\,I^-(aq) \longrightarrow HgI_2(aq)$	$\beta_2 = \dfrac{[HgI_2(aq)]}{[Hg^{2+}][I^-]^2}$	24.3	2×10^{24}	(3)
$HgI_2(aq) + I^-(aq) \longrightarrow HgI_3^-(aq)$	$K_3 = \dfrac{[HgI_3^-]}{[HgI_2(aq)][I^-]}$	3.7	5×10^3	(4)
$HgI_3^-(aq) + I^-(aq) \longrightarrow HgI_4^{2-}(aq)$	$K_4 = \dfrac{[HgI_4^{2-}]}{[HgI_3^-][I^-]}$	2.2	1.6×10^2	(5)
$Hg^{2+}(aq) + 4\,I^-(aq) \longrightarrow HgI_4^{2-}(aq)$	$\beta_4 = \dfrac{[HgI_4^{2-}]}{[Hg^{2+}][I^-]^4}$	30.2	1.6×10^{30}	(6)

†The (aq) notation, which is dropped for convenience in the case of other species, is retained for the HgI_2(aq) to differentiate clearly between the species in solution and the solid HgI_2 that precipitates.

$$\beta_4 = K_1K_2K_3K_4$$

$HgI_2(s) \longrightarrow HgI_2(aq)$ $K_{s2} = [HgI_2(aq)]$ -4.0 1×10^{-4} (7)
in equilibrium with
solid HgI_2

$HgI_2(s) + I^-(aq) \longrightarrow HgI_3^-(aq)$ $K_{s3} = \dfrac{[HgI_3^-]}{[I^-]}$ -0.3 0.5 (8)

$HgI_2(s) + 2\,I^-(aq) \longrightarrow HgI_4^{2-}(aq)$ $K_{s4} = \dfrac{[HgI_4^{2-}]}{[I^-]^2}$ 1.9 80 (9)

Note that K_{s3} is related to K_{s2} and K_3 as follows:

$$K_{s2} \times K_3 = \cancel{[HgI_2(aq)]} \times \frac{[HgI_3^-]}{\cancel{[HgI_2(aq)]}\,[I^-]} = K_{s3}$$

$$\therefore \log K_{s3} = \log K_{s2} + \log K_3 = 3.7 - 4.0 = -0.3$$

Also note that

$$K_{s4} = K_{s3} \times K_4$$

$$\therefore \log K_{s4} = -0.3 + 2.2 = 1.9$$

The figure plots the percentage of Hg^{2+} that is present in the solution as each of the species indicated versus $-\log[I^-]$. Values are calculated from the listed stability constants. Table 1 summarizes the same data. For a long range of concentration, $HgI_2(aq)$ is the dominant species, and the maximum concentration for this species in

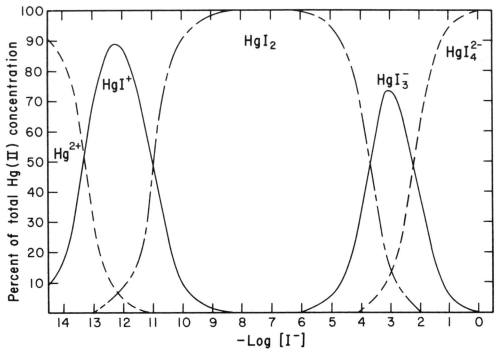

Mercury(II)/iodide ion species distribution.

Table 1. Percentage Distribution of Hg^{2+}/I^- Species

$[I^-]$	Hg^{2+}	HgI^+	$HgI_2(aq)$	HgI_3^-	HgI_4^{2-}
10^{-14}	83.3%	16.7%	—	—	—
10^{-13}	33.1%	66.2%	0.7%	—	—
10^{-12}	4.3%	87%	8.7%	—	—
10^{-11}	0.02%	49.9%	49.9%	—	—
10^{-10}	—	10%	90%	—	—
10^{-9}	—	1%	99%	—	—
10^{-8}	—	0.1%	99.9%	—	—
10^{-7}	—	—	99.95%	0.05%	—
10^{-6}	—	—	99.5%	0.5%	—
10^{-5}	—	—	95%	5%	—
10^{-4}	—	—	66.3%	33.1%	0.6%
10^{-3}	—	—	14.6%	73%	12.9%
10^{-2}	—	—	0.7%	36.8%	62.5%
10^{-1}	—	—	—	5.6%	94.4%
10^{0}	—	—	—	0.6%	99.4%

solution is 1×10^{-4}M. Solid HgI_2 precipitates when $HgI_2(aq)$ exceeds 1.0×10^{-4}M, and solid HgI_2 dissolves when the molar concentration of $HgI_2(aq)$ falls below 1.0×10^{-4}M. This threshold is the key to the interpretation of the data obtained from the quantitative titration of Hg^{2+}(aq) with I^-(aq).

The initial precipitation of HgI_2(s) should occur after sufficient I^- has been added to the solution to raise the concentration of $HgI_2(aq)$ to 1.0×10^{-4}M. The initial solution contains 3.5×10^{-3} moles of Hg^{2+}. Therefore, 3.5×10^{-3} moles of I^- would be required to convert Hg^{2+} into HgI^+. Then, sufficient I^- would have to be added to make $[HgI_2(aq)]$ in 3.5 liters equal 1.0×10^{-4}M (3.5×10^{-4} moles). Therefore, the minimum volume of 1.0M KI required is 3.85 ml (3.5×10^{-3} + $3.5 \times 10^{-4} = 3.85 \times 10^{-3}$ moles). An amount in excess of this must be added to produce a visible precipitate. Ordinarily, precipitation is observed after the addition of 4.0 ml of 1M KI, which allows for a very satisfactory interpretation.

The solubility of HgI_2 in water at 25°C is 0.059 g/liter, which is equivalent to 1.3×10^{-4} moles/liter, a reasonably good agreement with the value of 1×10^{-4} for K_{s2}. Seidell [3] provides the solubilities of HgI_2 in various concentrations of KI at 25°C, but only to one or two significant figures (see Table 2). However, even by casual inspection, the data show a 2:1 mole ratio of KI to HgI_2. This ratio is consistent with the following solution reaction:

$$HgI_2(s) + 2\,I^-(aq) \longrightarrow HgI_4^{2-}(aq) \tag{10}$$

Nevertheless, the addition of one equivalent of I^- dissolves the HgI_2, which suggests that the equation representing the solution process is

$$HgI_2(s) + I^-(aq) \longrightarrow HgI_3^-(aq) \tag{11}$$

Of course, both processes take place to some degree, but it is worth asking whether the data obtained from the experiment shed any light on the identity of the mercury iodide species without reference to the species diagram already presented.

The data obtained can be analyzed by starting with an endpoint at which the HgI_2

Table 2. Solubility of HgI_2 in KI Solutions at 25°C in Moles per Liter [3]

[KI]	[HgI_2]
0.05	0.025
0.10	0.05
0.20	0.10
0.50	0.25
1.0	0.50
1.5	0.75
2.0	1.0
2.5	1.25

just dissolved. On the one hand, assume that the soluble species is HgI_3^- ($\bar{n} = 3$) and calculate the free iodide ion concentration and value for K_{s3}. On the other hand, assume $\bar{n} = 4$, or that the soluble species is HgI_4^{2-}, and calculate $[I^-]$ and K_{s4}. The symbol $\bar{n}$ represents the ratio of the total number of moles of bound ligand to the total number of moles of metal ions in solution, or the average number of bound ligands per metal ion. The calculation that yields the most nearly constant values for K_{sn} and a monotonic increase in I^- as the concentration of total Hg^{2+} increases can be assumed to represent best the behavior of the system.

The following is a sample calculation taken from an actual experiment:

Volume of 0.10M $Hg(NO_3)_2$ solution added = 55.0 ml; contains 5.5×10^{-3} moles

Volume of 1.0M KI solution added = 27.5 ml; contains 2.75×10^{-2} moles
Total volume $\approx$ 3500 ml

Moles of I^- used if $\bar{n} = 3$: $3 \times 5.5 \times 10^{-3} = 1.65 \times 10^{-2}$
Moles of I^- in solution unbound : $(2.75 - 1.65) \times 10^{-2} = 1.1 \times 10^{-2}$

$$[I^-] = \frac{11 \times 10^{-3}}{3.5} = 3.14 \times 10^{-3} \text{M}$$

$$K_{s3} = \frac{[HgI_3^-]}{[I^-]} = \frac{1.57 \times 10^{-3}}{3.14 \times 10^{-3}} = 0.50$$

Table 3 contains calculations for five endpoints observed in a single run. In each case, the endpoint was the disappearance of the last solid as excess KI solution was added. Note that the value of K_{s3} is constant except for the first point, whereas K_{s4} shows significantly greater spread. Also, $[I^-(aq)]$ increases monotonically for $\bar{n} = 3$ but not for $\bar{n} = 4$, where there is an overall decrease. Since the value of $[I^-]$ *must* increase, the experimental data favor equation 11 over equation 10 to represent the change occurring in the system when $HgI_2(s)$ dissolves in excess KI solution.

In an equilibrium system, the triiodo complex cannot exist alone, but equation 11 fits the data best. The values presented in references 1 and 2 that are used to obtain reasonable internal consistency in K and β values are $K_{s3} \cong 0.50$ and $K_{s4} \cong 80$. The

Table 3. Values for K_{eq} as a Function of $\bar{n}$[a]

Vol. (ml)[b] Hg(NO$_3$)$_2$	Vol. (ml)[c] KI	Moles Hg^{2+} added ($\times 10^3$)	Moles KI added ($\times 10^2$)	[Hg^{2+}]$_T$ ($\times 10^3$)	$\bar{n} = 3$ (assumed) [I$^-$(aq)] ($\times 10^3$)	K_{s3}	$\bar{n} = 4$ (assumed) [I$^-$(aq)] ($\times 10^3$)	K_{s4}
35.0	21.0	3.5	2.1	1.0	3.0	0.33	2.0	250
55.0	27.5	5.5	2.75	1.57	3.14	0.50	1.57	638
59.0	29.0	5.9	2.9	1.69	3.23	0.52	1.54	713
63.0	31.0	6.3	3.1	1.80	3.46	0.52	1.66	652
66.0	32.0	6.6	3.2	1.89	3.49	0.54	1.60	738

[a]Volumes were measured from burets. The endpoints used are for the solution of the precipitate. Side lighting was used.
[b][Hg(NO$_3$)$_2$] in stock solution = 0.10M.
[c][KI] in stock solution = 1.0M.

Table 4. Values for K_{eq} as a Function of $\bar{n}$[a]

Vol. (ml) Hg(NO$_3$)$_2$	Vol. (ml) KI	Moles Hg^{2+} added ($\times 10^3$)	Moles KI added ($\times 10^2$)	[Hg^{2+}]$_T$ ($\times 10^3$)	$\bar{n} = 3$ (assumed) [I$^-$(aq)] ($\times 10^3$)	K_{s3}	$\bar{n} = 4$ (assumed) [I$^-$(aq)] ($\times 10^3$)	K_{s4}
51.0	25.0	5.10	2.50	1.46	2.77	0.53	1.31	849
60.0	29.0	6.0	2.9	1.71	3.14	0.54	1.43	838

[a]The same conditions apply as in Table 3, except that precipitation endpoints were used.

agreement with K_{s3} and the lack of agreement with K_{s4} are convincing that the dominant species in solution when all the HgI$_2$ has just dissolved is HgI$_3^-$(aq). This species is also dominant just prior to the precipitation of HgI$_2$(s) by the addition of Hg^{2+} to an iodide-rich solution.

Table 4 shows the results obtained using the formation of the first precipitate as the endpoint. Both K_{s3} and K_{s4} values are consistently larger than those shown in equations 8 and 9.

Solid mercuric iodide exists in two crystalline forms, red and yellow [4]. At room temperature the red form is the more stable, but almost all first precipitations of HgI$_2$ produce the higher-energy yellow form. Apparently, the yellow form is both more rapidly formed than the red and is more rapidly dissolved in the presence of excess iodide ion.

In the red form, the Hg^{2+} ions are in tetrahedral holes in a nearly undistorted, face-centered, cubic array of iodide ions. The lack of distortion implies no unique covalent bonding to any one or more iodide ions. In the yellow form, the Hg^{2+} ions are in octahedral coordination in which two of the iodide-to-Hg^{2+} distances are shorter than the other four. This structure appears reasonable in light of the large decrease in stability constants from K_2 to K_3. In coordinating with halide ions, Hg^{2+} consistently displays this tight binding of two ligands and significantly weaker binding of the third and fourth ligands. To some degree, the color changes observed during the demonstration can be interpreted in terms of these facts.

REFERENCES

1. Sillén, L. G.; Martell, A. E., Eds. "Stability Constants of Metal-Ion Complexes," Special Publication No. 17; The Chemical Society: London, 1964; p 341.
2. Sillén, L. G.; Martell, A. E., Eds. "Stability Constants of Metal-Ion Complexes," Supplement No. 1, Special Publication No. 25; The Chemical Society: London, 1971; p 220.
3. Seidell, A. "Solubilities of Inorganic and Metal Organic Compounds," 3rd ed.; Van Nostrand: New York, 1940; Vol. I, pp 640, 641.
4. Wyckoff, R. W. G. "Crystal Structures," 2nd ed.; Interscience Publishers, John Wiley and Sons: New York, 1963; Vol. I, p 309.

4.2

Chloro and Thiocyanato Complexes of Cobalt(II)

Anhydrous cobalt(II) chloride is blue, but the hexahydrate is pink. These colors are evident in aqueous solutions of Co(II) containing chloride ions or thiocyanate ions under certain conditions. Several procedures illustrate the conditions that favor the blue color over the pink.

MATERIALS FOR PROCEDURE A

100 ml 0.1M cobalt(II) chloride solution, $CoCl_2$ (To prepare 1 liter of 0.1M stock solution, dissolve 23.8 g $CoCl_2 \cdot 6H_2O$ in distilled water and dilute to 1 liter.)

1 liter concentrated (12M) hydrochloric acid, HCl

2 liters distilled water

1-liter beaker

stirring rod

2-liter beaker

MATERIALS FOR PROCEDURE B

200 ml 0.1M cobalt(II) chloride solution, $CoCl_2$ (For preparation, see Materials for Procedure A.)

200 ml concentrated (12M) hydrochloric acid, HCl

500 ml distilled water

ice-salt bath at least 15 cm deep

600-ml beaker

stirring rod

2 400-ml beakers

hot plate

test tube, 25 mm $\times$ 200 mm (optional)

MATERIALS FOR PROCEDURE C

1.0 g cobalt(II) chloride hexahydrate, $CoCl_2 \cdot 6H_2O$

25 ml 1-butanol, C_4H_9OH

25 ml distilled water

12 ml concentrated (12M) hydrochloric acid, HCl

2.0 g potassium thiocyanate, KSCN

10 ml 0.1M silver nitrate solution, $AgNO_3$ (optional) (To prepare 1 liter of 0.1M stock solution, dissolve 17 g of $AgNO_3$ in distilled water and dilute to 1 liter.)

500 ml ice-salt bath (optional)

test tube, 25 mm × 200 mm, with stopper

2 600-ml beakers (optional)

10-ml pipette (optional)

hot plate (optional)

MATERIALS FOR PROCEDURE D

1.0 g cobalt(II) chloride hexahydrate, $CoCl_2 \cdot 6H_2O$

25 ml 2-propanol, C_3H_7OH

10 ml distilled water

500 ml ice-salt mixture

test tube, 25 mm × 200 mm, with stopper

2 600-ml beakers

hot plate

PROCEDURE A

Place 100 ml of 0.1M $CoCl_2$ in a 1-liter beaker. While stirring the $CoCl_2$ solution, slowly add concentrated hydrochloric acid until the solution turns purple and then blue. You will need slightly more HCl than the volume of $CoCl_2$ solution used.

When the solution is satisfactorily blue, add distilled water with stirring until the solution turns pink.

Still stirring, add concentrated hydrochloric acid until the solution once more turns blue.

Repeat this cycle, trying to end up with the 1-liter beaker nearly full of blue solution. Then pour the blue solution into the remaining water in the 2-liter beaker.

PROCEDURE B

Place 200 ml of 0.1M $CoCl_2$ in a 600-beaker and add 200 ml of concentrated hydrochloric acid with stirring. The solution should be purple. If the solution is blue, add distilled water slowly until it becomes purple. You may wish to practice this demonstration to achieve the desired shade of purple.

Place half the solution in a 400-ml beaker and heat it on a hot plate. Place the remaining solution in a 400-ml beaker in an ice-salt bath and chill to 0°C. The heated solution turns blue while the chilled solution turns pink.

If the hot, blue solution is poured into a 25 × 200 mm test tube placed in an ice bath covering only the lower half of the tube, the lower half of the solution turns pink while the upper half remains blue.

PROCEDURE C

Place approximately 1 g of $CoCl_2 \cdot 6H_2O$ in a 25 × 200 mm test tube and add 25 ml of 1-butanol. Stopper the test tube and shake it to dissolve most of the solid. The solution will be blue.

Add 25 ml of distilled water and again stopper and shake the test tube to mix the phases. Avoid shaking the mixture too long or too vigorously since it sometimes forms an emulsion. Allow the two liquid layers to separate. The aqueous layer (bottom) will be pink and the butanol layer colorless.

Gently pour 10–12 ml of concentrated hydrochloric acid into the butanol layer without stirring. The butanol layer turns blue while the aqueous layer remains pink. Mix the phases by shaking. The blue color disappears from the butanol layer.

Add to the test tube 1 g of potassium thiocyanate, KSCN, and shake the tube until the solid dissolves. The aqueous layer remains pink, but the butanol layer becomes bright blue.

To emphasize the effect of temperature on the color change, add a second gram of potassium thiocyanate, dissolve it, and heat the test tube in a beaker of boiling water (ca. 500 ml). The aqueous layer becomes intensely purple, thus diminishing greatly the contrast between it and the blue butanol layer. Chilling the test tube in an ice bath restores the pink color to the aqueous phase.

To generate a red, white, and blue test tube, add about 10 ml of 0.1M silver nitrate solution to the bottom of the aqueous layer, using a 10-ml pipette and creating as little agitation as possible. The white precipitate formed will settle into a white layer relatively quickly.

PROCEDURE D

Place approximately 1 g of $CoCl_2 \cdot 6H_2O$ in a dry 25 × 200 mm test tube and add 25 ml of 2-propanol (isopropanol). Stopper and shake to dissolve the solid. The solution will be blue. Carefully add distilled water to the test tube, with mixing, until the homogeneous solution is a light purple. Normally, this takes between 4 and 5 ml of water.

Place the test tube in a beaker of boiling water (ca. 500 ml); the solution turns blue. Place the test tube in an ice-salt bath; the solution turns pink.

HAZARDS

Concentrated hydrochloric acid can cause severe burns. Hydrochloric acid vapors are extremely irritating to the skin, eyes, and respiratory system.

The alcohols, 1-butanol and 2-propanol, are flammable, and their vapors are irritating to the eyes and respiratory system.

Silver nitrate solutions are irritating to the skin and eyes. Also, they may stain skin or clothing an unsightly brown to black color. If a spill is recognized when it occurs, rinse the spot with some sodium thiosulfate solution followed by water. If a black spot develops later, try removing it as follows. Prepare a 10% solution of potassium ferricyanide in water and a 10% solution of sodium thiosulfate in 1% ammonia water. Mix equal volumes of these two solutions and scrub the black stain thoroughly with the resulting solution. Rinse well with water.

DISPOSAL

Since the solutions produced are highly acidic, they should be neutralized with sodium bicarbonate or soda ash and flushed down the drain.

DISCUSSION

The nature of the species in many of the solutions in this demonstration is not fully known. In the dilute aqueous solutions, the predominant species is $Co(H_2O)_6{}^{2+}$. However, the replacement of one of the water molecules by a chloride ion or a thiocyanate ion only slightly changes the observed pink color, which makes it impossible to know by observation when significant coordination begins. The other extreme, that is, complete coordination to Cl^- and SCN^-, is also well established. Compounds having the composition B_2CoCl_4 and $B_2Co(SCN)_4$, where B is a large, univalent metal ion or a large, substituted ammonium ion, have been isolated from organic solvents. These solids have the intense blue color apparent in some of the solutions in this demonstration, and the cobalt(II) in these solids is tetrahedrally coordinated to the four ligands.

The existence of the blue color, however, does not necessarily mean that the full coordination to Cl^- or SCN^- has been reached. Evidence exists that the shift to tetrahedral coordination and the blue color occurs when the second chloride ion enters the coordination sphere [1]. Although the evidence is not as clear for thiocyanate coordination, it suggests that the shift to tetrahedral coordination occurs later in the replacement of waters by thiocyanates [2]. In the totally aqueous system where the color change is induced simply by the addition of concentrated hydrochloric acid, the relationship between the various species has been reported by Zeltman et al. as follows [1]:

Octahedral (pink)	Tetrahedral (blue)
$Co(H_2O)_6{}^{2+}$	$CoCl_2(H_2O)_2$
	$\beta_2 = 1.7 \times 10^{-3}$
$CoCl(H_2O)_5{}^+$	$CoCl_3H_2O^-$
$K_1 = 0.17$	$\beta_3 = 3.1 \times 10^{-3}$
	$CoCl_4{}^{2-}$
	$\beta_4 = 8.8 \times 10^{-3}$

The reaction responsible for the color change is

$$CoCl(H_2O)_5{}^+(aq) + Cl^-(aq) \longrightarrow CoCl_2(H_2O)_2(aq) + 3 H_2O(l) \quad \beta_2 = 1.7 \times 10^{-3}$$

In mixed solvents, one or more of the water molecules may be replaced by an organic solvent molecule.

Although definitive equations are hard to write, the explanation for the change in color follows directly from the basic arguments of crystal field or ligand field theory. One expects the splitting of the d orbitals to be greater for an octahedral coordination than for a tetrahedral one, because in the octahedral case two of the d orbitals point directly at the six ligands, whereas in the tetrahedral case there are only four ligands and no orbitals point directly at them. A d-d transition in the octahedral case requires a higher photon energy and occurs at a shorter wavelength than the corresponding transition in a tetrahedral configuration. The shift from pink to blue indicates a shift from absorption in the middle of the visible spectrum to absorption in the red (see figure). Note the difference in molar absorptivities, which will be discussed later.

As observed in this demonstration, increasing the concentration of chloride ions and inducing more displacement of water molecules by chloride ions ultimately favors a change from octahedral to tetrahedral geometry. The change is favored by an increase in temperature, which means it is an endothermic process, and it is favored by the use of organic solvents instead of water.

Actually, this change in geometry is not unusual in the chemistry of transition metal ions [3]. In every case the change is accompanied by a positive change in enthalpy amounting to 42–54 kJ/mole. This significant enthalpy change requires that the reaction also be accompanied by a significant increase in entropy.

The knowledge that the formation of the blue species is endothermic can help avoid a possible misconception. In Procedure A, when concentrated hydrochloric acid is poured into the reaction mixture, heat is generated by the dilution of concentrated hydrochloric acid. This increase in temperature favors the formation of the blue species.

Evidence also shows that these tetrahedral complexes are favored by less polar solvents [4]. When $CoCl_2 \cdot 6H_2O$ dissolves in either 2-propanol or 1-butanol, the blue complex is formed even though the average number of chlorides per cobalt must be only two. The blue complex forms with the addition of less hydrochloric acid to a solution of cobalt(II) chloride in a 1:1 mixture of 2-propanol and water than is required in aqueous solution. The alcohol certainly enters the coordination sphere of the co-

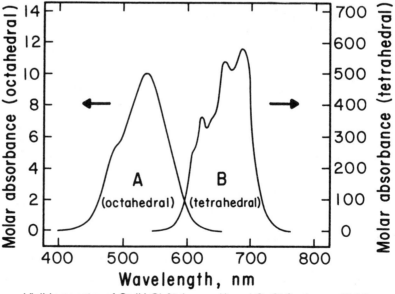

Visible spectra of $Co(H_2O)_6^{2+}$ (curve A) and $CoCl_4^{2-}$ (curve B) [4].

balt(II) in some cases, but the decrease in the dielectric constant is probably another significant factor in promoting the change.

In Procedure C, a large amount of hydrochloric acid is added to the solution, which temporarily turns the butanol layer blue but does not maintain the blue color after equilibrating with the water layer. In contrast, the addition of a smaller molar quantity of solid potassium thiocyanate produces a distinctly blue butanol layer. This result may be caused in part by the fact that cobalt(II) forms somewhat more stable complexes with thiocyanate ions than with chloride ions, and it may also be caused by the higher distribution coefficient for the thiocyanato complex than for the chloro complex.

Earlier, we noted that in water the molar absorptivity of the tetrahedral chloro complex of cobalt(II) ranged from 50–100 times greater than the molar absorptivity of the octahedral complex. This great difference helps make the color changes much sharper than if the absorptivities were equal. To make the blue color completely dominant, we need to change less than 10% of the cobalt complex from the octahedral to the tetrahedral form. Procedure C demonstrated this situation when the butanol layer is blue and the water layer is pink. When both the blue and pink complexes are present in the same solution, only the blue color is seen. An explanation of the increase in absorptivity with the formation of the tetrahedral complex is beyond the scope of this discussion, but its nature can be indicated. All the d-d transitions observed are LaPorte forbidden, so the absorptivity depends on the degree to which the selection rules are relaxed in a given case. In the octahedral case, the selection rule is relaxed primarily by vibrational contributions. In the tetrahedral case, the LaPorte selection rule is automatically relaxed because of the different symmetry, and it may be further relaxed by the mixing of d and p orbitals. In any case, the marked difference in absorptivity is of great benefit in this demonstration.

For other procedures which use acetone instead of an alcohol, see reference 5.

REFERENCES

1. Zeltman, A. H.; Matwiyoff, N. A.; Morgan, L. L. *J. Phys. Chem.* **1968**, *72*, 121.
2. Tribalat, S.; Zeller, C. *Bull. Soc. Chim. France* **1962**, 2041.
3. Lever, A. B. P. *J. Chem. Educ.* **1974**, *51*, 612.
4. Cotton, F. A.; Wilkinson, G. "Advanced Inorganic Chemistry: A Comprehensive Text," 4th ed.; Interscience Publishers, John Wiley and Sons: New York, 1980; p 772.
5. Ophardt, C. E. *J. Chem. Educ.* **1980**, *57*, 453.

4.3

Precipitates and Complexes of Lead(II)

To a lead(II) nitrate solution in a 2-liter beaker, solutions of chloride, iodide, carbonate, chromate, and hydroxide ions are added in succession. The relative solubilities and colors of various lead salts in water and in excess hydroxide ion solutions are observed. The insolubility of other salts such as sulfide, phosphate, or higher oxides can be observed by precipitation from the basic solution.

MATERIALS

2 liters distilled water

25 ml 1.0M lead nitrate, $Pb(NO_3)_2$ (To prepare 1 liter of 1.0M stock solution, dissolve 332 g $Pb(NO_3)_2$ in distilled water and dilute to 1 liter.)

50 ml 1.0M potassium chloride, KCl (To prepare 1 liter of 1.0M stock solution, dissolve 75 g KCl in distilled water and dilute to 1 liter.)

30 ml 1.0M potassium iodide, KI (To prepare 1 liter of 1.0M stock solution, dissolve 166 g KI in distilled water and dilute to 1 liter.)

2.0 g sodium bicarbonate, $NaHCO_3$

25 ml 1.0M potassium chromate, K_2CrO_4 (To prepare 1 liter of 1.0M stock solution, dissolve 194 g K_2CrO_4 in distilled water and dilute to 1 liter.)

30 ml 1.0M sodium hydroxide, NaOH (To prepare 1 liter of 1.0M stock solution, dissolve 40 g NaOH in distilled water and dilute to 1 liter.)

30 g solid sodium hydroxide, NaOH

25 ml commercial bleach (5% NaOCl solution)

8 g sodium acetate trihydrate, $NaC_2H_3O_2 \cdot 3H_2O$

250–300 ml crushed ice or ice cubes

ca. 2 ml 0.1M sodium sulfide, Na_2S (To prepare 1 liter of 0.1M stock solution, dissolve 24 g $Na_2S \cdot 9H_2O$ in distilled water and dilute to 1 liter.)

ca. 1 g sodium dihydrogen phosphate, NaH_2PO_4

10 ml 6M nitric acid, HNO_3, or 10 ml 6M acetic acid, $HC_2H_3O_2$

10 g sodium carbonate, Na_2CO_3

2 2-liter beakers

100-ml graduated cylinder

10–12 50-ml beakers

24 test tube labels

hot plate, with magnetic stirrer

10–12 test tubes, 25mm $\times$ 200 mm, with stoppers

test tube rack

small dipper and mittens or towel, or other means of removing ca. 50-ml samples from hot beaker and pouring into test tubes

Meker burner

thermometer, $-10°C$ to $+110°C$

4 400-ml beakers

PROCEDURE

To prepare for the demonstration, heat about 2 liters of distilled water to boiling in a 2-liter beaker.

Measure into labelled, 50-ml beakers each of the following:

20 ml of 1.0M $Pb(NO_3)_2$
50 ml of 1.0M KCl
30 ml of 1.0M KI
2.0 g of $NaHCO_3$, dissolved in 25 ml of distilled water
20 ml of 1.0M K_2CrO_4
30 ml of 1.0M NaOH
30 g of NaOH pellets
25 ml of commercial bleach
4 g of $NaC_2H_3O_2·3H_2O$
4 g of $NaC_2H_3O_2·3H_2O$

Prepare the following labels for test tubes: lead chloride, lead iodide (3 labels), lead carbonate, lead chromate, basic lead chromate, and sodium plumbite.

To begin the demonstration, put a large stirring bar in the second 2-liter beaker and place the beaker on a hot plate with a magnetic stirrer. Add 20 ml of 1.0M $Pb(NO_3)_2$ solution and 200 ml of cold distilled water. Stir gently and, if necessary, add a few drops of dilute nitric acid to maintain a clear solution.

To this clear solution of lead nitrate, add 50 ml of 1.0M KCl while continuing stirring. A white precipitate of $PbCl_2$ will form slowly. When this precipitate becomes visible, heat the beaker. The lead chloride will dissolve, yielding a clear, colorless solution. Transfer approximately 50 ml of this solution to one of the test tubes, label it, and allow it to cool.

To the rest of the solution in the beaker, add 30 ml of 1.0M potassium iodide. A dense, yellow precipitate of PbI_2 will form. Additional iodide ion will make the lead iodide difficult to dissolve. Continue heating this suspension and dilute it to approximately 1800 ml with hot, preferably boiling, distilled water. Stir and heat to dissolve the precipitate. Try to obtain a clear, colorless solution, but if the precipitate does not dissolve, continue the demonstration. Transfer 50-ml portions of the solution (or suspension) to each of three labelled test tubes, and remove the beaker from the hot plate.

Let the solution cool somewhat before the next addition is made. If the lead iodide was not completely dissolved in the samples transferred to the test tubes, heat these over the burner until solution is complete. Dilute if necessary. Complete solution and slow cooling produce a more aesthetic precipitate.

Add 250–300 ml of ice to the beaker to adjust the temperature to 50–60°C. With stirring, add 2.0 g of sodium bicarbonate dissolved in 25 ml of distilled water. A white precipitate forms, and sometimes the gas evolution is visible as the bicarbonate solution is added. Transfer a representative 50-ml sample of this suspension to a test tube and label the tube.

With vigorous stirring but without heating, add 20 ml of 1.0M potassium chromate to the large beaker. The white suspension changes fairly rapidly to a yellow suspension. As soon as this occurs, transfer 50 ml of the suspension to a test tube. Add a few drops of dilute nitric acid to the test tube to prevent the precipitation of basic lead chromate.

While continuing to stir the suspension of lead chromate in the large beaker, add 30 ml of 1.0M NaOH slowly enough to observe the color change at the region of addition. The yellow precipitate becomes orange-brown as the chromate is converted to the basic chromate. Place a 50-ml sample of this suspension in a test tube and label it.

Gradually add solid NaOH to the large beaker with stirring, giving the pellets time to dissolve before making subsequent additions. The suspension clears to a yellow chromate–colored solution as the last 5 g or so of the NaOH dissolve. Transfer 50 ml of this solution to a test tube and label it. The remaining solution may be used in the supplementary procedures (below).

The demonstration to this point illustrates many simple reactions of Pb(II), and the materials in the test tubes allow review of the reactions. By the end of the demonstration, the $PbCl_2$ should have reprecipitated as white needles, and the PbI_2 should have come out of solution in the form of golden platelets.

Supplementary Procedures

We offer the following procedures as suggestions for further exploration of this demonstration:

1. Pour 250 ml of the alkaline solution of lead salts into each of four 400-ml beakers. To one of these, add a stirring bar and 25 ml of commercial hypochlorite bleach. Heat and stir until a precipitate forms (PbO_2 or Pb_3O_4). To a second sample, add 0.1M sodium sulfide solution. Black lead sulfide precipitates. To a third, add about 1 g of NaH_2PO_4 dissolved in a minimum of water. A white precipitate of lead phosphate forms. To a fourth sample, add enough strong acid to acidify the solution. This demonstrates the rich mixture of substances present and thereby illustrates how dominant each one of the equilibria had to be to produce a single product. Usually, the most obvious change is the oxidation of iodide ions to molecular iodine. The use of acetic acid in place of nitric acid sometimes produces other changes.

2. To one of the test tubes containing PbI_2, add 4.0 g of $NaC_2H_3O_2 \cdot 3H_2O$ and shake to dissolve. The golden precipitate disappears as the acetate complexes of Pb^{2+} form. Sometimes a new precipitate (basic lead iodide) forms. If so, add 4.0 g of $NaC_2H_3O_2 \cdot 3H_2O$ to a second test tube containing PbI_2; then add a few drops of acetic acid to prevent formation of the basic salt.

3. To the test tube containing the lead carbonate, add 4.0 g of $NaC_2H_3O_2 \cdot 3H_2O$ and shake. Although the suspension clears somewhat, not all the lead carbonate dissolves.

4. The conversion of lead chromate to lead carbonate shows the reversal of one of the reactions observed in the demonstration. Add 2 ml of 1.0M $Pb(NO_3)_2$ and 2 ml of 1.0M K_2CrO_4 to a test tube and add about 20 ml of distilled water. The suspension

($PbCrO_4$) should be bright yellow. Heat the test tube over the burner. Usually this produces basic lead chromate. Add solid Na_2CO_3 in small amounts while continuing to heat gently. The solid may darken at first but eventually turns almost white, while the solution becomes yellow. This process illustrates the reversal of the reaction first observed and illustrates the procedure frequently employed in separating cations from anions in a qualitative analysis scheme.

HAZARDS

Lead nitrate is harmful if taken internally. The dust from lead salts should not be inhaled. The effects of exposure to small concentrations can be cumulative, causing loss of appetite and anemia.

Chromates are irritating to the skin and eyes and are poisonous. Hexavalent chromium compounds are cancer-suspect agents.

Sodium hydroxide can cause severe burns of the eyes and skin. Dust from solid sodium hydroxide is very caustic.

Sodium hypochlorite solution (commercial bleach) can cause burns and is irritating to the skin and eyes.

Nitric acid is a strong acid and a powerful oxidizing agent. Contact with combustible material can cause violent and explosive reactions. The liquid can cause severe burns to the skin and eyes. The vapor is irritating to the eyes and to the respiratory system.

Acetic acid is a skin and eye irritant.

DISPOSAL

Lead salts can be converted to lead sulfide by treatment with sodium or ammonium sulfide in basic solution. The collected lead sulfide should be buried in a landfill designed for heavy metals.

Alternatively, small volumes may be discarded by flushing down the drain with large volumes of water. Local regulations should be consulted.

DISCUSSION

The following list presents the net ionic equations and equilibrium constants for reactions as they occur in this demonstration.

$$Pb(NO_3)_2(s) + H_2O(l) \longrightarrow Pb^{2+}(aq) + 2\,NO_3^-(aq) \tag{1}$$

$$Pb^{2+}(aq) + 2\,Cl^-(aq) \longrightarrow PbCl_2(s) \qquad K_{eq} = 1.0 \times 10^5 \tag{2}$$

$$PbCl_2(s) + 2\,I^-(aq) \longrightarrow PbI_2(s) + 2\,Cl^-(aq) \qquad K_{eq} = 4 \times 10^2 \tag{3}$$

$$PbI_2(s) + 2\,HCO_3^-(aq) \longrightarrow PbCO_3(s) + H_2CO_3(aq) + 2\,I^-(aq)$$
$$K_{eq} = 3.3 \times 10^{-1} \tag{4}$$

$$PbCO_3(s) + CrO_4^{2-}(aq) \longrightarrow PbCrO_4(s) + CO_3^{2-}(aq) \qquad K_{eq} = 36 \tag{5}$$

$2\,PbCrO_4(s) + 2\,OH^-(aq) \longrightarrow Pb(OH)_2\!\cdot\!PbCrO_4(s) + CrO_4^{2-}(aq)$

$$K_{eq} = 4.3 \times 10^6 \qquad (6)$$

$Pb(OH)_2\!\cdot\!PbCrO_4(s) + 4\,OH^-(aq) \longrightarrow 2\,Pb(OH)_3^-(aq) + CrO_4^{2-}(aq)$

$$K_{eq} = 4.6 \times 10^{-5} \qquad (7)$$

$Pb(OH)_3^-(aq) + S^{2-}(aq) \longrightarrow PbS(s) + 3\,OH^-(aq) \qquad K_{eq} = 2.0 \times 10^{14} \qquad (8)$

$3\,Pb(OH)_3^-(aq) + 2\,PO_4^{3-}(aq) \longrightarrow Pb_3(PO_4)_2(s) + 9\,OH^-(aq)$

$$K_{eq} = 2.5 \times 10^2 \qquad (9)$$

$Pb(OH)_3^-(aq) + ClO^-(aq) \longrightarrow PbO_2(s) + Cl^-(aq) + H_2O(l) + OH^-(aq) \qquad (10)$

(Undoubtedly, there are other redox reactions taking place in this mixture, but this one is the observable one. If Pb_3O_4 is to be considered the product, it may be considered as $PbO_2\!\cdot\!2PbO$.)

$PbI_2(s) + 4\,C_2H_3O_2^-(aq) \longrightarrow Pb(C_2H_3O_2)_4^{2-}(aq) + 2\,I^-(aq)$

$$K_{eq} = 6.3 \times 10^{-5} \qquad (11)$$

The values of equilibrium constants were calculated from the primary equilibrium constants in Table 1 and from $\log \beta_3$ for $Pb(OH)_3^-$ which is 13.7. In this demonstration either $PbBr_2$ or $Pb(SCN)_2$ could be substituted for $PbCl_2$, and $PbSO_4$ might be substituted for $PbCO_3$, but none of these substitutions provides any advantage. Because all the precipitates are white, visual demonstration of conversion from one to the other is not possible.

Table 1. Equilibrium Constants for Lead(II) Reactions [1]

Substance	$\log K_{s0}$	K_{s0}
$PbCl_2(s)$	-5.0	1.0×10^{-5}
$PbI_2(s)$	-7.61	2.5×10^{-8}
$PbCO_3(s)$	-11.0	1.0×10^{-11}
$PbCO_3(s)$ in saturated CO_2	ca. -10.0	1.0×10^{-10}
$PbCrO_4(s)$	-12.55	2.8×10^{-13}
$Pb(OH)_2\!\cdot\!PbCrO_4(s)$	-31.74	1.8×10^{-32}
$PbS(s)$	-28.0	1.0×10^{-28}
$Pb_3(PO_4)_2(s)$	-43.53	3.0×10^{-44}
$PbBr_2(s)$	-5.67	2.1×10^{-6}
$Pb(SCN)_2(s)$	-4.70	2.0×10^{-5}
$PbSO_4(s)$	-7.80	1.6×10^{-8}

Consider the reactions listed above and the sequence in which they are carried out. Reactions 1, 2, and 3 proceed readily. In reaction 2, an excess of Cl^- is employed to favor formation of the relatively soluble $PbCl_2$, whereas in reaction 3, an excess of I^- should be avoided because it would hinder the formation of a complete solution of the lead iodide.

Reaction 4 is favored by heating because the product H_2CO_3 is lost as CO_2 (see Demonstration 4.10). This loss of CO_2 raises the pH of the solution and favors the precipitation of basic lead chromate instead of $PbCrO_4$ in reaction 5. Addition of nitric acid lowers the pH and allows the precipitation of the yellow $PbCrO_4$.

The formation of the basic lead chromate, reaction 6, proceeds without difficulty, frequently upon simply heating the suspension of $PbCrO_4$. The dissolving of basic lead chromate in excess sodium hydroxide almost certainly is misrepresented by equation 7. As noted previously, a variety of hydroxo complexes form, and the equilibrium constant for the solution process would be considerably larger than 4.6×10^{-5} if all the soluble complexes were considered. We chose the complex $Pb(OH)_3^-$ because it is the most stable mononuclear complex. The oxidation of Pb(II) to Pb(IV) by hypochlorite, reaction 10, is a slow reaction and takes place primarily on the bottom of the beaker over several minutes.

Sometimes the basicity of the sodium acetate solution is sufficient to promote the precipitation of basic lead iodide when the acetate is added to dissolve the lead iodide, as in reaction 11. That product is slightly off-white in color and crystallizes as needles. Addition of a small quantity of acetic acid normally inhibits the precipitation or dissolves the precipitate.

The conversion of relatively insoluble salts to carbonates by flooding the system with carbonate ion solution and then heating is a standard method of freeing anions for qualitative analysis and of converting one salt to another, which would be extremely difficult otherwise. The conversion of lead chromate to the carbonate uses this technique.

Lead(II) forms complexes with halide ions and with acetate ions, but these complexes are colorless and soluble and therefore not readily observed in this demonstration. Lead(II) also forms a series of complexes with hydroxide ions, which is further complicated by the formation of a series of polynuclear species, such as $[Pb_4(OH)_4]^{4+}$ and $[Pb_6(OH)_8]^{4+}$, some of which have been well characterized [2]. Table 2 presents the equilibrium constants for the formation of many of these complexes.

Table 2. Logarithms of Stability Constants for Lead(II) Complexes[a] [1]

Ion	β_1	β_2	β_3	β_4	β_5
Cl^-	1.17	1.3	1.4	1.2	—
Br^-	1.1	1.8	2.2	2.0	2.4
I^-	1.26	2.8	3.4	3.9	—
$C_2H_3O_2^-$	2.1	2.17	3.18	3.4	—
OH^-	6.3	10.9	13.7	—	—
$\beta_{21} = 7.9$; $\beta_{34} = 33.8$					
$\beta_{44} = 37.5$; $\beta_{68} = 71.3$					

[a]These values indicate that the solution of PbX_2 is not a solution of $Pb^{2+}(aq)$ and $X^-(aq)$ as we normally assume for ionic substances, but rather a mixture of complex species such as $PbX^+(aq)$, $PbX_2(aq)$, and $PbX_3^-(aq)$.

Although this demonstration does not directly include lead(II) oxide or hydroxide, the oxide precipitate has three forms, each with its own solubility product constant:

	log K_{s0}
$Pb_2O(OH)_2(s)$ as $[Pb^{2+}][OH^-]^2$	−14.9
$PbO(s)$(yellow) as $[Pb^{2+}][OH^-]^2$	−15.1
$PbO(s)$(red) as $[Pb^{2+}][OH^-]^2$	−15.3

REFERENCES

1. Smith, R. M.; Martell, A. E., Eds. "Critical Stability Constants," Vol. IV, Inorganic Complexes; Plenum Press: New York, 1976.
2. Trotman-Dickenson, A. F., Executive Ed. "Comprehensive Inorganic Chemistry"; Pergamon Press: Oxford, 1973.

4.4

Iodo and Silver(I) Complexes of Silver Iodide

In each of three beakers, approximately the same amount of silver iodide is precipitated with continuous stirring. Separate solutions are then added to each beaker: 3M KI to the first beaker, 3M $AgNO_3$ to the second, and 3M KNO_3 to the third. In the first two beakers but not the third, the silver iodide dissolves before 100 ml of solution has been added. Addition of water to the clear solutions reprecipitates silver iodide.

MATERIALS

500 ml distilled water

500 ml 3.0M silver nitrate, $AgNO_3$ (To prepare 1 liter of 3.0M stock solution, dissolve 510 g $AgNO_3$ in distilled water and dilute to 1 liter.)

500 ml 3.0M potassium iodide, KI (To prepare 1 liter of 3.0M stock solution, dissolve 500 g KI in distilled water and dilute to 1 liter. This solution is easily air-oxidized and will turn yellow if exposed to air.)

500 ml 3.0M potassium nitrate, KNO_3 (To prepare 1 liter of 3.0M stock solution, dissolve 303 g KNO_3 in distilled water and dilute to 1 liter.)

3 magnetic stirrers with stirring bars

3 600-ml beakers, preferably tall-form

2 10-ml graduated cylinders or 2 droppers marked to deliver 1.0 ml

4 100-ml graduated cylinders

1500- or 2000-ml beaker

PROCEDURE

Set up three magnetic stirrers. Place on each a 600-ml beaker containing a stirring bar. To each add 100 ml of distilled water and adjust the stirring rate to provide efficient but nonturbulent stirring. Add to each beaker 1.0 ml of 3.0M $AgNO_3$ and 1.0 ml of 3.0M KI to produce a fairly dense precipitate of yellow AgI.

Fill three 100-ml graduated cylinders, one with 3.0M KI, one with 3.0M $AgNO_3$, and one with 3.0M KNO_3 solution. To the first beaker, slowly add the 3.0M KI solution until the precipitate dissolves. Note the volume of KI solution used.

To the second beaker, slowly add the 3.0M $AgNO_3$ solution until the precipitate dissolves. Do not add more $AgNO_3$ than necessary, even if the dissolution process seems slow. Note the volume of $AgNO_3$ solution required.

To the third beaker, add 3.0M KNO_3 solution. Add the entire 100 ml, demonstrating that the precipitate does not dissolve even in a quantity of the KNO_3 solution greater than either the volume of KI or $AgNO_3$ required to dissolve the AgI.

Carefully add distilled water to the first two beakers until each develops a convincing turbidity, showing the reprecipitation of AgI upon dilution.

Add 3.0M KI solution to the first beaker and 3.0M $AgNO_3$ solution to the second beaker until the precipitate redissolves. If desired, repeat the additions of water and 3M reagents to demonstrate the continuing reversibility of the system.

To terminate the demonstration, transfer the contents of the three beakers into a single large beaker, demonstrating graphically the amount of KI and $AgNO_3$ required to hold a relatively small amount of AgI in solution. The suspension of AgI becomes as thick as pudding, and the beaker becomes markedly warmer. This combination of solutions also serves as the first step in the recovery of silver from the demonstration.

HAZARDS

Silver nitrate and its solutions are irritating to the skin and eyes and can cause burns. Also, they may stain skin or clothing an unsightly brown to black color. If a spill is recognized when it occurs, rinse the spot with some sodium thiosulfate solution followed by water. If a black spot develops later, try removing it as follows. Prepare a 10% solution of potassium ferricyanide in water and a 10% solution of sodium thiosulfate in 1% ammonia water. Mix equal volumes of these two solutions and scrub the black stain thoroughly with the resulting solution. Rinse well with water.

Mixtures of potassium nitrate and combustible materials are readily ignited. If finely divided, such mixtures react explosively when ignited.

DISPOSAL

The silver iodide suspension should be allowed to settle, after which the supernatant liquid can be decanted and flushed down the drain. The solid silver iodide should be collected for later recovery of the silver. The silver can be recovered by procedures described in reference 1.

DISCUSSION

This demonstration shows that silver iodide dissolves in an excess of either silver nitrate or potassium iodide and that in either case the concentration of the dissolving species must be quite high. The fact that silver iodide does not dissolve upon the addition of potassium nitrate demonstrates that the phenomenon is not caused simply by a change in ionic strength.

To explain these observations requires the postulation of complex formation in the presence of either excess Ag^+ or excess I^-. Figure 1 presents the formulae of the various complexes postulated to account for quantitative measurements. These formulae are arranged in a manner intended to illustrate the relationships between the various species and to identify the equilibrium constants relating these species to one another. Table 1 presents numerical values for the equilibrium constants, and Table 2 presents a

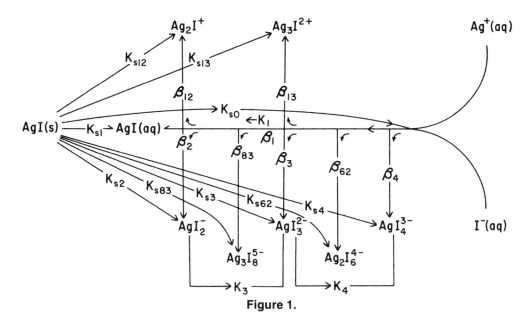

Figure 1.

number of equilibrium constants derived from those in Table 1. The complexes formed in excess iodide ions look quite normal down to the polynuclear species $Ag_2I_6^{4-}$ and $Ag_3I_8^{5-}$. Such polynuclear species are encountered relatively often, but the positively charged polynuclear species formed in excess Ag^+ are less common.

This demonstration, which shows that AgI is soluble in excess Ag^+ or excess I^-, is likely to upset some preconceived notions about the silver iodide system. Silver halides often are used to illustrate the principles of solubility and the application of

Table 1. Equilibrium Constants for the Ag^+/I^- System [2]

Constant	Equilibrium expression	log K_{eq}	K_{eq}
K_{s0}	$[Ag^+][I^-]$	-16.35	4.5×10^{-17}
$K_1 = \beta_1$	$\dfrac{[AgI(aq)]}{[Ag^+][I^-]}$	8.1	1.3×10^8
β_2	$\dfrac{[AgI_2^-]}{[Ag^+][I^-]^2}$	11.0	1.0×10^{11}
β_3	$\dfrac{[AgI_3^{2-}]}{[Ag^+][I^-]^3}$	13.8	6.3×10^{13}
β_4	$\dfrac{[AgI_4^{3-}]}{[Ag^+][I^-]^4}$	14.3	2.0×10^{14}
β_{62}	$\dfrac{[Ag_2I_6^{4-}]}{[Ag^+]^2[I^-]^6}$	29.7	5.0×10^{29}
β_{83}	$\dfrac{[Ag_3I_8^{5-}]}{[Ag^+]^3[I^-]^8}$	46.4	2.5×10^{46}
K_{s12}	$\dfrac{[Ag_2I^+]}{[Ag^+]}$	-6.0	1.0×10^{-6}
K_{s13}	$\dfrac{[Ag_3I^{2+}]}{[Ag^+]^2}$	-3.15	7×10^{-4}

Table 2. Equilibrium Constants Derived from Tabulated Values

Constant	log K	K
K_2	$\log\beta_2 - \log K_1 = 2.9$	8×10^2
K_3	$\log\beta_3 - \log\beta_2 = 2.8$	6.3×10^2
K_4	$\log\beta_4 - \log\beta_3 = 0.5$	3
K_{s2}	$\log\beta_2 + \log K_{s0} = -5.35$	4.5×10^{-6}
K_{s3}	$\log\beta_2 + \log K_{s0} = -2.55$	2.8×10^{-3}
K_{s4}	$\log\beta_4 + \log K_{s0} = -2.05$	8.9×10^{-3}
K_{s62}	$\log\beta_{62} + 2\log K_{s0} = 29.7 - 32.7 = -3.0$	1×10^{-3}
K_{s83}	$\log\beta_{83} + 3\log K_{s0} = 46.4 - 49.05 = -2.65$	2.2×10^{-3}
K for reaction $2\,AgI_3^{2-} \longrightarrow Ag_2I_6^{4-}$		
	$\log\beta_{62} - 2\log\beta_3 = 29.7 - 27.6 = 2.1$	1.3×10^2

solubility products to quantitative analysis. The behavior of silver iodide, which is the least soluble of these halides, is treated as the most ideal in terms of correspondence to theoretical principles. Because the value of K_{s0} for AgI is 4.5×10^{-17}, in a saturated solution of pure AgI,

$$[Ag^+] = [I^-] = \sqrt{45} \times 10^{-9} = 6.7 \times 10^{-9} M$$

Because β_1 is 1.3×10^8, the concentration of AgI(aq) is

$$(1.3 \times 10^8) \times (4.5 \times 10^{-17}) = 5.9 \times 10^{-9} M$$

The law of mass action requires that the addition of a small excess of either Ag^+ or I^- precipitate solid AgI. Nephelometric studies of the AgI system indicate that it behaves relatively ideally in the region of the equivalence point. A small excess of silver(I) produces the same turbidity in a saturated solution of AgI as does exactly the same small excess of iodide ions. Thus, the system does behave ideally in this sense.

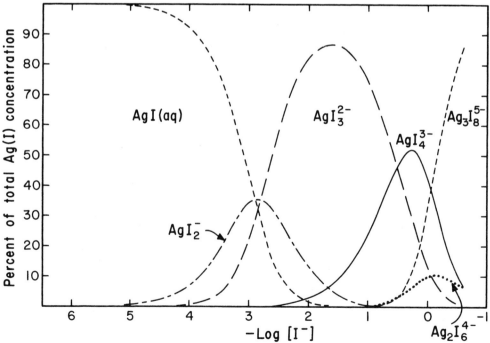

Figure 2. Species distribution of AgI in excess I^-.

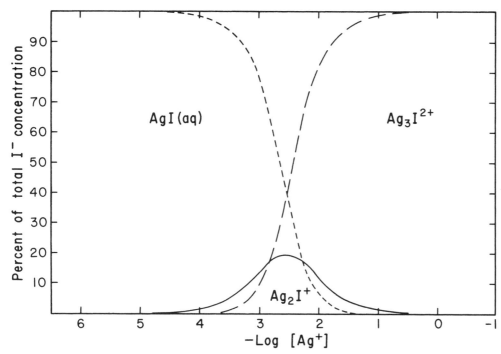

Figure 3. Species distribution for AgI in excess Ag^+.

Figures 2 and 3 present the species distribution diagrams for AgI in excess I^- and excess Ag^+, respectively. They show that in either case the free ion concentration must reach approximately $1 \times 10^{-5}M$ before the first complex approaches 1% of the concentration of AgI(aq), which is constant at $5.9 \times 10^{-9}M$. Thus, on either side of the equivalence point, there are approximately 4 decades of ion concentration in which complexing is insignificant.

At Ag^+(aq) concentrations above $10^{-5}M$, the complexes Ag_2I^+(aq) and Ag_3I^{2+}(aq) form, but the Ag_3I^{2+}(aq) soon begins to dominate (Figure 3). When $[Ag^+]$ is about $3 \times 10^{-3}M$, the concentration of Ag_3I^{2+}(aq) equals the concentration of AgI(aq). By $[Ag^+] = 3 \times 10^{-2}M$, the concentration of Ag_3I^{2+}(aq) has become 100 times the concentration of AgI(aq), and its concentration rises sharply thereafter until it becomes high enough to dissolve all the AgI(s). This dissolution appears to occur when $[Ag^+]$ is somewhat less than 1M.

When excess iodide is added to the suspension of solid AgI, complex formation becomes significant at $[I^-] = 10^{-5}M$ (Figure 2). When $[I^-]$ is slightly greater than $10^{-3}M$, the concentrations of AgI(aq), AgI_2^-(aq), and AgI_3^{2-} are approximately the same, i.e., about $5.9 \times 10^{-9}M$. The concentration of AgI_3^{2-} continues to increase until it maximizes at an I^- concentration of about $2.5 \times 10^{-2}M$. At this point, [AgI] is still about 1% of the total, so the AgI in this demonstration will not dissolve. The AgI(s) dissolves as $[I^-]$ approaches 1M. In this concentration range, AgI_4^{3-} is the most important species, but AgI_3^{2-}, $Ag_2I_6^{4-}$, and $Ag_3I_8^{5-}$ are all present in significant concentrations. Although the system of complexes is more complicated on the iodide-rich side of the equivalence point than on the silver-rich side, the net effect of adding excess ions is approximately the same on both sides, and the initially precipitated AgI dissolves as the concentration of free ions approaches 1M.

When a solution of AgI in excess Ag^+ is diluted with distilled water, the free silver ion concentration is reduced and, in effect, the system moves to the left in the diagram in Figure 3. The percentage of the total iodide present as AgI(aq) increases. When the actual concentration of that species exceeds 5.9×10^{-9}M, silver iodide precipitates. A similar argument explains the precipitation of AgI from its solution in an excess of iodide ions.

REFERENCES

1. Bush, K. J.; Diehl, H. *J. Chem. Educ.* **1979**, *56,* 54.
2. Smith, R. M.; Martell, A. E., Eds. "Critical Stability Constants," Vol. IV, Inorganic Complexes; Plenum Press: New York, 1976.

4.5

Precipitates and Complexes of Nickel(II)

Reagents are added in a specified order to a large beaker containing an aqueous solution of nickel(II) sulfate [1]. Complex ions of nickel(II) are prepared by using aqueous solutions of ammonia, ethylenediamine, dimethylglyoxime, and cyanide ions. As each complex ion is formed, its color and solubility are observed.

MATERIALS FOR PROCEDURE A

300 ml distilled water

20 ml 1M nickel sulfate, $NiSO_4$ (To prepare 1 liter of 1M stock solution, dissolve 263 g $NiSO_4 \cdot 6H_2O$ in distilled water and dilute to 1 liter.)

40 ml 5M aqueous ammonia, NH_3 (To prepare 1 liter of 5M stock solution, dilute 333 ml of concentrated ammonia solution, 15M, to 1 liter with distilled water.)

20 ml 25% ethylenediamine, $H_2NCH_2CH_2NH_2$ (To prepare 1 liter of 25% stock solution, dilute 250 ml ethylenediamine to 1 liter with distilled water.)

25 ml 1% dimethylglyoxime, $C_4H_8N_2O_2$ (To prepare 1 liter of 1% stock solution, dissolve 8 g dimethylglyoxime in 1 liter ethanol.)

200 ml 1M potassium cyanide, KCN (To prepare 1 liter of 1M stock solution, dissolve 65 g KCN in distilled water and dilute to 1 liter.) **(See Hazards section before handling cyanide salts.)**

1-liter beaker

magnetic stirrer with stirring bar

MATERIALS FOR PROCEDURE B

75 ml 1M nickel sulfate, $NiSO_4$ (For preparation, see Materials for Procedure A.)

1125 ml distilled water

100 ml 15M aqueous ammonia, NH_3

30 ml 25% ethylenediamine, $H_2NCH_2CH_2NH_2$ (For preparation, see Materials for Procedure A.)

30 ml 1% dimethylglyoxime, $C_4H_8N_2O_2$ (For preparation, see Materials for Procedure A.)

200 ml 1M potassium cyanide, KCN (For preparation, see Materials for Procedure A.) **(See Hazards section before handling cyanide salts.)**

299

2-liter beaker, preferably with approximate graduations

4 1-liter beakers, preferably with approximate graduations

4 glass stirring rods to fit 1-liter beakers

MATERIALS FOR PROCEDURE C

2100 ml distilled water

140 ml 1M nickel sulfate, $NiSO_4$ (For preparation, see Materials for Procedure A.)

40 ml 5M aqueous ammonia, NH_3 (For preparation, see Materials for Procedure A.)

30 ml 25% ethylenediamine, $H_2NCH_2CH_2NH_2$ (For preparation, see Materials for Procedure A.)

25 ml 1% dimethylglyoxime, $C_4H_8N_2O_2$ (For preparation, see Materials for Procedure A.)

80 ml 1M potassium cyanide, KCN (For preparation, see Materials for Procedure A.) **(See Hazards section before handling cyanide salts.)**

7 600-ml beakers

6 glass stirring rods to fit 600-ml beakers

PROCEDURE A

Place the 1-liter beaker containing 300 ml of distilled water and the stirring bar on the magnetic stirrer, and adjust the stirring rate to mix efficiently without much turbulence. Add 20 ml of 1M $NiSO_4$ solution, noting the color and its intensity. In sequence, add 40 ml of 5M NH_3 solution, 20 ml of 25% ethylenediamine solution, 25 ml of 1% dimethylglyoxime solution, and 200 ml of 1M KCN solution. During and after each addition, note the color and general appearance of the reaction mixture.

PROCEDURE B

Place the beakers in a row in front of the audience, with the 2-liter beaker at one end. To the 2-liter beaker add 75 ml of 1M $NiSO_4$ and dilute to 1200 ml with distilled water. Pour approximately 800 ml of the resulting solution into the first of the 1-liter beakers, then add to that beaker 100 ml of 15M NH_3 solution with stirring.

Pour about 650 ml of the resulting solution into the second 1-liter beaker, and add 30 ml of 25% ethylenediamine solution with stirring.

Pour about 350 ml of the resulting solution into the third 1-liter beaker, and add 30 ml of 1% dimethylglyoxime solution with stirring.

Finally, pour about 100 ml of the resulting suspension into the fourth 1-liter beaker, and add 200 ml of 1M KCN solution with stirring.

This procedure allows the viewer to observe changes which occur upon the addition of each reagent and to contrast the colors of all five complex ions.

PROCEDURE C

Place the seven beakers in a row in front of the audience. To each beaker add 300 ml of distilled water and 20 ml of 1M $NiSO_4$. Place a stirring rod in each beaker except the first one. To the second beaker, add 40 ml of 5M NH_3 solution and stir. Note the color change throughout the addition and the final color. Add the reagents slowly, with stirring, so that any intermediate changes can be seen.

To the third beaker, add 5 ml of 25% ethylenediamine solution, stir, and note the color change.

To the fourth beaker, add 10 ml of 25% ethylenediamine solution with stirring and note the color changes.

To the fifth beaker, add 15 ml of 25% ethylenediamine solution with stirring and note the color changes.

To the sixth beaker, add 25 ml of 1% dimethylglyoxime solution, stir, and note the changes.

To the seventh beaker, add 80 ml of 1M KCN solution, stir, and note all changes. As the KCN solution is added, a precipitate will form and subsequently redissolve.

At the end of this procedure, you have beaker contents similar to those obtained from Procedure B plus two beakers in which smaller amounts of ethylenediamine have been added to the $NiSO_4$ solution. This variation allows observation of intermediate ethylenediamine complexes of Ni(II).

HAZARDS

Cyanide salts, their solutions, and hydrogen cyanide gas produced by the reaction of cyanides with acids are all extremely poisonous. Hydrogen cyanide is among the most toxic and rapidly acting of all poisons. The solutions and the gas can be absorbed through the skin. Solutions are irritating to the skin, nose, and eyes. Cyanide compounds and acids must not be stored or transported together. An open bottle of potassium cyanide can generate HCN in moist air.

Early symptoms of cyanide poisoning are weakness, difficult breathing, headache, dizziness, nausea, and vomiting; these may be followed by unconsciousness, cessation of breathing, and death.

Anyone exposed to hydrogen cyanide should be removed from the contaminated atmosphere immediately. Amyl nitrite should be held under the person's nose for not more than 15 seconds per minute, and oxygen should be administered in the intervals. If the person is not breathing, artificial resuscitation by the Silvester method (not mouth to mouth) should be attempted immediately.

Nickel salts and their solutions will irritate the eyes upon contact. Dust from solid nickel salts is harmful, and the compounds are assumed to be poisonous if ingested. Some nickel compounds, although not nickel sulfate or the complexes produced, are on the Category I list of "suspected carcinogens" published by the U.S. Occupational Safety and Health Administration.

Concentrated aqueous ammonia solution can cause burns and is irritating to the skin, eyes, and respiratory system. Like ammonia, ethylenediamine (1,2-diaminoethane) is caustic and has similar toxic properties.

DISPOSAL

Solutions containing cyanide ions should be mixed with an excess of sodium hydroxide plus sodium hypochlorite solution (household bleach) and allowed to stand for a few hours. Flush the drain with water to eliminate any residual acid and then flush the sodium hydroxide–bleach mixture down the drain with excess water.

The other solutions should be flushed down the drain.

DISCUSSION

This demonstration acquaints the observer with the colors of several nickel(II) complexes and with the insolubility of at least two nickel(II) compounds. The table presents the stability constants and solubility product constants for the species involved in the demonstration. As described in the introduction to this chapter, β_n stands for the stability constant of ML_n that is formed from the hydrated metal ion and the free ligand. For example,

$$Ni(H_2O)_6{}^{2+}(aq) + 6\ NH_3(aq) \longrightarrow Ni(NH_3)_6{}^{2+}(aq) + 6\ H_2O(l)$$

$$\beta_6 = \frac{[Ni(NH_3)_6{}^{2+}]}{[Ni^{2+}][NH_3]^6}$$

Throughout the discussion, we use these standard abbreviations:

en   represents ethylenediamine, $H_2NCH_2CH_2NH_2$

Hdmg   represents dimethylglyoxime, CH_3—C—C—CH_3

$$\overset{\displaystyle \|\quad\|}{\underset{\displaystyle \underset{\textstyle H—O\ \ O—H}{|\quad|}}{N\ \ N}}$$

This demonstration has many aspects capable of exploitation. Your immediate purpose will dictate your choice of procedure and the nature of your discussion.

Procedure A and Procedure B reveal the relative stabilities of the various complexes. The net ionic equations and their equilibrium constants derived from tabulated equilibrium constants [2] are listed on the following page.

Equilibrium Constants for Ni(II) Species [2]

Symbol of K_{eq}	NH_3	en	dmg^-	CN^-	S^{2-}
β_1	6.5×10^2	3.0×10^7	—	1.07×10^7	—
β_2	1.2×10^5	6.6×10^{13}	1.7×10^{17}	—	—
β_3	7.1×10^6	1.35×10^{18}	—	—	—
β_4	1.3×10^8	—	—	1.15×10^{31}	—
β_5	8.5×10^8	—	—	—	—
β_6	1.2×10^9	—	—	—	—
K_{s0}	—	—	—	—	[a]$\alpha\ 4.0 \times 10^{-20}$
					$\beta\ 1.3 \times 10^{-25}$
					$\gamma\ 2.5 \times 10^{-27}$
K_{s2}	—	—	9.3×10^{-7}	—	—

[a]Three forms of NiS are identified by the symbols α, β, and γ. The aging of a nickel sulfide precipitate forms the most stable (least soluble) form.

$$Ni(H_2O)_6^{2+}(aq) + 6\,NH_3(aq) \longrightarrow Ni(NH_3)_6^{2+}(aq) + 6\,H_2O(l) \quad K_{eq} = 1.2 \times 10^9$$

$$Ni(NH_3)_6^{2+}(aq) + 3\,en(aq) \longrightarrow Ni(en)_3^{2+}(aq) + 6\,NH_3(aq) \qquad K_{eq} = 1.1 \times 10^9$$

$$Ni(en)_3^{2+}(aq) + 2\,Hdmg(aq) \longrightarrow Ni(dmg)_2(aq) + 3\,en(aq) + 2\,H^+(aq)$$
$$K_{eq} = 1.3 \times 10^{-1}$$

$$Ni(en)_3^{2+}(aq) + 2\,Hdmg(aq) \longrightarrow Ni(dmg)_2(s) + 3\,en(aq) + 2\,H^+(aq)$$
$$K_{eq} = 1.35 \times 10^5$$

$$Ni(dmg)_2(s) + 4\,CN^-(aq) \longrightarrow Ni(CN)_4^{2-}(aq) + 2\,dmg^-(aq) \quad K_{eq} = 6.3 \times 10^7$$

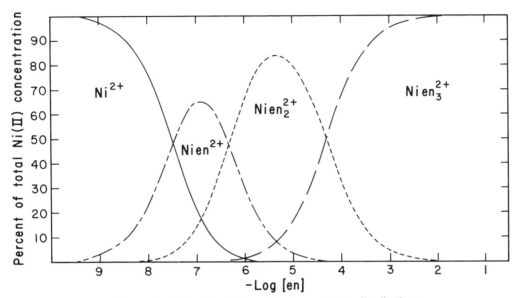

Figure 1. Nickel(II)/ethylenediamine species distribution.

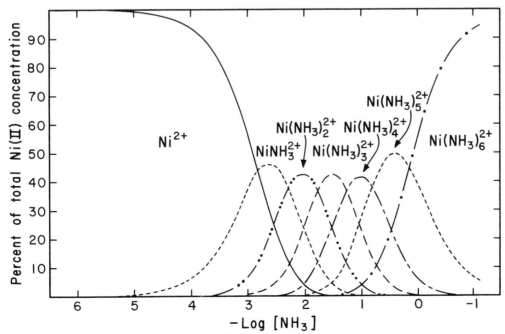

Figure 2. Nickel(II)/ammonia species distribution.

Procedure C allows observation of the formation by steps of the Ni^{2+}/en complexes. This observation requires that each complex have a unique color and that each complex be the dominant species in solution over some finite range of ligand concentration. Figure 1 and Figure 2 are the species distribution diagrams for the Ni^{2+}/en and Ni^{2+}/NH_3 systems, respectively. In the former case, each intermediate complex represents more than 60% of the total Ni^{2+} in solution at some concentration of ethylenediamine. This percentage allows each to dominate the color of the solution in that ligand concentration range so that the individual complex formation can be seen. In the Ni^{2+}/NH_3 system, no intermediate complex ever represents more than 50% of the total Ni^{2+}, so that the only clearly detectable complex is the terminal one, $Ni(NH_3)_6^{2+}$. The separation of the species depends on the relative values of successive formation constants. If these values differ by several powers of 10, as they do in the Ni^{2+}/en case, intermediate species become dominant at some point. If the differences between successive formation constants are small, the intermediate complexes overlap, as in the Ni^{2+}/NH_3 system.

In the table no value appears for the classical solubility product of nickel(II) glyoximate. Values are given both for β_2, the formation constant for $Ni(dmg)_2(aq)$, and for K_{s2}, the equilibrium constant for the solution of $Ni(dmg)_2(s)$ to form $Ni(dmg)_2(aq)$. These values are related to the classical solubility product, K_{s0}, as follows:

$$Ni^{2+}(aq) + 2\ dmg^-(aq) \xrightleftharpoons{\beta_2} Ni(dmg)_2(aq)$$

$$K_{s0} \searrow \qquad\qquad \nearrow\!\!\!\!\!/\ K_{s2}$$

$$Ni(dmg)_2(s)$$

$$K_{s0} = [Ni^{2+}] \cdot [dmg^-]^2 = \frac{K_{s2}}{\beta_2} = \frac{[Ni(dmg)_2(aq)] \times [Ni^{2+}][dmg^-]^2}{[Ni(dmg)_2(aq)]}$$

$$\therefore K_{s0} = \frac{9.3 \times 10^{-7}}{1.7 \times 10^{17}} = 5.5 \times 10^{-24}$$

The solubility can be dealt with in terms of the classical solubility product, but such a treatment neglects the finite solubility of the neutral $Ni(dmg)_2$.

While the geometries of the complexes are not apparent in the observations, both octahedral and square planar geometries are exhibited. The color and geometry of each complex are summarized below:

Formula	Color	Geometry
$Ni(H_2O)_6^{2+}$	green	octahedral
$Ni(NH_3)_6^{2+}$	deep blue	octahedral
$[Ni(H_2O)_4(en)]^{2+}$	light blue	octahedral
$[Ni(H_2O)_2(en)_2]^{2+}$	blue ~	octahedral
$Ni(en)_3^{2+}$	purple	octahedral
$Ni(dmg)_2$	red	square planar
$Ni(CN)_4^{2-}$	yellow	square planar

While the drastic changes in color are associated with drastic changes in geometry, the relationship is not a simple one and cannot be explained easily.

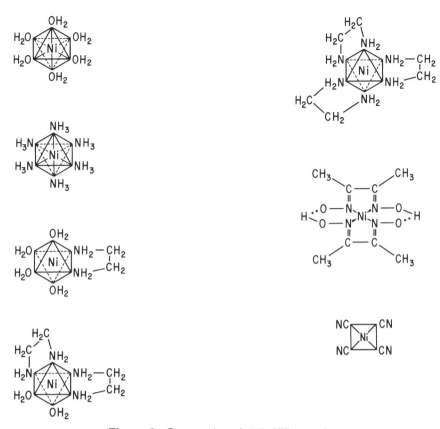

Figure 3. Geometries of nickel(II) complexes.

Figure 3 presents the geometries of the complexes of nickel(II) considered in this demonstration. The normal octahedral coordination of first row, $2+$, transition metal ions is displayed in the aquo, ammine, and ethylenediamine complexes. Possibly the geometric restraints imposed by the dimethylglyoxime strongly favor the square planar configuration of the complex with that ligand. The cyanide ion is small enough to coordinate in the octahedral, tetrahedral, or square planar geometry. The square planar arrangement is plainly caused by preferential bonding.

The square planar structure is invoked by the strong-field ligand, cyanide ion. Ni(II) is a d^8 ion, which presents the possibility of stabilizing a complex in the square planar configuration and of elevating the energy of the unoccupied d_{z^2} orbital [3]. The antibonding orbitals on the carbon atom of the cyanides can accept electron density from the appropriate occupied d orbitals of the Ni(II), thereby strengthening the bonding in the square planar arrangement.

The relative stabilities of the complexes of Ni(II) are apparent in the demonstration. Essentially, the stabilities are consistent with the positions of the ligands in the spectrochemical series. Ligands are arranged in the spectrochemical series according to the degree to which they split the degeneracy of the d orbitals in octahedral complexes. While the assignment is made on the basis of spectral data, the ability to split the d orbitals' degeneracy is closely correlated with the strength of bonding. Consequently, one expects a positive correlation between the relative stabilities of complexes and the position of the ligand in the spectrochemical series. The series varies somewhat depending on which ligands are included. In the following version of the series [4], ligands are listed in order of increasing ability to split the d orbital degeneracy (* labels the donor atom):

$$I^- < Br^- < Cl^- < \overset{*}{S}CN^- < F^- \sim urea < OH^- < acetate < C_2O_4^{2-} <$$

$$H_2O < \overset{*}{N}CS^- < glycine < pyridine \sim NH_3 < en < S\overset{*}{O}_3^{2-} <$$

$$\alpha,\alpha\text{-dipyridyl} \sim o\text{-phenanthroline} < \overset{*}{N}O_2^- < \overset{*}{C}N^-$$

The stability constants for some of the complexes of Ni(II) are presented in the table.

The differences between the stability constants for the ammonia and ethylenediamine complexes illustrate the stabilization due to chelation. In both cases, the ligand atoms are essentially ammonia nitrogens, but in the ethylenediamine complex these are bridged by CH_2–CH_2 groups, whereas in the ammonia complex, each nitrogen is a free entity. The value of β_2 for the NH_3 complex is 1.2×10^5, and the value of β_1 for the ethylenediamine complex is 3.0×10^7, indicating that the ethylenediamine complex is roughly 200 times more stable than the corresponding complex of ammonia. This effect is very common. Polydentate ligands (ligands with two or more appropriately spaced atoms capable of coordination) tend to form more stable complexes than do their monodentate analogues. One contributing factor to this increased stability is that the formation of a chelate complex releases more molecules or ions than it binds. Consequently, the product system has more degrees of freedom and greater entropy; for example,

$$Ni(H_2O)_6^{2+} + 6\ NH_3(aq) \longrightarrow Ni(NH_3)_6^{2+} + 6\ H_2O(l)$$

$$Ni(H_2O)_6^{2+} + 3\ en(aq) \longrightarrow Ni(en)_3^{2+} + 6\ H_2O(l)$$

The total number of chemical species in the formation of $Ni(NH_3)_6^{2+}$ from $Ni(H_2O)_6^{2+}$ does not change. In the formation of $Ni(en)_3^{2+}$ from $Ni(H_2O)_6^{2+}$, the total number of chemical species increases by 3.

REFERENCES

1. Shakhashiri, B. Z.; Dirreen, G. E.; Juergens, F. *J. Chem. Educ.* **1980**, *57,* 900.
2. Smith, R. M.; Martell, A. E., Eds. "Critical Stability Constants," Vol. IV, Inorganic Complexes; Plenum Press: New York, 1976.
3. Cotton, F. A.; Wilkinson, G. "Advanced Inorganic Chemistry," 4th ed.; Interscience Publishers, John Wiley and Sons, Inc.: New York, 1980.
4. Purcell, K. F.; Kotz, J. C. "An Introduction to Inorganic Chemistry"; W. B. Saunders Company: Philadelphia, 1980; p 336.

4.6

Precipitates and Complexes of Silver(I)

Several clear, colorless solutions are added in sequence to a stirred solution on a magnetic stirrer [1]. Each addition changes the appearance of the contents of the beaker: a precipitate forms or dissolves, or the mixture changes color. The precipitates are Ag_2CO_3, Ag_2O, $AgCl$, $AgBr$, AgI, and Ag_2S; the complexes are $Ag(NH_3)_2^+$, $Ag(S_2O_3)_2^{3-}$, and $Ag(CN)_2^-$.

MATERIALS FOR PROCEDURE A

200 ml distilled water

10 ml 0.1M silver nitrate, $AgNO_3$ (To prepare 1 liter of 0.1M stock solution, dissolve 17.0 g $AgNO_3$ in distilled water and dilute to 1 liter.)

2 ml 0.1M sodium bicarbonate, $NaHCO_3$ (To prepare 1 liter of 0.1M stock solution, dissolve 8.4 g $NaHCO_3$ in distilled water and dilute to 1 liter.)

10 ml 0.1M sodium hydroxide, $NaOH$ (To prepare 1 liter of 0.1M stock solution, dissolve 4.0 g $NaOH$ in distilled water and dilute to 1 liter.)

30 ml 0.1M sodium chloride, $NaCl$ (To prepare 1 liter of 0.1M stock solution, dissolve 5.9 g $NaCl$ in distilled water and dilute to 1 liter.)

35 ml 5.0M aqueous ammonia, NH_3 (To prepare 1 liter of 5M stock solution, dilute 330 ml of concentrated, 15M, ammonia solution to 1 liter with distilled water.)

10 ml 0.1M sodium bromide, $NaBr$ (To prepare 1 liter of 0.1M stock solution, dissolve 10.3 g $NaBr$ in distilled water and dilute to 1 liter.)

50 ml 0.1M sodium thiosulfate, $Na_2S_2O_3$ (To prepare 1 liter of 0.1M stock solution, dissolve 24.8 g $Na_2S_2O_3 \cdot 5H_2O$ in distilled water and dilute to 1 liter.)

10 ml 0.1M potassium iodide, KI (To prepare 1 liter of 0.1M stock solution, dissolve 16.6 g KI in distilled water and dilute to 1 liter.)

20 ml 0.1M potassium cyanide, KCN (To prepare 1 liter of 0.1M stock solution, dissolve 6.5 g KCN in distilled water and dilute to 1 liter.) **(See Hazards section before handling cyanide salts.)**

10 ml 0.1M sodium sulfide, Na_2S (To prepare 1 liter of 0.1M stock solution, dissolve 24.0 g $Na_2S \cdot 9H_2O$ in distilled water and dilute to 1 liter.)

250-ml graduated cylinder

600-ml beaker

magnetic stirrer with stirring bar

10-ml graduated cylinder

50-ml graduated cylinder

MATERIALS FOR PROCEDURE B

6 ml 0.1M sodium bicarbonate, $NaHCO_3$ (For preparation, see Materials for Procedure A.)

40 ml 0.1M silver nitrate, $AgNO_3$ (For preparation, see Materials for Procedure A.)

35 ml 0.1M sodium hydroxide, NaOH (For preparation, see Materials for Procedure A.)

60 ml 0.1M sodium chloride, NaCl (For preparation, see Materials for Procedure A.)

20 ml 5.0M aqueous ammonia, NH_3 (For preparation, see Materials for Procedure A.)

25 ml 0.1M sodium bromide, NaBr (For preparation, see Materials for Procedure A.)

100 ml 0.1M sodium thiosulfate, $Na_2S_2O_3$ (For preparation, see Materials for Procedure A.)

15 ml 0.1M potassium iodide, KI (For preparation, see Materials for Procedure A.)

30 ml 0.1M potassium cyanide, KCN (For preparation, see Materials for Procedure A.) **(See Hazards section before handling cyanide salts.)**

10 ml 0.1M sodium sulfide, Na_2S (For preparation, see Materials for Procedure A.)

800 ml distilled water

9 150-ml beakers

10 glass stirring rods

10 100-ml beakers

10 labels for beakers

10-ml graduated cylinder

25-ml graduated cylinder

50-ml graduated cylinder

100-ml graduated cylinder

1-liter beaker, preferably with approximate graduations

PROCEDURE A

Place the 600-ml beaker containing 200 ml of distilled water and the stirring bar on the magnetic stirrer, and adjust the rate of stirring so it is vigorous but not turbulent.

Add 10 ml of 0.1M $AgNO_3$ solution and allow time for complete mixing. Add 2 ml of 0.1M $NaHCO_3$. A white to yellow precipitate of Ag_2CO_3 will form. (If Na_2CO_3 is used in place of $NaHCO_3$, the carbonate solution should be added before the silver nitrate solution to help prevent the precipitation of silver oxide.)

Add 10 ml of 0.1M NaOH solution. The light-colored, silver carbonate precipitate darkens to brown because of the conversion to silver oxide, Ag_2O.

Add in sequence the remaining specified solutions, NaCl, NH_3, NaBr, $Na_2S_2O_3$, KI, KCN, and Na_2S, and note the changes. Do not use HCl or any other acidic reagents as a source of Cl^- (see Hazards section). Aside from creating a hazard with HCN, acids counteract the effect of ammonia and induce the decomposition of thiosulfate. Photochemical decomposition results from prolonged exposure of the insoluble silver(I) compounds to light.

This procedure is a modification of the procedure developed by Schwenck [2].

PROCEDURE B

This procedure permits the same observations as in Procedure A but allows also for the preservation of a sample of each suspension or solution to assist in recapitulation.

Place the nine 150-ml beakers in a row and put a stirring rod in each beaker.

Measure the specified amounts of the ten reagents into 100-ml beakers and place each, except the silver nitrate solution, beside one of the 150-ml beakers in the sequence in which the reagents will be added. Labelling these beakers is helpful.

Place 800 ml of distilled water in the 1-liter beaker. Add 6 ml of 0.1M $NaHCO_3$ solution and stir thoroughly. With continued stirring, add 40 ml of 0.1M $AgNO_3$ solution to the same beaker. Pour approximately 100 ml of the resulting suspension into the first 150-ml beaker.

To the suspension in the 1-liter beaker, add 35 ml of 0.1M NaOH solution and stir vigorously. Transfer approximately 100 ml of the resulting dark suspension to the second 150-ml beaker.

Add 60 ml of 0.1M NaCl solution to the 1-liter beaker and stir vigorously. When the change in the suspension is complete, transfer 100 ml of the suspension to the next 150-ml beaker.

Continue in this fashion, adding a new reagent, allowing time for the reaction, and pouring off a 100-ml sample of the solution or suspension to be referred to later, until all the reagents have been added.

HAZARDS

Cyanide salts, their solutions, and hydrogen cyanide gas produced by the reaction of cyanides with acids are all extremely poisonous. Hydrogen cyanide is among the most toxic and rapidly acting of all poisons. The solutions and the gas can be absorbed through the skin. Solutions are irritating to the skin, nose, and eyes. Cyanide compounds and acids must not be stored or transported together. An open bottle of potassium cyanide can generate HCN in moist air.

Early symptoms of cyanide poisoning are weakness, difficult breathing, headache, dizziness, nausea, and vomiting; these may be followed by unconsciousness, cessation of breathing, and death.

Anyone exposed to hydrogen cyanide should be removed from the contaminated atmosphere immediately. Amyl nitrite should be held under the person's nose for not more than 15 seconds per minute, and oxygen should be administered in the intervals. If the person is not breathing, artificial resuscitation by the Silvester method (not mouth to mouth) should be attempted immediately.

Sodium sulfide solutions must not be acidified becuse toxic hydrogen sulfide gas will be produced. The solid or solution can cause severe burns of the eyes and skin.

Silver nitrate and its solutions are irritating to the skin and eyes and can cause burns. Also, they may stain skin or clothing an unsightly brown to black color. If a spill is recognized when it occurs, rinse the spot with some sodium thiosulfate solution followed by water. If a black spot develops later, try removing it as follows. Prepare a 10% solution of potassium ferricyanide in water and a 10% solution of sodium thiosulfate in 1% ammonia water. Mix equal volumes of these two solutions and scrub the black stain thoroughly with the resulting solution. Rinse well with water.

Mixtures of silver nitrate and ammonia solutions must not be stored because silver nitride, which is dangerously shock-sensitive, may be produced.

Sodium hydroxide can cause severe burns of the eyes and skin. Dust from solid sodium hydroxide is very caustic.

Concentrated aqueous ammonia solution can cause burns and is irritating to the skin, eyes, and respiratory system.

DISPOSAL

At the end of the demonstration, precipitate any silver still in solution by adding excess sodium sulfide solution. Collect the silver-containing precipitates to be reclaimed later [3].

Solutions containing cyanide ions should be mixed with an excess of sodium hydroxide plus sodium hypochlorite solution (household bleach) and allowed to stand for a few hours. Flush the drain with water to eliminate any residual acid, then flush the sodium hydroxide–bleach mixture down the drain with excess water.

After flushing the drain with water to remove residual acids, any remaining solutions should be flushed down the drain.

DISCUSSION

This demonstration is especially attractive because of the number of steps required before reaching a material that is so insoluble as to defy easy solution. It illustrates simple solubility and competitive equilibrium between two insoluble substances and between insoluble compounds and complexes. Most of these substances have 1:1 stoichiometries and thus are easily handled by students, although enough have 2:1 stoichiometries to illustrate clearly the effect of such stoichiometry on the relationship between solubility and the solubility product.

Solubility Product and Formation Constants of Ag(I) Species [4]

Ag(I) species	Symbol of K_{eq}	Log K_{eq}	K_{eq}
Ag_2CO_3	K_{s0}	-11.09	8.1×10^{-12}
Ag_2O as			
AgOH	K_{s0}	-7.37	4.3×10^{-8}
AgCl	K_{s0}	-9.74	1.8×10^{-10}
AgBr	K_{s0}	-12.10	8.0×10^{-13}
AgI	K_{s0}	-16.35	4.5×10^{-17}
AgCN	K_{s0}	-15.4	4.0×10^{-16}
Ag_2S	K_{s0}	-49.7	2.0×10^{-50}
$Ag(NH_3)_2^+$	β_2	7.20	1.6×10^7
$Ag(S_2O_3)_2^{3-}$	β_2	12.63	4.3×10^{12}
$Ag(CN)_2^-$	β_2	20.0	1.0×10^{20}

The table lists the equilibrium constants relevant to the demonstration. From these values, the equilibrium constant for each of the reactions observed in the reaction sequence can be calculated. The net ionic equations for those reactions along with the calculated values for the equilibrium constants are

$2 \, Ag^+(aq) + CO_3^{2-}(aq) \longrightarrow Ag_2CO_3(s)$ $K_{eq} = 1.2 \times 10^{11}$

$Ag_2CO_3(s) + 2 \, OH^-(aq) \longrightarrow Ag_2O \cdot H_2O(s) + CO_3^{2-}(aq)$ $K_{eq} = 4.4 \times 10^3$
   or $2 \, AgOH(s)$

$AgOH(s) + Cl^-(aq) \longrightarrow AgCl(s) + OH^-(aq)$ $K_{eq} = 2.4 \times 10^2$

$AgCl(s) + 2 \, NH_3(aq) \longrightarrow Ag(NH_3)_2^+(aq) + Cl^-(aq)$ $K_{eq} = 2.9 \times 10^{-3}$

$Ag(NH_3)_2^+(aq) + Br^-(aq) \longrightarrow AgBr(s) + 2 \, NH_3(aq)$ $K_{eq} = 7.8 \times 10^4$

$AgBr(s) + 2 \, S_2O_3^{2-}(aq) \longrightarrow Ag(S_2O_3)_2^{3-}(aq) + Br^-(aq)$ $K_{eq} = 3.4$

$Ag(S_2O_3)_2^{3-}(aq) + I^-(aq) \longrightarrow AgI(s) + 2 \, S_2O_3^{2-}(aq)$ $K_{eq} = 5.2 \times 10^3$

$AgI(s) + 2 \, CN^-(aq) \longrightarrow Ag(CN)_2^-(aq) + I^-(aq)$ $K_{eq} = 4.5 \times 10^3$

$2 \, Ag(CN)_2^-(aq) + S^{2-}(aq) \longrightarrow Ag_2S(s) + 4 \, CN^-(aq)$ $K_{eq} = 5.0 \times 10^9$

The combination of equilibrium constants is accomplished relatively easily by treating the reaction as a competition between two ligands for the same metal ion for which only one concentration in the solution exists. This single concentration must satisfy both equilibrium constants. Consider the conversion of Ag_2O to AgCl. According to custom, Ag_2O is dealt with as AgOH to simplify the treatment of its solubility. For a single concentration of Ag^+ to satisfy the solubility constants of both AgCl and AgOH, we can write:

$$\frac{1.8 \times 10^{-10}}{[Cl^-]} = [Ag^+] = \frac{4.3 \times 10^{-8}}{[OH^-]} \tag{1}$$

The observed reaction is $AgOH + Cl^- \longrightarrow AgCl + OH^-$, so the expression for its equilibrium constant is

$$K_{eq} = \frac{[OH^-]}{[Cl^-]} \tag{2}$$

Expression 1 is easily rearranged to provide a numerical value for this ratio:

$$\frac{[OH^-]}{[Cl^-]} = \frac{4.3 \times 10^{-8}}{1.8 \times 10^{-10}} = 2.4 \times 10^2 \tag{3}$$

The custom of dealing with Ag_2O as if it were $AgOH$ is quite justifiable. For the solubility of Ag_2O, we write:

$$Ag_2O + H_2O \longrightarrow 2\,Ag^+(aq) + 2\,OH^-(aq)$$

$$K_{s0} = [OH^-]^2[Ag^+]^2$$

Taking the square root of both sides, we get:

$$\sqrt{K_{s0}} = [Ag^+]\,[OH^-]$$

Thus, the K_{s0} value of 4.3×10^{-8} that we use is simply the square root of the value we would use if we dealt with Ag_2O each time.

The calculation for the conversion of silver carbonate to silver oxide involves a squared term, but it is essentially the same as for the conversion of the oxide to the chloride. The observed reaction is represented by the following equation:

$$Ag_2CO_3(s) + 2\,OH^-(aq) \longrightarrow 2\,AgOH(s) + CO_3^{2-}(aq) \tag{4}$$

Thus, the equilibrium constant has the form:

$$K_{eq} = \frac{[CO_3^{2-}]}{[OH^-]^2} \tag{5}$$

Only one $[Ag^+]$ can satisfy both solubility products, but in the K_{s0} for silver carbonate this concentration appears squared. Therefore we square the entire expression for silver oxide to equate the $[Ag^+]^2$ terms.

$$[Ag^+] \cdot [OH^-] = 4.3 \times 10^{-8}$$

$$\therefore [Ag^+]^2[OH^-]^2 = (4.3)^2 \times 10^{-16} = 1.85 \times 10^{-15}$$

$$\frac{8.1 \times 10^{-12}}{[CO_3^{2-}]} = [Ag^+]^2 = \frac{1.85 \times 10^{-15}}{[OH^-]^2}$$

$$K_{eq}\,(4) = \frac{[CO_3^{2-}]}{[OH^-]^2} = \frac{8.1 \times 10^{-12}}{1.85 \times 10^{-15}} = 4.4 \times 10^3$$

To illustrate further the method of calculation, we present the value of K_{eq} for the precipitation of Ag_2S from $Ag(CN)_2^-$. The equation for the observed reaction is

$$2\,Ag(CN)_2^- + S^{2-} \longrightarrow Ag_2S + 4\,CN^-$$

K_{eq} has the form:

$$\frac{[CN^-]^4}{[Ag(CN)_2^-]^2[S^{2-}]}$$

Combining K_{s0} for Ag_2S with β_2 for $Ag(CN)_2^-$ yields:

$$\frac{[Ag(CN)_2^-]^2}{\beta_2^2[CN^-]^4} = [Ag^+] = \frac{K_{s0}}{[S^{2-}]}$$

Thus,

$$K_{eq} = \frac{1}{K_{s0} \cdot \beta_2^2} = \frac{1}{2 \times 10^{-50} \times 1 \times 10^{40}} = 5 \times 10^9$$

The series in this demonstration is initiated using sodium bicarbonate rather than sodium carbonate because carbonate ions hydrolyze extensively to produce $OH^-(aq)$ in sufficient concentration to bring about the precipitation of some silver oxide along with the silver carbonate. Using bicarbonate provides a high enough concentration of carbonate ions to precipitate Ag_2CO_3, but the solution starts with a pH of about 8.5 and gradually becomes more acidic as the reaction proceeds:

$$2\,HCO_3^-(aq) + 2\,Ag^+(aq) \longrightarrow Ag_2CO_3(s) + H_2CO_3(aq)$$

Students can be asked to calculate values for the equilibrium constants for many other potential reactions. One can also specify conditions for any one of a number of systems and ask students to calculate what they predict would happen. For example, they might be asked to calculate the solubility of silver bromide in 15M ammonia, or the concentration of thiosulfate ions required to hold 0.1 mole per liter of silver bromide in solution. All the traditional calculations involving solubility products can be initiated by this sequence.

REFERENCES

1. Shakhashiri, B. Z.; Dirreen, G. E.; Juergens, F. *J. Chem. Educ.* **1980**, *57*, 813.
2. Schwenck, J. R. *J. Chem. Educ.* **1959**, *56*, 45.
3. Bush, K. J.; Diehl, H. *J. Chem. Educ.* **1979**, *56*, 54.
4. Smith, R. M.; Martell, A. E., Eds. "Critical Stability Constants," Vol. IV, Inorganic Complexes; Plenum Press: New York, 1976.

4.7

Bromo Complexes of Copper(II)

When increments of concentrated hydrobromic acid solution are added to a dilute solution of copper(II) sulfate, the color of the solution changes from light blue to bluish green, then to yellowish green, then to brownish or greenish yellow, and finally to brown. Heating the yellowish green solution changes it to brown, while cooling a light brown solution changes it to brownish yellow.

MATERIALS

200 ml 0.10M copper(II) sulfate, $CuSO_4$ (To prepare 1 liter of 0.10M stock solution, dissolve 25.0 g $CuSO_4 \cdot 5H_2O$ in distilled water and dilute to 1 liter.)

150 ml distilled water

200 ml concentrated (8.8M) 48% hydrobromic acid, HBr

ice bath

3–5 g copper turnings, fine

3 600-ml beakers, preferably tall-form

2 100-ml graduated cylinders

4 glass stirring rods to fit beakers

hot plate

500-ml Erlenmeyer flask with stopper

300-ml tall-form beaker

dropper

PROCEDURE

Pour 150 ml of 0.10M $CuSO_4$ into one of the 600-ml beakers and add 150 ml of distilled water. Stir thoroughly and pour half the solution into a second 600-ml beaker.

Measure 100 ml of concentrated hydrobromic acid solution in a 100-ml graduated cylinder. Add this solution in 10-ml increments to one of the copper(II) sulfate solutions with stirring. Use the solution in the other beaker for comparison purposes. Adding the HBr solution in 10-ml increments avoids adding it too rapidly and provides useful time intervals for observing color changes.

After the addition of the first 10-ml aliquot of HBr solution, the color is detectably greener than the Cu(II) solution. After the second 10-ml aliquot of HBr solution, the color is definitely green. After the third 10-ml aliquot, the solution is clear and yellowish green.

Transfer half the yellowish green solution to another 600-ml beaker and heat on the hot plate. The solution gradually turns brown. Stop heating and place the beaker in an ice bath. Upon cooling, the color returns to the initial yellowish green, and the solutions can be recombined.

Alternatively, add 5 ml of concentrated HBr solution to both the cold (yellowish green) solutions and to the hot (brown) solution. The cold solution remains green while the hot solution becomes more intensely brown. Cooling the brown solution restores the yellowish green color, and the solutions can be recombined.

When a fifth 10-ml aliquot of concentrated HBr solution is added to the recombined solutions, the color becomes noticeably brown. If a portion of the solution is chilled in an ice bath, the brown tinge is removed and the solution becomes yellow to yellowish green.

The heating and cooling steps are not necessary but demonstrate the effect of temperature on the color of the solution.

Addition of a sixth 10-ml aliquot of concentrated HBr solution intensifies the brown color significantly. Further additions of concentrated HBr solution render the solution opaque before the full 100 ml of HBr has been added.

The true color of the complex formed in the highest concentration of HBr is not brown but violet, which can be demonstrated in either of two ways, both of which are pedagogically useful. After adding the full 100 ml of concentrated HBr solution to the 150 ml of 0.050M Cu(II), use a hot plate to heat the solution almost to boiling. The solution appears to be almost black. Pour this solution into a 500-ml Erlenmeyer flask containing 3–5 g of fine copper turnings. Shake the flask to heat the air in it, then stopper it and shake. The intensity of the color diminishes markedly for about 5 minutes, and the violet color becomes visible.

The same effect can also be demonstrated by measuring 100 ml of concentrated HBr solution into a 300-ml tall-form beaker and then adding 0.10M $CuSO_4$ to it by drops. One drop produces a noticeable color change even when the HBr solution is colored, as it frequently is. Four drops produce a fairly intense, clearly observable, violet color. Ten drops make the solution opaque.

The color changes can be produced in reverse order by diluting either opaque solution with water. However, dilution of the mixture of the concentrated HBr with 10 drops of 0.10M $CuSO_4$ solution produces pale colors after the brown color disappears.

HAZARDS

Hydrobromic acid gives off an acrid vapor, and both the liquid and the vapor can cause burns and are irritating to the skin, eyes, and respiratory system.

Copper compounds are harmful if taken internally. Dust from copper compounds can irritate mucous membranes.

DISPOSAL

The acidic solutions should be neutralized with sodium bicarbonate or soda ash and flushed down the drain with excess water.

DISCUSSION

This demonstration is an expansion of the procedure described by Burke [1]. The literature describing the bromo complexes of Cu(II) is limited and sometimes apparently conflicting. This uncertain status is reflected by the inclusion in "Critical Stability Constants" [2] of only a value for K_1 in spite of the obvious formation of higher complexes [3]. Probably the monobromo complex is $[CuBr(H_2O)_5]^+$, with a coordination number of 6 rather than 4:

$$Cu(H_2O)_6^{2+} + Br^- \longrightarrow [Cu(H_2O)_5Br]^+ + H_2O \qquad K_1 = 0.93$$

Also, the two axial ligands are probably farther from the Cu(II) than are the equatorial ligands. Copper(II) is a d^9 ion, subject to significant distortion because of the Jahn-Teller effect. Such distortion helps to account for the difficulties encountered in studying Cu(II) solutions spectroscopically, namely, the presence of many more transitions than one would expect for a symmetrical d^9 case.

The high concentrations of HBr solution required to form the bromo complexes in this demonstration make it obvious that the formation constants are small. The broad range of concentrations over which the ultimate bromo complex is formed also makes it apparent that the formation constant for that species is very small. Such low formation constants in aqueous solution are in accord with the position of Br^- in the spectrochemical series relative to water (see Discussion section of Demonstration 4.5).

The violet species formed in the highest concentration of HBr solution is almost certainly $CuBr_4^{2-}$, in which the bromide ions are in a distorted tetrahedral arrangement around the copper [4, 5]. As expected, the tetrabromocuprate(II) ions form more readily in organic solvents than in water, and salts such as Cs_2CuBr_4, $((CH_3)_4N)_2CuBr_4$, $((C_2H_5)_4N)_2CuBr_4$, and Li_2CuBr_4 have been isolated from such solvents. The absorption spectra of these salts support the conclusion that the violet species in concentrated aqueous HBr is the same. The tetrabromocuprate(II) has been collected on an ion exchange resin, where it behaved as a simple monomeric species [4].

The observations in this demonstration make it obvious that the molar absorbance coefficient for the violet complex, $CuBr_4^{2-}$, is much larger than those of the other complexes formed in aqueous solution. The absorptions of those other species are due to d-d transitions, so their molar absorbance coefficients are not large. Some of these d-d transitions are apparent in the spectrum of $CuBr_4^{2-}$, but the main absorption, responsible for the intense color of the ion, has been assigned to charge transfer from the ligand Br^- to the Cu(II) ion [4, 5]. Such charge transfers are apparent in the other complexes of Cu(II), such as $[Cu(NH_3)_5X]^+$, where X is Cl^-, Br^-, or I^-. In such complexes, the charge transfers occur far enough in the ultraviolet so as not to be noticeable visually. In $CuBr_4^{2-}$, the charge-transfer bands occur near enough to the visible region of the spectrum to dominate the visual observations of the solution. Spectrophotometric studies of solutions of $CuBr_4^{2-}$ are complicated by the complete transfer of an electron to Cu(II), which results in the formation of Cu(I) complexes [4].

The generation of the intensely colored species when the solution is heated indicates that the formation of $CuBr_4^{2-}$ is endothermic, as is the formation of $CoCl_4^{2-}$ (see Demonstration 4.2). As with $CoCl_4^{2-}$, the formation of the product requires a decrease in the coordination number of the central metal ion from 6 to 4 and a change from octahedral to tetrahedral geometry. Also, in both systems the tetracoordinate species is favored by less polar solvents. The systems differ, however, in the reason for the high intensity of color of the four coordinate species.

The compound $(NH_4)_2CuBr_4 \cdot 2H_2O$ has been isolated, and its structure has been determined to contain the following unit [6]:

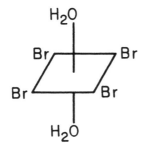

This compound shows typical d-d transitions in the visible and near ultraviolet regions. Clearly, the shift to the nearly tetrahedral geometry in the violet species, $CuBr_4^{2-}$, increases the ease of transferring charge from the bromide to the copper(II), possibly because of the shorter internuclear distances and increased orbital overlap.

REFERENCES

1. Burke, Barbara A. *J. Chem. Educ.* **1977**, *54*, 29.
2. Smith, R. M.; Martell, A. E. "Critical Stability Constants," Vol. IV, Inorganic Complexes; Plenum Press: New York, 1976; p 115.
3. Andrew, S. N.; Khaldin, V. G.; Andreeva, M. V.; Smirnova, M. F. *Zhur. neorg. Khim.* **1967**, *12*, 1791.
4. Barnes, J. C.; Hume, D. N. *Inorg. Chem.* **1963**, *2*, 444.
5. Braterman, P. S. *Inorg. Chem.* **1963**, *2*, 448.
6. "Table of Interatomic Distances in Molecules and Ions," Special Publication No. 11; The Chemical Society: London, 1958.

4.8

Precipitates and Complexes of Copper(II)

Different reagents are added sequentially to two beakers containing aqueous solutions of $CuSO_4$ and $Cu(NO_3)_2$. The color and solubility of a series of Cu species including $Cu(H_2O)_6{}^{2+}$, $Cu(NH_3)_4{}^{2+}$, $Cu(en)_2{}^{2+}$, $Cu(EDTA)^{2-}$, $Cu(CN)_3{}^{2-}$, $Cu(OH)_2$, CuO, $CuSO_4 \cdot 3Cu(OH)_2$, and CuS are observed.

MATERIALS

500 ml distilled water

50 ml 0.1M copper sulfate, $CuSO_4$ (To prepare 1 liter of 0.1M stock solution, dissolve 25 g $CuSO_4 \cdot 5H_2O$ in distilled water and dilute to 1 liter.)

50 ml 0.1M copper nitrate, $Cu(NO_3)_2$ (To prepare 1 liter of 0.1M stock solution, dissolve 24.2 g $Cu(NO_3)_2 \cdot 3H_2O$ in distilled water and dilute to 1 liter.)

ca. 5 g copper sulfate pentahydrate, $CuSO_4 \cdot 5H_2O$

1 ml 1M sulfuric acid, H_2SO_4

ca. 5 g copper sulfate, anhydrous, $CuSO_4$ (optional)

50 ml 5M aqueous ammonia, NH_3 (To prepare 1 liter of 5M stock solution, dilute 333 ml concentrated ammonia, 15M, to 1 liter with distilled water.)

50 ml 25% ethylenediamine, $H_2NCH_2CH_2NH_2$ (To prepare 1 liter of 25% stock solution, dilute 250 ml $C_2H_8N_2$ to 1 liter with distilled water.)

50 ml 0.1M disodiumdihydrogenethylenediaminetetraacetate, Na_2H_2EDTA. (To prepare 1 liter of 0.1M stock solution, dissolve 37.2 g $Na_2H_2EDTA \cdot 2H_2O$ in distilled water and dilute to 1 liter.)

25 ml 1.0M potassium cyanide, KCN (To prepare 1 liter of 1.0M stock solution, dissolve 65 g KCN in distilled water and dilute to 1 liter.) **(See Hazards section before handling cyanide salts.)**

25 ml 0.1M sodium sulfide, Na_2S (To prepare 1 liter of 0.1M stock solution, dissolve 24 g $Na_2S \cdot 9H_2O$ in distilled water and dilute to 1 liter.)

2 hot plates with magnetic stirrers and stirring bars

3 600-ml beakers, preferably tall-form

3 100-ml graduated cylinders

casserole, 3–4 inches in diameter

Meker burner

ring stand

iron ring to hold casserole

heat-protective glove

2 droppers

100-ml beaker

2 10-ml graduated cylinders

PROCEDURE

Set up two hot plates with magnetic stirrers and place on each a 600-ml beaker containing a stirring bar. Add to each beaker 75 ml of distilled water, turn on the stirrers and adjust them to a moderate stirring rate. To one beaker add 25 ml of 0.10M $CuSO_4$, and to the other add 25 ml of 0.1M $Cu(NO_3)_2$. Turn the hot plates to "high" to heat the solutions rapidly.

While the solutions are heating, you can demonstrate that the blue color characteristic of Cu(II) solutions depends on the presence of the water. Place about 5 g of ground $CuSO_4 \cdot 5H_2O$ crystals in a casserole and heat over a Meker burner. Occasionally show observers the contents, so they can follow the color changes. In a few minutes, the solid becomes almost white. Let it cool slightly, then add the solid to a 100-ml beaker containing 75 ml of distilled water and a couple of drops of 1M sulfuric acid. The solid turns blue on contact with the solution and then dissolves to yield a blue solution. You can conclude that the two original solutions contain the species that is represented as $Cu^{2+}(aq)$, regardless of its coordination number. Commercial anhydrous copper sulfate can be used to demonstrate the generation of the blue color on contact with water.

When the two solutions are boiling moderately, add 5M ammonia by drops to each and note the precipitation of a light blue solid. Continue the addition until the first purplish color of the copper-ammonia complex appears in the copper sulfate beaker and until the precipitate in the copper nitrate beaker turns brownish black. This takes approximately 3 medicine droppers of 5M ammonia in each case, often slightly less for the copper sulfate solution and slightly more for the copper nitrate solution. Cool the solutions, or allow them to cool.

To the cooled suspensions, add 10 ml of 5M ammonia and stir. The bluish green precipitate in the copper sulfate beaker dissolves, but the brownish black precipitate in the copper nitrate beaker does not. Let the suspension with the dark precipitate stir while you proceed with the solution in the copper sulfate beaker. In all likelihood, the color in that beaker is so intense that it cannot be perceived well. Discard about one half to three quarters of the solution and dilute with distilled water until the purple of the $Cu(NH_3)_4^{2+}$ can be seen. You may wish to add a little more ammonia.

To this purple solution, add about 2 ml of 25% ethylenediamine solution. If the solution does not become redder, add more ethylenediamine in small increments until it does. Be careful not to add a large excess of ethylenediamine. By inference, the color change indicates the displacement of ammonia from the inner coordination sphere of Cu(II) by ethylenediamine.

When that color change is complete, begin adding 2–3 ml portions of 0.10M EDTA solution, allowing time for reactions to occur after each addition. The solution changes to a light blue, reminiscent of the $Cu^{2+}(aq)$ color but distinguishable from it.

Returning to the copper nitrate beaker, note that the brownish black precipitate has not dissolved completely in the 5M ammonia. Some color may develop in the solution, but the solid is highly resistant to attack.

Since the addition of ethylenediamine to the ammoniacal solution of Cu(II) in the $CuSO_4$ beaker resulted in a reaction, add about 2 ml of the 25% ethylenediamine to the stirred suspension in the $Cu(NO_3)_2$ beaker. Given a little time, it dissolves the precipitate. From then on, the two solutions behave the same, and the ethylenediamine complex can be converted to the ethylenediaminetetraacetato complex.

This system can be confusing, partly because the relative stabilities of the ethylenediamine and EDTA complexes are nearly the same, and partly because the EDTA solution reacts more slowly with the Cu(II) oxide than does ethylenediamine. Sometimes, when the EDTA solution is added to the suspension of Cu(II) oxide, no reaction is apparent for some time, but then, upon the addition of ethylenediamine, the precipitate dissolves, only to yield the light blue of the EDTA complex. These reactions depend on the balance of reagents, since one complex can be converted to the other simply by adding a sufficient excess of the appropriate ligand.

You now have two solutions with essentially the same appearance, that of the EDTA complex of Cu(II). Stir the solutions and add 3–4 ml of 1.0M KCN solution to one of them. Watch closely because the initial darkening of the solution color is often quite transitory. The solution then becomes essentially colorless. If you want a second look at this phenomenon, add KCN solution to the second beaker of EDTA complex.

Alternatively, you can add to the second solution 3–4 ml of 0.10M sodium sulfide solution to create a black precipitate of CuS. If you choose this route, follow the addition of the sulfide solution by adding 3–4 ml of 1.0M KCN, which dissolves black copper sulfide. You can then return to the beaker previously treated with KCN and add Na_2S solution. As the preceding observation implies, no sulfide precipitate forms.

HAZARDS

Cyanide salts, their solutions, and hydrogen cyanide gas produced by the reaction of cyanides with acids are all extremely poisonous. Hydrogen cyanide is among the most toxic and rapidly acting of all poisons. The solutions and the gas can be absorbed through the skin. Solutions are irritating to the skin, nose, and eyes. Cyanide compounds and acids must not be stored or transported together. An open bottle of potassium cyanide can generate HCN in moist air.

Early symptoms of cyanide poisoning are weakness, difficult breathing, headache, dizziness, nausea, and vomiting; these may be followed by unconsciousness, cessation of breathing, and death.

Anyone exposed to hydrogen cyanide should be removed from the contaminated atmosphere immediately. Amyl nitrite should be held under the person's nose for not more than 15 seconds per minute, and oxygen should be administered in the intervals. If the person is not breathing, artificial resuscitation by the Silvester method (not mouth to mouth) should be attempted immediately.

Sodium sulfide solutions must not be acidified since toxic hydrogen sulfide gas will be produced. The solid or solution can cause severe burns of the eyes and skin.

Copper compounds are harmful if taken internally. Dust from copper compounds can irritate mucous membranes.

Concentrated aqueous ammonia solution can cause burns and is irritating to the skin, eyes, and respiratory system. Like ammonia, ethylenediamine (1,2-diamino-ethane) is caustic and has similar toxic properties.

Because sulfuric acid is a strong acid and a powerful dehydrating agent, it can cause burns. Spills should be neutralized with an appropriate agent, such as sodium bicarbonate, and then rinsed clean.

DISPOSAL

Solutions containing cyanide ions should be mixed with an excess of sodium hydroxide solution plus sodium hypochlorite solution (household bleach) and allowed to stand for a few hours. Flush the drain with water to eliminate any residual acid and then flush the sodium hydroxide–bleach mixture down the drain with excess water.

After flushing the drain with water to remove residual acids, any remaining solutions should be flushed down the drain.

DISCUSSION

Aside from the dehydration and rehydration of copper sulfate, this demonstration consists of a series of displacement reactions in which the copper(II) is bound ever more tightly, either in an insoluble substance or in a soluble complex. In the following net ionic equations for the reactions involved, en = ethylenediamine, $H_2NCH_2CH_2NH_2$, and EDTA = H_4Y = ethylenediaminetetraacetic acid, $(HOOCCH_2)_2NCH_2CH_2N$-$(CH_2COOH)_2$:

$$CuSO_4 \cdot 5H_2O(s) \xrightarrow{\Delta} CuSO_4(s) + 5 H_2O(g) \tag{1}$$

(Some well-defined hydrates occur between these two limits, but they are of no significance if the heating is vigorous.)

$$CuSO_4(s) + H_2O(l) \longrightarrow CuSO_4 \cdot xH_2O(s) \longrightarrow$$
$$Cu(H_2O)_6{}^{2+}(aq) + SO_4{}^{2-}(aq) \tag{2}$$

$$4 Cu^{2+}(aq) + SO_4{}^{2-}(aq) + 6 H_2O(l) + 6 NH_3(aq) \longrightarrow$$
$$CuSO_4 \cdot 3Cu(OH)_2(s) + 6 NH_4{}^+(aq) \tag{3}$$

$$Cu^{2+}(aq) + 2 NH_3(aq) + 2 H_2O(l) \longrightarrow Cu(OH)_2(s) + 2 NH_4{}^+(aq)$$
$$K_{eq} = 1.1 \times 10^9 \tag{4}$$

$$Cu(OH)_2(s) + 4 NH_3(aq) \longrightarrow Cu(NH_3)_4{}^{2+}(aq) + 2 OH^-(aq)$$
$$K_{eq} = 3 \times 10^{-6} \tag{5}$$

$$CuSO_4 \cdot 3Cu(OH)_2(s) + 16 NH_3(aq) \longrightarrow$$
$$4 Cu(NH_3)_4{}^{2+}(aq) + 6 OH^-(aq) + SO_4{}^{2-}(aq) \tag{6}$$

$$Cu(NH_3)_4{}^{2+}(aq) + 2 en(aq) \longrightarrow Cu(en)_2{}^{2+}(aq) + 4 NH_3(aq)$$
$$K_{eq} = 4 \times 10^6 \tag{7}$$

$$CuO(s) + H_2O(l) + 2 en(aq) \longrightarrow Cu(en)_2{}^{2+}(aq) + 2 OH^-(aq)$$
$$K_{eq} = 1.2 \tag{8}$$

$$Cu(en)_2{}^{2+}(aq) + Y^{4-}(aq) \longrightarrow CuY^{2-}(aq) + 2 en(aq) \qquad K_{eq} = 0.12 \tag{9}$$

$$2\,CuY^{2-}(aq) + 8\,CN^-(aq) \longrightarrow 2\,Cu(CN)_3{}^{2-}(aq) + (CN)_2(aq) + 2\,Y^{4-}(aq) \tag{10}$$

$$CuY^{2-}(aq) + S^{2-}(aq) \longrightarrow CuS(s) + Y^{4-}(aq) \qquad K_{eq} = 2.5 \times 10^{17} \tag{11}$$

$$CuS(s) + 8\,CN^-(aq) \longrightarrow 2\,Cu(CN)_3{}^{2-}(aq) + (CN)_2(aq) + 2\,S^{2-}(aq) \tag{12}$$

The table contains the relevant stability constants and solubility product constants. The series of displacement reactions (equations 1 and 2) shows that the blue species characteristic of aqueous solutions is hydrated Cu(II), presumably $Cu(H_2O)_6{}^{2+}$. The structure of solid $CuSO_4 \cdot 5H_2O$ shows the Cu(II) in a distorted octahedral coordination with 6 oxygens [2]. The Cu(II) ions in concentrated solutions of $CuCl_2$ in aqueous HCl are octahedrally coordinated [3].

The systems obtained when aqueous NH_3 is added to boiling solutions of $CuSO_4$ and $Cu(NO_3)_2$ are complex, and all of the species involved probably cannot be identified. However, the addition of aqueous NH_3 to a boiling solution of $CuSO_4$ is the standard method of preparing the basic sulfate: $CuSO_4 \cdot 3Cu(OH)_2$. This basic sulfate is bluish green. If washed and dissolved in dilute HCl and then treated with $BaCl_2$ solution, it produces a precipitate of $BaSO_4$.

No basic salt is formed in the solution starting with $Cu(NO_3)_2$, so the initial precipitate is $Cu(OH)_2$, as in the $CuSO_4$ solution. When the suspension of $Cu(OH)_2$ is heated in the presence of some excess OH^-, the $Cu(OH)_2$ dehydrates, gradually turning into black CuO. This dehydration has been studied carefully [4], and apparently no solid phases are identifiable by x-ray diffraction other than $Cu(OH)_2 \cdot xH_2O$ and CuO. The greater stability of the CuO is demonstrated by the spontaneous conversion of the $Cu(OH)_2$ to CuO and by the difference in the solubility products of a factor of 10.

What controls the solution or nonsolution of the various precipitates is unclear. In the normal formation of $Cu(NH_3)_4{}^{2+}$ by the addition of fairly concentrated aqueous ammonia to a solution containing Cu(II)(aq), $Cu(OH)_2$ precipitates and then redissolves in spite of the low equilibrium constant for the formation of $Cu(NH_3)_4{}^{2+}$. However, CuO, which has a solubility product only 1 power of 10 smaller than $Cu(OH)_2$, fails to dissolve under the same conditions. The controlling factor appears to be kinetic. The basic sulfate, which forms in preference to $Cu(OH)_2$, also dissolves rapidly in the less concentrated ammonia used in this demonstration. In all likelihood, the failure of the CuO to dissolve in the EDTA solution, even though it dissolves fairly readily in the ethylenediamine solution, is a kinetic phenomenon as well, since the two soluble complexes have virtually the same stability constants. Whatever the explanation or the sequence of steps employed, both solutions can be brought to the same composition

Equilibrium Constants for Cu(II) Species [1]

K_{eq}	NH_3	en	EDTA	OH^-	CN^- Cu(I)	S^{2-}
β_1	1.75×10^4	3.5×10^{10}	5.0×10^{18}	2.0×10^6	—	—
β_2	6.8×10^7	4.0×10^{19}	—	6.3×10^{12}	1.6×10^{16}	—
β_3	6.3×10^{10}	—	—	3×10^{14}	4.0×10^{21}	—
β_4	1.0×10^{13}	—	—	4.0×10^{15}	1.3×10^{23}	—
β_5	2.7×10^{12}	—	—	—	—	—
β_{22}	—	—	—	1.9×10^{17}	—	—
K_{s0} (Cu(OH)$_2$)	—	—	—	3×10^{-19}	—	—
K_{s0} (CuO)	—	—	—	3×10^{-20}	—	—
K_{s0}	—	—	—	—	—	8×10^{-37}

containing predominantly the light blue EDTA complex. Ultimately, one can form the EDTA complex regardless of the path taken.

The insolubility of copper(II) sulfide is so great that it will precipitate from almost any complex. However, it will not precipitate from a cyanide solution of copper, and it will dissolve in a solution of potassium cyanide.

When excess KCN solution is added to the light blue solution containing Cu(II)-EDTA, the mixture initially darkens and then the color fades rapidly. The transient purple is due to the formation of tetracyanocuprate(II), which is unstable at room temperature, and decomposes to give cyanogen and cyanide complexes of Cu(I) [5]. At least three of these complexes are known, and more can be isolated from the solutions by the appropriate choices of cation and conditions of isolation [6]. The dominant species under the conditions of this demonstration is probably $Cu(CN)_3^{2-}$. The oxidation-reduction reactions between Cu(II) and CN^- are discussed in Demonstration 2.5.

REFERENCES

1. Smith, R. M.; Martell, A. E., Eds. "Critical Stability Constants," Vol. IV, Inorganic Complexes; Plenum Press: New York, 1976.
2. Wyckoff, R. W. G., Ed. "Crystal Structures," 2nd ed.; Interscience Publishers, John Wiley and Sons: New York, 1965; Vol. III, p 766.
3. Wertz, D. L.; Tyvoll, J. L. *J. Inorg. Nucl. Chem.* **1974**, *36*, 3713.
4. Trotman-Dickenson, A. F., Executive Ed. "Comprehensive Inorganic Chemistry"; Pergamon Press: Oxford, 1973; p 46.
5. Trotman-Dickenson, A. F., Executive Ed. "Comprehensive Inorganic Chemistry"; Pergamon Press: Oxford, 1973; p 45.
6. Trotman-Dickenson, A. F., Executive Ed. "Comprehensive Inorganic Chemistry"; Pergamon Press: Oxford, 1973; p 28.

4.9

Reactions Between Antimony(III) and Chloride Ions

Solid antimony trichloride dissolves readily in dilute hydrochloric acid solution. Addition of water causes the precipitation of a white solid, which redissolves upon the addition of concentrated hydrochloric acid solution. The precipitate can be dissolved using either sulfuric acid or saturated ammonium chloride solution, although neither is as effective as hydrochloric acid.

MATERIALS

100 ml of 4M hydrochloric acid, HCl (To prepare 1 liter of 4M stock solution, dilute 333 ml of concentrated, 12M, HCl with distilled water to 1 liter.)

10 g of antimony trichloride, $SbCl_3$ ($SbCl_3$ is very hygroscopic; keep in tightly closed vial until used.)

2 liters distilled water

200 ml concentrated (12M) hydrochloric acid, HCl

100 ml 9M sulfuric acid, H_2SO_4 (To prepare 1 liter of 9M stock solution, slowly and carefully add 500 ml of concentrated, 18M, H_2SO_4 to 500 ml of distilled water.)

100 g ammonium chloride, $NH_4Cl(s)$

2 1-liter beakers

2 magnetic stirrers with stirring bars

2 100-ml graduated cylinders

black background for beakers

PROCEDURE

Place a 1-liter beaker and a stirring bar on a magnetic stirrer. Add 100 ml of 4M HCl and about 10 g of solid $SbCl_3$. Start gentle stirring, wait until the solid dissolves, and then establish a moderate rate of stirring. The initial solution can also be prepared by dissolving antimony(III) oxide in HCl, although any residue must be removed by filtration.

Add distilled water from a 100-ml graduated cylinder to the solution until the precipitate formed becomes dense and does not redissolve with continued stirring. Note the volume of water used. Up to 150 ml may be needed.

Using a 100-ml graduated cylinder, add concentrated (12M) HCl to the suspension until the precipitate dissolves. Note the volume of HCl solution used. A total of 25 ml is ordinarily sufficient. If 5 ml of concentrated HCl is added initially, the solution becomes clear, but it slowly clouds up with the formation of a more granular precipitate. Dissolving requires much more HCl.

Add distilled water until the precipitate reforms. Addition of water causes temporary cloudiness which disappears with mixing. A large volume of water is needed to make that cloudiness permanent. However, if 150–200 ml of H_2O are added and the solution is allowed to stir for a few minutes, the precipitate does form.

Add concentrated HCl to dissolve the precipitate (25 ml should be enough). This step should demonstrate the reversibility of the phenomenon, but the cycle can be repeated.

When the solution is clear, pour half of it into a second 1-liter beaker, preferably also equipped with a stirring bar and magnetic stirrer. Reprecipitate the solid in both beakers by adding distilled water. (If this step is taken when the total volume is approximately 500 ml, the addition of 100 ml of H_2O to each beaker slowly produces a fairly large amount of granular precipitate.)

When the precipitation is convincingly great, add 9M H_2SO_4 to one beaker. About 50 ml of this acid yields a clear solution.

To the other beaker, add 60–100 g solid NH_4Cl as rapidly as it dissolves. Mild heating helps counteract the cooling caused by the positive ΔH of solution of NH_4Cl. The precipitate dissolves slowly.

Either of the clear solutions produces a precipitate when diluted further with water, although the volume required may be relatively large.

HAZARDS

Antimony trichloride and all soluble antimony compounds are poisonous. Most are irritating to skin, eyes, and respiratory system. Stibine, an extremely poisonous gas, may be formed by the action of acidic reducing agents on antimony compounds.

Hydrochloric acid may cause severe burns. Hydrochloric acid vapors are extremely irritating to the skin, eyes, and respiratory system.

Since sulfuric acid is a strong acid and a powerful dehydrating agent, it can cause burns. Spills should be neutralized with an appropriate agent, such as sodium bicarbonate, and then rinsed clean.

DISPOSAL

The acidic solution should be cautiously neutralized with sodium bicarbonate or soda ash and washed down the drain with excess water.

DISCUSSION

The reversibility of the Sb(III)/Cl$^-$ system makes it especially useful for the purpose of displaying the hydrolysis reaction of metal ions in aqueous solution (see Demonstration 4.11). The initial precipitate obtained when HCl solutions of $SbCl_3$ are diluted is quite probably a complicated mixture of basic chlorides. Equilibrium is reached slowly in the system, and the literature contains reports of compounds that seem improbable in the light of more complex studies [1]. Investigators agree that the basic chlorides SbOCl and $Sb_4O_5Cl_2$ exist, with the former precipitating from more concentrated solutions and the latter from more dilute solutions [1]. They also agree that on extreme dilution the hydrolysis may proceed all the way to Sb_2O_3 [1]. Under the circumstances, discussing the observations in terms of solubility products is not useful. Instead, we will discuss the observations in less specific terms.

The precipitation of a basic chloride by diluting a solution of antimony(III) chloride has been described by the equation:

$$Sb^{3+}(aq) + H_2O(l) + Cl^-(aq) \longrightarrow SbOCl(s) + 2 H^+(aq) \qquad (1)$$

The equilibrium constant expression for reaction 1 is

$$K_{eq} = \frac{[H^+]^2}{[Sb^{3+}][Cl^-]}$$

However, if one dilutes a solution containing Sb^{3+}, Cl^-, and H^+, the value of the expression above remains constant, so that equation 1 by itself does not account for the precipitation of the basic chloride.

If we assume that Sb(III) in hydrochloric acid solution exists as a chloro complex containing more than one chloride ion, we can account for the precipitation. For the purpose of illustration, we assume that the complex is $SbCl_4^-$:

$$SbCl_4^-(aq) + H_2O(l) \longrightarrow SbOCl(s) + 2 H^+(aq) + 3 Cl^-(aq) \qquad (2)$$

$$K_{eq} = \frac{[H^+]^2[Cl^-]^3}{[SbCl_4^-]}$$

The effect of dilution is clearly magnified in the numerator of this expression, so that the numerical value decreases rapidly, thus leading to precipitation.

This explanation of the observations also implies that either H^+ or Cl^- should be effective in dissolving the precipitate, and that hydrochloric acid is doubly effective because it adds both species to the solution simultaneously. The final steps of the demonstration illustrate that the precipitate is dissolved by a strong acid, such as concentrated sulfuric acid or 60% perchloric acid ($HClO_4$), or by sufficient concentration of chloride ions provided by ammonium chloride. When the quantities of these agents required to dissolve the precipitate are compared to the quantity of HCl solution required to do so, the special effectiveness of hydrochloric acid is readily revealed. If HCl solution is diluted to the same molarity as 9M H_2SO_4 (or 60% $HClO_4$) and all other factors are equal, a greater volume of the H_2SO_4 (or $HClO_4$) solution than the HCl is needed to redissolve the hydrolysis product.

In an experiment designed to show the solubility of $SbCl_3$ in different solutions, 11.4 g of $SbCl_3$ was added to the following solutions with the stated results:

(a) 50 ml of 5M HCl. Solution clear at all times, and all $SbCl_3$ dissolved rapidly.

(b) 50 ml of 60% $HClO_4$. Solution clear at all times, and all $SbCl_3$ dissolved over a period of 3 hours.

(c) 50 ml of NH_4Cl solution saturated at room temperature. Cloudy solution produced, which might have cleared spontaneously but cleared rapidly when warmed and did not reprecipitate any solid after standing for more than 24 hours.

(d) 50 ml of KCl solution saturated at room temperature. Cloudy solution produced, which cleared with heating but which deposited a precipitate upon cooling. This solid is not characterized but is believed [*1*] to be the granular hydrolysis product assigned the formula of $Sb_4O_5Cl_2$.

The hydrolysis of Sb(III) produces hydroxyl complexes. The presence of hydroxyl complexes in the solution prior to dilution reduces the effectiveness of acid in reversing the precipitation. Consider this hypothetical reaction:

$$SbOH^{2+}(aq) + Cl^-(aq) \longrightarrow SbOCl(s) + H^+(aq) \qquad (3)$$

$$K_{eq} = \frac{[H^+]}{[Cl^-][SbOH^{2+}]}$$

Since the equilibrium constant expression involves $[H^+]$ only to the first power instead of the second power, as in equation 2, the effect of altering $[H^+]$ is diminished.

Although the formation constants for the hydroxyl complexes of Sb(III) are quite large and the formation constants for the chloro complexes are relatively small, the presence of HCl simultaneously decreases the concentration of the hydroxyl complexes and increases the concentration of the chloro complexes. The table presents some of the relevant stability constants.

Stability Constants for Cl^- and OH^- Complexes of Sb(III) [2]

Chloro complexes	K_{eq}	Hydroxo complexes	K_{eq}
β_1	2.0×10^2	$\dfrac{[Sb(OH)_2{}^+]}{[SbOH^{2+}][OH^-]}$	3×10^{15}
β_2	3.0×10^3	$\dfrac{[Sb(OH)_3]}{[Sb(OH)_2{}^+][OH^-]}$	6×10^{12}
β_3	1.6×10^4		
β_4	5.0×10^4	$\dfrac{[Sb(OH)_4{}^-]}{[Sb(OH)_3][OH^-]}$	1.6×10^2
β_5	5.0×10^4	$\dfrac{[Sb_2(OH)_2{}^{4+}]}{[Sb(OH)_2{}^+]^2[H^+]^2}$	5
β_6	1.26×10^4	$\dfrac{[Sb(OH)_2{}^+]}{[H^+]}$	8×10^{-4}

The species distribution diagram for the chloro complexes (see figure) shows that, in the high concentrations of chloride ions used in this demonstration, the concentrations of $SbCl_4{}^-$, $SbCl_5{}^{2-}$, and $SbCl_6{}^{3-}$ differ only slightly. This result agrees with the observation that the solids which may be crystallized from these concentrated chloride solutions depend upon the nature of the cation chosen. Salts of each of the anions listed may be formed as well as salts of polynuclear species, such as $Sb_2Cl_9{}^{3-}$.

Additional supplementary procedures can be employed to help understand the behavior of the aqueous Sb(III)/Cl^- system. If a solution of $SbCl_3$ in HCl is treated with a solution of tris(ethylenediamine)cobalt(III) chloride, $[Co(en)_3]Cl_3$, a precipitate with the characteristic color of the $[Co(en)_3]^{3+}$ ion is obtained. If this product is collected by filtration and treated with water, the white precipitate characteristic of the hydrolysis of $SbCl_3$ is observed, which demonstrates the presence of Sb^{3+} in an anionic form. The compound formed is $[Co(en)_3][SbCl_6]$. A similar compound forms with $[Co(NH_3)_6]^{3+}$.

In another procedure, a solution can be simultaneously saturated with NH_4Cl and $SbCl_3$ at elevated temperature. When cooled, this solution deposits crystals which are

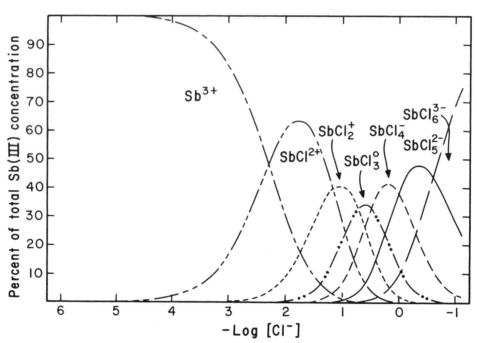

Antimony(III)/chloride ion species distribution.

different from either starting material. The literature [3] indicates that the composition is $(NH_4)_2SbCl_5$.

If one desires to show the effect of complex formation on solubility, tartaric acid can be used. With $SbCl_3$ in solution, add a solution of tartaric acid, then dilute with distilled water. No precipitate will form. The soluble species is apparently a tartrate complex of the partially hydrolyzed Sb^{3+}.

REFERENCES

1. Mellor, J. W. "A Comprehensive Treatise on Inorganic and Theoretical Chemistry," Vol. IX; Longmans, Green and Co.: London, 1929.
2. Smith, R. M.; Martell, A. E., Eds. "Critical Stability Constants," Vol. IV, Inorganic Substances; Plenum Press: New York, 1976.
3. Trotman-Dickenson, A. F., Executive Ed., "Comprehensive Inorganic Chemistry," Vol. II; Pergamon Press: Oxford, 1973.

4.10

Reactions Between Carbon Dioxide and Limewater

Carbon dioxide is bubbled through a clear, colorless solution in a glass vessel. A relatively dense precipitate forms fairly rapidly and then, upon continued bubbling, eventually redissolves. If the solution is then heated, the precipitate reforms. A person exhaling through a sample of the same solution produces the precipitate but is not able to redissolve it.

MATERIALS

saturated calcium hydroxide solution, $Ca(OH)_2$ (To prepare, boil 2 liters of distilled water, cover, and allow to cool overnight. Add 3.5 grams of solid $Ca(OH)_2$. Shake well and allow to settle. If cloudy at time of use, filter.)

source of $CO_2(g)$, either solid CO_2 (dry ice) or CO_2 cylinder with valve and rubber tubing

1 ml bromthymol blue indicator (To prepare 1 liter of stock solution, dissolve 0.4 g of bromthymol blue in 64 ml of 0.01M NaOH and dilute to 1 liter with distilled water.)

50 ml 0.1M disodiumdihydrogenethylenediaminetetraacetate solution, Na_2H_2EDTA (To prepare 1 liter of 0.1M stock solution, dissolve 37.2 g $Na_2H_2EDTA \cdot 2H_2O$ in distilled water and dilute to 1 liter.)

5 500-ml beakers

3 pilsner glasses (optional)

3 straws or 20–25 cm lengths of 6–8 mm glass tubing, fire polished

gloves or towel suitable for handling dry ice

hot plate with magnetic stirrer and stirring bar

PROCEDURE

Place 150–250 ml of clear, saturated $Ca(OH)_2$ solution in each of two containers. Have a volunteer blow gently through a straw or glass tube into the solution in one container. Caution: Because $Ca(OH)_2$ is a strong base, warn the volunteer not to get the liquid into the mouth (see Hazards section). Bubble pure CO_2 into the other sample of $Ca(OH)_2$ solution, either by dropping a piece of solid CO_2 into the container or by using a cylinder of CO_2. A precipitate of calcium carbonate, $CaCO_3$, forms in both containers. The precipitate then dissolves in the container through which pure CO_2 is bubbled because of the formation of soluble calcium bicarbonate, $Ca(HCO_3)_2$. The volunteer is unable to dissolve the precipitate in the other container. Caution the volunteer not to hyperventilate.

The effectiveness of this demonstration can be enhanced by having two volunteers in gentle competition with each other as well as with the source of pure CO_2. In this case, the pilsner glasses make very effective display vessels.

Place the solution of $Ca(HCO_3)_2$ in a beaker, and place the beaker on the combination hot plate and magnetic stirrer. Heat and stir the solution. The solution degases and, as it approaches the boiling temperature, a solid begins to settle out of solution. This precipitate will not appear to be as dense as the initial precipitate. When the solution is allowed to cool, more solid precipitates. Further saturation of this solution with pure CO_2 does not redissolve all the precipitate.

Supplementary Procedures

1. Place approximately 200 ml of saturated $Ca(OH)_2$ solution in a 500-ml beaker and add a few drops of bromthymol blue indicator. Bubble CO_2 through the solution. A white precipitate forms and redissolves while the bromthymol blue is undergoing its color change, thereby establishing pH 7 to pH 6 as the range in which complete solution of the initial precipitate occurs.

2. Place approximately 200 ml of saturated $Ca(OH)_2$ solution in a 500-ml beaker on a hot plate and heat almost to boiling. Bubble pure CO_2 into the hot solution, either from a cylinder or a piece of dry ice. A white precipitate forms but does not redissolve upon further bubbling of CO_2.

3. Place approximately 200 ml of saturated $Ca(OH)_2$ solution in a 500-ml beaker and add 25 ml of 0.1M Na_2H_2EDTA solution. Bubble CO_2 into the resultant solution. No precipitate forms.

HAZARDS

Saturated calcium hydroxide solution (limewater) is a base with pH = 12.4. It can irritate the skin and eyes. In case of accidental ingestion, give copious drinks of water to dilute the limewater in the stomach, and seek medical advice.

Solid CO_2, dry ice, has a temperature of $-78°C$ and can cause frostbite. Thermal protection in the form of gloves or a towel should be used when handling dry ice.

DISPOSAL

Dry ice can be left to evaporate in a well-ventilated area. Solutions and suspensions should be flushed down the drain with water.

DISCUSSION

This system is useful in illustrating some geological phenomena since CO_2 dissolved in surface water leaches $CaCO_3$ from the soil and transports it. The $CaCO_3$ deposits in caves result from the redeposition of the $CaCO_3$ when the CO_2 and water evaporate. Carbon dioxide is one of the few substances significantly involved in both biological and geological cycles and present in the free or combined state in the atmosphere, hydrosphere, lithosphere, and biosphere. An understanding of the carbonic acid system is essential in almost every field of applied chemistry. For that reason, we will discuss some aspects of this system before considering the details of the demonstration.

Table 1. Equilibrium Constants
for the Ca^{2+}, OH^-, CO_2 System [1]

Reaction	K_{eq}
$Ca(OH)_2(s) \longrightarrow Ca^{2+}(aq) + 2\,OH^-(aq)$	6.46×10^{-6}
$CaCO_3(s) \longrightarrow Ca^{2+}(aq) + CO_3^{2-}(aq)$	
For calcite	4.5×10^{-9}
For aragonite	6.0×10^{-9}
$CO_2(g) + H_2O(l) \longrightarrow H_2CO_3(aq)$	0.035
$H_2CO_3(aq) \longrightarrow H^+(aq) + HCO_3^-(aq)$	4.5×10^{-7}
$HCO_3^-(aq) \longrightarrow H^+(aq) + CO_3^{2-}(aq)$	4.7×10^{-11}
$Ca^{2+}(aq) + EDTA^{4-} \longrightarrow Ca(EDTA)^{2-}$	4×10^{10}

We follow the customary practice of treating the H_2O/CO_2 system as if all dissolved CO_2 is converted to carbonic acid or to one of its ions. The first dissociation constant of carbonic acid in Table 1 is based upon this practice. However, dissolved CO_2 is principally just that and should be represented as $CO_2(aq)$, not as $H_2CO_3(aq)$. The ratio of $CO_2(aq)$ to $H_2CO_3(aq)$ is approximately 600, which means that the concentration of $H_2CO_3(aq)$ in water saturated with CO_2 at 25°C is approximately $3.5 \times 10^{-2} \div 600$, or 6×10^{-5}M [2]. Thus, carbonic acid, H_2CO_3, is a much stronger acid than the tabulated K_1 (4.5×10^{-7}) would imply, and it has a true ionization constant of approximately 2.7×10^{-4}:

$$K_1 = \frac{[H^+][HCO_3^-]}{[H_2CO_3]}$$

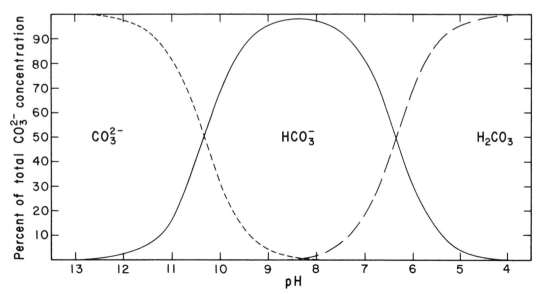

Figure 1. Carbonic acid species distribution.

If one assumes that $[H_2CO_3] = 0.035$, the value of K_1 is the tabulated value of 4.5×10^{-7}. However, if one assumes that $[H_2CO_3] = \dfrac{0.035}{600}$, then K_1 must equal $600 \times 4.5 \times 10^{-7} \simeq 2.7 \times 10^{-4}$. The concentrations of H^+ and HCO_3^- are unaffected by this assumption. In considering the equilibrium between $CO_2(g)$ and the various carbonic acid species in solution, it is usually convenient to use the tabulated value of K_1 and to overlook the distribution of dissolved CO_2 between $CO_2(aq)$ and $H_2CO_3(aq)$. The species distribution diagram in Figure 1 is calculated on the assumption that all dissolved CO_2 is present as H_2CO_3 or its derivative ions.

The reaction of $CO_2(aq)$ with either water or hydroxide ions is slow, so this system should be given adequate time to equilibrate. The dehydration of carbonic acid is also slow, so slow that it must be catalyzed in the lungs to permit the release of CO_2. The catalyst, appropriately called carbonic anhydrase, is found in many organisms in which the interconversion between $CO_2(aq)$ and $H_2CO_3(aq)$ is necessary.

The net ionic equations and calculated values of equilibrium constants for the reactions occurring in this demonstration are

$$2\,OH^-(aq) + CO_2(g) \longrightarrow H_2O(l) + CO_3^{2-}(aq) \qquad K_{eq} = 7.4 \times 10^9 \quad (1)$$

$$Ca^{2+}(aq) + CO_3^{2-}(aq) \longrightarrow CaCO_3(s) \qquad K_{eq} = 2.2 \times 10^8 \quad (2)$$

$$CO_2(g) + H_2O(l) + CaCO_3(s) \longrightarrow Ca^{2+}(aq) + 2\,HCO_3^-(aq)$$
$$K_{eq} = 1.5 \times 10^{-6} \quad (3)$$

$$H_2CO_3(aq) + CaCO_3(s) \longrightarrow Ca^{2+}(aq) + 2\,HCO_3^-(aq) \qquad K_{eq} = 4.3 \times 10^{-5} \quad (4)$$

$$Ca^{2+}(aq) + HCO_3^-(aq) \longrightarrow CaHCO_3^+(aq) \qquad K_{eq} = 10 \qquad (5)$$

$$Ca^{2+}(aq) + 2\,HCO_3^-(aq) \longrightarrow CaCO_3(s) + CO_2(g) + H_2O(l)$$
$$K_{eq} = 6.7 \times 10^5 \quad (6)$$

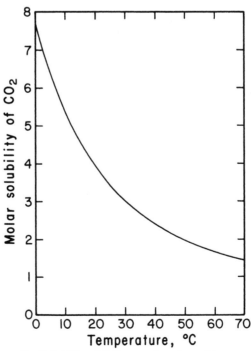

Figure 2. Solubility of CO_2 as a function of temperature.

The sum of equations 1 and 2 represents the precipitation of $CaCO_3$. Equations 3 and 4 represent the dissolution of freshly precipitated $CaCO_3$ in excess CO_2. Equation 5 represents a reaction that aids in the dissolution of freshly precipitated $CaCO_3$. Equation 6 represents the reprecipitation of $CaCO_3$ from a solution of calcium bicarbonate, essentially the reverse of equation 3. Carbon dioxide is removed from the solution by heating. Figure 2 and Table 2 show the solubility of CO_2 as a function of temperature.

Table 2. Solubility of CO_2 in Water as a Function of Temperature [3]

Temperature t(°C)	q^a	β^b	Moles[c] per 1000 ml H_2O $\times 10^2$
0	0.335	1.713	7.64
5	0.277	1.424	6.35
10	0.231	1.194	5.33
12.5	—	(1.075)	—
15	0.197	1.019 (1.000)	4.57
20	0.169	0.878 (0.865)	3.92
25	0.145	0.759 (0.754)	3.39
30	0.126	0.664	2.96
35	—	(0.595)	2.67
40	0.097	0.530 (0.533)	2.37
45	—	(0.480)	2.16
50	0.076	(0.438)	1.95
55	—	(0.397)	1.77
60	—	(0.365)	1.63
70	—	(0.319)	1.42

[a]q = grams of CO_2 dissolved by 100 g H_2O at total pressure equal to 1 atmosphere.

[b]β = ml of CO_2 (at 760 mm and 0°C) dissolved in 1 ml of water when P_{CO_2} = 760 mm. Values in parentheses are from a second set of measurements.

[c]These values are calculated from β values and therefore imply that P_{CO_2} = 760 mm of Hg.

Although equations 1–6 represent the observed reactions, a quantitative account of the observations using the tabulated values for the equilibrium constants is impossible. Direct application of these constants to the problem leads to the conclusion that not all the precipitated $CaCO_3$ should dissolve in the solution saturated with CO_2. In the following calculation, the solution process can be conceptualized as a competition between $H_2CO_3(aq)$ and $Ca^{2+}(aq)$ for the carbonate ion:

$$H_2CO_3(aq) + CO_3^{2-}(aq) \longrightarrow 2\,HCO_3^-(aq) \tag{7}$$

$$K_{eq} = K_1/K_2$$

$$Ca^{2+}(aq) + CO_3^{2-}(aq) \longrightarrow CaCO_3(s) \tag{2}$$

$$K_{eq} = 1/K_{s0}$$

We combine these reactions to get the reaction:

$$CaCO_3(s) + H_2CO_3(aq) \longrightarrow Ca^{2+}(aq) + 2\,HCO_3^-(aq) \tag{4}$$

For this reaction,

$$K_{eq} = K_{s0} \times \frac{K_1}{K_2}$$

$$= [Ca^{2+}][\cancel{CO_3^{2-}}] \times \frac{[\cancel{H^+}][HCO_3^-]}{[H_2CO_3]} \times \frac{[HCO_3^-]}{[\cancel{H^+}][\cancel{CO_3^{2-}}]} = \frac{[Ca^{2+}][HCO_3^-]^2}{[H_2CO_3]}$$

So $K_{eq} = \dfrac{4.5 \times 10^{-9} \times 4.5 \times 10^{-7}}{4.7 \times 10^{-11}} = 4.3 \times 10^{-5}$

With a numerical value for this equilibrium constant, we can substitute known values for ionic concentrations and test the applicability of the constant to the system at hand.

When carbon dioxide reacts completely with hydroxide ions to form bicarbonate ions, one bicarbonate ion is formed for every initial hydroxide ion:

$$CO_2(g) + OH^-(aq) \longrightarrow HCO_3^-(aq) \qquad (8)$$

Therefore, the bicarbonate ion concentration in a solution of calcium hydroxide that has been converted to calcium bicarbonate should be the same as the concentration of hydroxide ions in the saturated calcium hydroxide solution. Consequently, we can write the following equality to describe that solution:

$$[Ca^{2+}][HCO_3^-]^2 = K_{s0} \text{ of } Ca(OH)_2 = 6.46 \times 10^{-6}$$

We can then substitute this value in equation 4, the expression for the equilibrium constant for the solution process:

$$4.3 \times 10^{-5} = \frac{6.46 \times 10^{-6}}{[H_2CO_3]}$$

Therefore,

$$[H_2CO_3] = \frac{6.46 \times 10^{-6}}{4.3 \times 10^{-5}} = 0.15M$$

However, at 25°C the maximum solubility of CO_2 in water is 0.035M. According to the equilibrium constants, therefore, the H_2CO_3 concentration cannot be high enough to dissolve all the $CaCO_3$. And yet the precipitate dissolves.

Direct measurements of the solubility of calcium hydroxide in water and of calcium carbonate (calcite) in water that is equilibrated with various partial pressures of carbon dioxide show clearly that the amount of calcium carbonate formed from a saturated solution of calcium hydroxide should not redissolve when that solution is saturated with carbon dioxide at 1 atmosphere. Figure 3 shows the solubility of calcite in water as a function of partial pressure of carbon dioxide [4]. Point A is the solubility required if all the $CaCO_3$ from a saturated solution of CO_2 were to redissolve $(2.2 \times 10^{-2}$ moles/liter). Thus, if the precipitate were calcite of macroscopic size and if the original solution of calcium hydroxide were truly saturated, only about half the calcite should redissolve.

Three factors may contribute to a plausible explanation for the discrepancy between the experimental observations and the calculated results. First, a complex may be formed between $Ca^{2+}(aq)$ and $HCO_3^-(aq)$ (equation 5) with a formation constant value of 10. The formation of such a complex would favor dissolution of $CaCO_3$ and would account for the discrepancy between the observations and the equilibrium calculations. Calculation of the magnitude of this effect shows that it is inadequate to account for the observation, although it reduces the required concentration of H_2CO_3 to just under 0.1M.

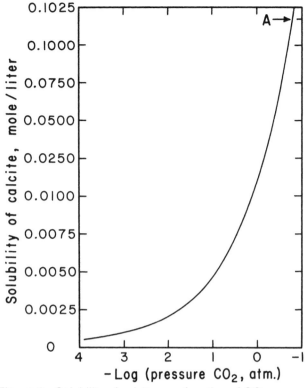

Figure 3. Solubility of calcite as a function of CO₂ pressure.

Second, whenever we use concentrations in an equilibrium calculation, we recognize that we should be using activities. The discrepancy between observation and theroretical prediction is large enough that activity coefficients might make the difference. Standard calculations from ionic strengths show that the introduction of activity coefficients decreases the discrepancy but does not account for it entirely.

The third factor is that the observation may not be a result of the equilibria we assumed in making our calculations. Calcium carbonate exists in nature in two stable forms, calcite and aragonite. Since calcite is the less soluble material and the more stable, we used its solubility product for the above calculations. If we assume that the precipitate is aragonite and make similar calculations using its solubility product, we do not find sufficient decrease in the discrepancy between prediction and observation. However, the solid formed in the initial precipitation may be neither calcite nor aragonite. The precipitation of calcium carbonate has been studied extensively because of its geological importance, and researchers know that rapid precipitation of $CaCO_3$ produces little or no calcite, some aragonite, and mostly a hydrated calcium carbonate that is more soluble than either aragonite or calcite [5]. Digestion of precipitates to enhance their ease of filtration and to improve their purity is a common procedure aimed at forming large particles that are less soluble than small fine particles. In this demonstration the initial precipitate of $CaCO_3$ is a very fine solid. In addition, some freshly precipitated solids are distinctly more soluble and more reactive than aged precipitates.

All three factors mentioned may contribute to the observed phenomena. The third factor may be dominant, since the precipitate formed by homogeneous precipitation when the solutions of $Ca(HCO_3)_2$ are heated does not redissolve when the cooled solution is resaturated with carbon dioxide. However, this observation may simply represent another case of slow equilibration. Henry's law, which states that the solubility of a gas is proportional to the partial pressure of that gas, explains the fact that exhaling through the solution fails to redissolve the $CaCO_3$. Carbon dioxide constitutes only about 4% of exhaled breath, which means that its partial pressure is approximately 0.04 atmospheres. Consequently, the concentration of CO_2 is never high enough in that solution to dissolve the freshly precipitated calcium carbonate.

The supplementary demonstration, in which the calcium carbonate precipitate fails to dissolve when the calcium hydroxide solution is heated as the CO_2 is added, illustrates that CO_2 is less soluble at higher temperatures than at lower temperatures and that the concentration of dissolved CO_2 is never high enough to dissolve the calcium carbonate. Table 2 and Figure 2 show the solubility of CO_2 as a function of temperature. Bringing about the precipitation in a hot solution may conceivably favor the formation of larger particles of a more stable form of $CaCO_3$, but the principal factor appears to be the insolubility of CO_2 in hot water.

The precipitation and redissolution of the $CaCO_3$ in the presence of bromthymol blue helps to emphasize the acid-base character of the reaction taking place. A rough calculation, based on the assumptions that all original OH^- is converted to HCO_3^- and that the solution is saturated with CO_2 at 25°C, provides an estimated pH of about 6.2. This value is near the lower edge of the bromthymol-blue color change from blue to yellow and is compatible with the observation that the solution process occurs while the indicator is changing color. The ratio of carbonic acid species at any pH can be read from Figure 1. A pH of 6.2 is very near the mid-point of the first buffer range.

Performing the addition of CO_2 in the presence of ethylenediaminetetraacetic acid (EDTA) demonstrates the stability of the Ca^{2+}/EDTA complex and shows how a complexing or sequestering agent prevents precipitation of a metal ion. This observation, coupled with the observation that heating a bicarbonate-rich solution containing Ca^{2+} leads to the precipitation of $CaCO_3$, may be useful in a discussion of water hardness and the problems it causes.

Many animals create shells by removing calcium ions and carbonate ions from solution. Plants remove CO_2 from solution during photosynthesis and may thereby cause the precipitation of $CaCO_3$ if the water source is saturated. Both combustion of fossil fuels and respiration increase the CO_2 content of the atmosphere, which has been increasing with only seasonal reversals since accurate measurements have been made. This increase has led to concern about the greenhouse effect, which is based on the observation that CO_2 in the atmosphere inhibits the reradiation of solar energy and on the conclusion that too high a concentration of CO_2 could lead to significant elevation of the average temperature of the atmosphere, with catastrophic results.

The carbonate system is also part of the buffer system maintaining optimum pH in biological systems. Hyperventilation, the rapid, shallow breathing often experienced by novices at high altitudes, decreases the CO_2 content of the body and makes the blood more alkaline than normal. The apprehension this generates, coupled with the involuntary muscle contractions experienced, makes this a frightening experience. Hyperventilation can be treated by having the victim breathe into and out of a paper bag, thereby recycling the CO_2 back into the victim's system.

REFERENCES

1. Smith, R. M.; Martell, A. E., Eds. "Critical Stability Constants," Vol. IV, Inorganic Substances; Plenum Press: New York, 1976.
2. Trotman-Dickenson, A. F., Executive Ed. "Comprehensive Inorganic Chemistry"; Pergamon Press: Oxford, 1973; Vol. I, p 1232.
3. Linke, W. F. "Solubilities of Inorganic and Metal Organic Compounds," 4th ed.; Van Nostrand: New York, 1958; Vol. I, p 459.
4. Linke, W. F. "Solubilities of Inorganic and Metal Organic Compounds," 4th ed.; Van Nostrand: New York, 1958; Vol. I, p 536 ff.
5. Bathurst, R. G. C. "Carbonate Sediments and Their Diagenesis," 2nd ed.; Elsevier Publishing Co.: New York, 1975; Ch. 6.

4.11

Precipitates and Complexes of Iron(III)

Reagents are added in a specified order to aqueous solutions of iron(III). The reagents are nitric acid, sodium hydroxide, sodium chloride, phosphoric acid, potassium thiocyanate, sodium fluoride, and potassium hexacyanoferrate(II). Two procedures reveal the colors of precipitates and complexes of iron(III).

MATERIALS FOR PROCEDURE A

1500 ml distilled water

50 ml 1M iron(III) nitrate, $Fe(NO_3)_3$ (To prepare 1 liter of 1M stock solution, dissolve 404 g of $Fe(NO_3)_3 \cdot 9H_2O$ in distilled water and dilute to 1 liter.)

35 ml 6M nitric acid, HNO_3 (To prepare 1 liter of 6M stock solution, dilute 385 ml of concentrated, 16M, HNO_3 solution to 1 liter with distilled water.)

10 ml 6M sodium hydroxide, NaOH (To prepare 1 liter of 6M stock solution, dissolve 240 g of NaOH in distilled water and dilute to 1 liter.)

100 ml 1M sodium chloride, NaCl (To prepare 1 liter of 1M stock solution, dissolve 58.4 g of NaCl in distilled water and dilute to 1 liter.)

15 ml 2M sodium dihydrogen phosphate, NaH_2PO_4 (To prepare 1 liter of 2M stock solution, dissolve 276 g of $NaH_2PO_4 \cdot H_2O$ in distilled water and dilute to 1 liter.)

70 ml 0.1M potassium thiocyanate, KSCN (To prepare 1 liter of stock solution, dissolve 9.7 g KSCN in distilled water and dilute to 1 liter.)

100 ml 1M sodium fluoride, NaF (To prepare 1 liter of stock solution, dissolve 42 g of NaF in distilled water and dilute to 1 liter.)

5 ml 0.1M potassium hexacyanoferrate(II), $K_4Fe(CN)_6$ (To prepare 1 liter of stock solution, dissolve 42.2 g of $K_4Fe(CN)_6 \cdot 3H_2O$ in distilled water and dilute to 1 liter.)

2-liter beaker, preferably with approximate graduations

8 1-liter beakers, preferably with approximate graduations

8 stirring rods

100-ml graduated cylinder

10-ml graduated cylinder

dropper

MATERIALS FOR PROCEDURE B

2 g iron(III) nitrate nonahydrate, $Fe(NO_3)_3 \cdot 9H_2O$

600 ml distilled water

3 ml 6M nitric acid, HNO_3 (For preparation, see Materials for Procedure A.)

4 g sodium chloride, NaCl

2.5 ml concentrated, 14.6M (85%), phosphoric acid, H_3PO_4

2 g potassium thiocyanate, KSCN

5 g sodium fluoride, NaF

0.4 g potassium hexacyanoferrate(II) trihydrate, $K_4Fe(CN)_6 \cdot 3H_2O$

1-liter beaker

magnetic stirrer with stirring bar

10-ml graduated cylinder

PROCEDURE A

This procedure allows the viewer to observe changes that occur upon addition of each successive reagent and to contrast the colors of nine iron(III) species. In each case, leave about 200 ml of each solution in the reaction vessel for comparison purposes.

Place the nine beakers in a row, with the 2-liter beaker third from the right. (From here on, the beakers to its right are referred to as #1 and #2, the 2-liter beaker is #3, and the beakers to its left are #4 through #9).

Pour 1500 ml of distilled water and 50 ml of 1M $Fe(NO_3)_3$ solution into beaker #3. Pour 900 ml of this yellowish orange solution into beaker #4 and 400 ml into beaker #2.

To beaker #4, add 15 ml of 6M HNO_3 solution and stir. Note the immediate disappearance of color.

To beaker #2, add 10 ml of 6M NaOH solution and note the color and appearance of the precipitate. Stir the mixture and add half to beaker #1.

To beaker #1, add 15–20 ml of 6M HNO_3 solution and stir occasionally during the next 10–15 minutes while the precipitate dissolves. Note the gradual changes in color until the solution becomes colorless. While this precipitate is dissolving, continue with the demonstration.

Next, pour 700 ml of the colorless solution in beaker #4 into beaker #5. With stirring, add enough 1M NaCl solution (ca. 100 ml) to produce a yellow color.

Pour about 600 ml of the yellow solution into beaker #6, remembering to leave about 200 ml in beaker #5. While stirring, add enough 2M NaH_2PO_4 solution (ca. 15 ml) to produce a colorless solution.

Pour about 400 ml from beaker #6 into beaker #7. With stirring, add enough 0.1M KSCN solution (ca. 70 ml) to produce a red color. Do not add more KSCN solution than necessary for this color change.

Pour about 270 ml from beaker #7 into beaker #8. With stirring, add enough 1M NaF solution (ca. 70 ml) to produce a colorless solution. Do not add more NaF solution than necessary to produce a colorless solution.

Finally, pour about 150 ml from beaker #8 into beaker #9. With stirring, use a dropper to add about 2 ml of 0.1M $K_4Fe(CN)_6$ solution. Note the solution colors and the precipitate formation.

PROCEDURE B

Place the 1-liter beaker containing the stirring bar on the magnetic stirrer. Add 2 g of $Fe(NO_3)_3 \cdot 9H_2O$, noting its color. Next, add 600 ml of distilled water and adjust the stirring rate to mix efficiently without too much turbulence. Continue stirring until the solid is all dissolved, and note the shade and intensity of the color of the solution. Add, in sequence, enough of each reagent to produce the indicated color change:

6M HNO_3 (ca. 3 ml)	colorless
solid NaCl (ca. 4 g)	yellow
85% H_3PO_4 (ca. 2.5 ml)	colorless
solid KSCN (ca. 2 g)	red
solid NaF (ca. 5 g)	colorless
powdered solid $K_4Fe(CN)_6 \cdot 3H_2O$ (ca. 0.4 g)	dark blue

During and after each addition, note the color and general appearance of each solution. Adding an excess of any reagent may interfere with subsequent observations. For example, too much KSCN will require addition of more sodium fluoride; a white precipitate may form and, subsequently, the $Fe_4[Fe(CN)_6]_3$ will not form.

HAZARDS

Nitric acid is a strong acid and a powerful oxidizing agent. Contact with combustible material can cause violent and explosive reactions. The liquid can cause severe burns to the skin and eyes. The vapor is irritating to the eyes and the respiratory system.

Sodium hydroxide can cause severe burns of the eyes and skin. Dust from solid sodium hydroxide is very caustic.

Sodium fluoride is very poisonous, and dust from the solid is a severe irritant to the skin, eyes, and respiratory system.

Potassium hexacyanoferrate(II) must not be mixed with hot or concentrated acids since extremely toxic hydrogen cyanide may be produced. Hydrogen cyanide is among the most toxic and rapidly acting of all poisons. The solutions and the gas can be absorbed through the skin. Solutions are irritating to the skin, nose, and eyes. Cyanide compounds and acids must not be stored or transported together.

Early symptoms of cyanide poisoning are weakness, difficult breathing, headache, dizziness, nausea, and vomiting; these may be followed by unconsciousness, cessation of breathing, and death.

Anyone exposed to hydrogen cyanide should be removed from the contaminated atmosphere immediately. Amyl nitrite should be held under the person's nose for not more than 15 seconds per minute, and oxygen should be administered in the intervals. If the person is not breathing, artificial resuscitation by the Silvester method (not mouth to mouth) should be attempted immediately.

Potassium thiocyanate is toxic if taken internally.

Phosphoric acid can cause burns to skin, eyes, and mucous membranes.

DISPOSAL

Solutions should be flushed down the drain with water.

DISCUSSION

All equilibrium constants are tabulated or calculated from tabulated values in "Critical Stability Constants" [1]. The table shows relevant stability constants for Fe(III) complexes.

Stability Constants[a] for Complexes of Fe(III) [1]

Ligand	β_1	β_2	β_3	Other
OH^-	1.23×10^{11}	2.9×10^{21}	—	$\beta_{22} = 7.9 \times 10^{24}$
				$\beta_{43} = 1 \times 10^{51}$
				$K_{s0} = 2.5 \times 10^{-39}$[b]
Cl^-	4.3	5.6	0.2	—
HPO_4^{2-}	2.0×10^8	—	—	—
$H_2PO_4^-$	2.95×10^3	—	—	—
SCN^-	1.3×10^2	1.6×10^3	1.0×10^5	—
F^-	1.51×10^5	1.17×10^9	1.26×10^{12}	—
$Fe(CN)_6^{4-}$	—	—	—	$K_{s0} = 1.85 \times 10^{-37}$[c]

[a]Values are for 1.0M ionic strength.
[b]This value is for freshly precipitated solid. Aged precipitates have lower solubilities.
[c]This value is calculated from the solubility given in Linke [2].

When the purple solid $Fe(NO_3)_3 \cdot 9\,H_2O$ dissolves in water, the solution becomes acidic and yellowish orange because of the hydrolysis reaction:

$$Fe(H_2O)_6^{3+} \longrightarrow [Fe(H_2O)_5OH]^{2+} + H^+ \qquad K_{eq} = 1.23 \times 10^{-3} \qquad (1)$$

   colorless   yellowish orange

The value of the equilibrium constant for this reaction indicates that $Fe(H_2O)_6^{3+}$ is a significantly stronger acid than acetic acid. The addition of nitric acid (or perchloric acid) reverses this hydrolysis reaction, and the color of the hydroxo complex disappears, indicating that $Fe(H_2O)_6^{3+}$ is colorless. (For another example of metal ion hydrolysis, see Demonstration 4.9.)

Addition of NaOH solution to solutions of either $Fe(H_2O)_6^{3+}$ or $[Fe(H_2O)_5OH]^{2+}$ produces a reddish brown precipitate of iron(III) hydroxide, $Fe(OH)_3$ or $Fe_2O_3 \cdot xH_2O$:

$$Fe(H_2O)_6^{3+} + 3\,OH^- \longrightarrow Fe(OH)_3(s) + 6\,H_2O \qquad K_{eq} = 4.0 \times 10^{38} \qquad (2)$$

Freshly precipitated iron(III) hydroxide dissolves slowly in nitric acid (or perchloric acid):

$$Fe(OH)_3(s) + 3\,H^+ + 3\,H_2O \longrightarrow Fe(H_3O)_6^{3+} \qquad K_{eq} = 2.5 \times 10^3 \qquad (3)$$

Aged iron(III) hydroxide, with the characteristic crystal structure of FeO(OH)(s) or Fe_2O_3(s), is less soluble than freshly precipitated iron(III) hydroxide and dissolves more slowly: for FeO(OH)(s), $K_{s0} = 8 \times 10^{-42}$; and for Fe_2O_3(s), $K_{s0} = 2 \times 10^{-43}$. In all cases, the solid dissolves to give polymeric hydroxo complexes in solution, which slowly depolymerize to the colorless $Fe(H_2O)_6^{3+}$ ion.

Addition of chloride ions in the form of sodium chloride produces $[Fe(H_2O)_5Cl]^{2+}$:

$$Fe(H_2O)_6^{3+} + Cl^- \longrightarrow [Fe(H_2O)_5Cl]^{2+} + H_2O \qquad K_{eq} = 4.3 \qquad (4)$$
$$\text{yellow}$$

Some $[Fe(H_2O)_4Cl_2]^+$ may also form, but, because the quantities of ligand added throughout this demonstration produce relatively low ligand concentrations in solution, the formation of the monosubstituted complex is assumed.

Addition of phosphoric acid ($K_1 = 2 \times 10^{-2}$) or NaH_2PO_4 solution to $[Fe(H_2O)_5Cl]^{2+}$ decolorizes the solution by converting the chloro complex to a colorless phosphato complex, as represented in equation 7, 8, or 9:

$$Fe(H_2O)_6^{3+} + H_3PO_4 \longrightarrow [Fe(H_2O)_5H_2PO_4]^{2+} + H^+ \qquad K_{eq} = 5.9 \times 10^1 \quad (5)$$

$$Fe(H_2O)_6^{3+} + H_3PO_4 \longrightarrow [Fe(H_2O)_5HPO_4]^+ + 2 H^+ \qquad K_{eq} = 1.6 \qquad (6)$$

$$[Fe(H_2O)_5Cl]^{2+} + H_3PO_4 \longrightarrow [Fe(H_2O)_5H_2PO_4]^{2+} + Cl^- + H^+$$
$$K_{eq} = 13.7 \qquad (7)$$

$$[Fe(H_2O)_5Cl]^{2+} + H_3PO_4 \longrightarrow [Fe(H_2O)_5HPO_4]^+ + Cl^- + 2 H^+$$
$$K_{eq} = 0.37 \qquad (8)$$

$$[Fe(H_2O)_5Cl]^{2+} + H_2PO_4^- \longrightarrow [Fe(H_2O)_5H_2PO_4]^{2+} + Cl^- \quad K_{eq} = 6.9 \times 10^2 \quad (9)$$

Although one cannot be sure which phosphate complex forms, the dihydrogen phosphate complex seems most likely, since the concentration of $H^+(aq)$ in the demonstration solution is approximately 0.1 molar. The equilibrium constants also appear to favor this species. Mixed chloro-phosphato complexes of Fe(III) may form, but these species have not yet been characterized.

Thiocyanate ions displace the phosphate ligand from the Fe(III) coordination sphere, as shown in equation 11 or 12:

$$[Fe(H_2O)_6]^{3+} + SCN^- \longrightarrow [Fe(H_2O)_5SCN]^{2+} + H_2O \qquad K_{eq} = 1.3 \times 10^2 \ (10)$$
$$\text{red}$$

$$[Fe(H_2O)_5H_2PO_4]^{2+} + SCN^- + H^+ \longrightarrow [Fe(H_2O)_5SCN]^{2+} + H_3PO_4$$
$$K_{eq} = 2.2 \qquad (11)$$

$$[Fe(H_2O)_5HPO_4]^+ + SCN^- + 2H^+ \longrightarrow [Fe(H_2O)_5SCN]^{2+} + H_3PO_4$$
$$K_{eq} = 8.1 \times 10^1 \ (12)$$

The thiocyanato complex is intensely colored, so this reaction need not go very far toward completion to produce a highly colored solution. In this case, the formation of the disubstituted complex is feasible, but, since its absorption of light is very similar to that of the monothiocyanato complex, its formation is not visually apparent. The addition of too much KSCN reagent will interfere with subsequent steps. At low pH, thiocyanate species decompose to yield sulfur, cyanogen, and other products. Because subsequent addition of $Fe(CN)_6^{4-}$ does not yield a blue species, we conclude that Fe^{3+} is reduced to Fe^{2+} by SCN^- or one of its oxidation products.

Addition of sodium fluoride brings about the displacement of the thiocyanate ligand by fluoride ions, as shown in equation 14:

$$Fe(H_2O)_6^{3+} + F^- \longrightarrow [Fe(H_2O)_5F]^{2+} + H_2O \qquad K_{eq} = 1.51 \times 10^5 \quad (13)$$

$$[Fe(H_2O)_5SCN]^{2+} + HF \longrightarrow [Fe(H_2O)_5F]^{2+} + SCN^- + H^+$$
$$K_{eq} = 7.0 \qquad (14)$$

Note that high acidity inhibits reaction 14, since HF is a weak acid ($K_{eq} = 1.1 \times 10^{-3}$). Note also that the absorption of the thiocyanato complex is due to an allowed charge transfer, so the molar absorbance coefficient of that species is high. Consequently, to turn the solution colorless, one must sometimes use additional NaF reagent. The concentration of Fe(III)/SCN$^-$ species must be reduced to a very low level for color to disappear. The amount of F$^-$ required may result in the formation of $[Fe(H_2O)_4F_2]^+$ and higher complexes, since these species are relatively stable.

The addition by drops of $Fe(CN)_6^{4-}$ solution to either $Fe(H_2O)_6^{3+}$ or $[Fe(H_2O)_5F]^{2+}$ solution produces different colors prior to the final deep blue color and the deep blue precipitate of $Fe_4[Fe(CN)_6]_3$:

$$4\ Fe(H_2O)_6^{3+} + 3\ Fe(CN)_6^{4-} \longrightarrow Fe_4[Fe(CN)_6]_3(s) + 24\ H_2O$$
$$\text{blue} \qquad K_{eq} = 1.85 \times 10^{-37} \quad (15)$$

$$4\ [Fe(H_2O)_5F]^{2+} + 3\ Fe\,(CN)_6^{4-} \longrightarrow Fe_4[Fe(CN)_6]_3(s) + 4\ F^- + 20\ H_2O$$
$$K_{eq} = \text{ca. } 1 \times 10^{16} \quad (16)$$

This deep blue precipitate has the same structure whether it is prepared by combining solutions of iron(III) with hexacyanoferrate(II) or solutions of iron(II) with hexacyanoferrate(III), so the distinction between Prussian blue and Turnbull's blue is an artificial one [3, 4].

REFERENCES

1. Smith, R. M.; Martell, A. E., Eds. "Critical Stability Constants," Vol. IV, Inorganic Complexes; Plenum Press: New York, 1976.
2. Linke, W. F. "Solubilities of Inorganic and Metal Organic Compounds," 4th ed.; Van Nostrand: New York, 1958; Vol. I.
3. Robin, M. B. *Inorg. Chem.* **1962**, *1*, 337.
4. Shriver, D. F.; Shriver, S. A.; Anderson, S. E. *Inorg. Chem.* **1965**, *4*, 725.

Future Volumes

The following demonstration topics will be included in future volumes:

acids, bases, and salts
atomic structure
chemical periodicity
chromatography
clock reactions
colloids and gels
corridor demonstrations and exhibits
cryogenics
electrochemistry
fluorescence and phosphorescence
kinetics and catalysis
lasers in chemistry
organic chemistry
oscillating reactions
overhead projector demonstrations
photochemistry
properties and behavior of gases
properties of liquids, solids, and solutions
radioactivity
spectroscopy and color